IN VITRO FERTILIZATION AND EMBRYO TRANSFER

A Manual of Basic Techniques

IN VITRO FERTILIZATION AND EMBRYO TRANSFER
A Manual of Basic Techniques

Edited by

Don P. Wolf
Oregon Regional Primate Research Center
Beaverton, Oregon

Associate Editors

Barry D. Bavister
Wisconsin Regional Primate Research Center
Madison, Wisconsin

Marybeth Gerrity
Evanston and Glenbrook Hospitals
Evanston, Illinois

and

Gregory S. Kopf
University of Pennsylvania
Philadelphia, Pennsylvania

PLENUM PRESS • NEW YORK AND LONDON

Library of Congress Cataloging in Publication Data

Workshop on Laboratory Techniques in In Vitro Fertilization and Embryo Culture
   (2nd: 1985: University of Wisconsin — Extension)
   In vitro fertilization and embryo transfer.

   "Based on the Second Annual Workshop on Laboratory Techniques in In Vitro Fertilization and Embryo Culture, sponsored by University of Wisconsin — Extension,
August 20–24, 1985" — T.p. verso.
   Includes bibliographies and index.
   1. Fertilization in vitro, Human — Congresses. 2. Fertilization in vitro — Congresses.
3. Human embryo — Transplantation — Congresses. 4. Embryo transplantation — Congresses. I. Wolf, Don P. II. University of Wisconsin — Extension. III. Title. [DNLM:. 1.
Embryo Transfer — methods — congresses. 2. Fertilization in Vitro — methods — congresses. WQ 205 W9256 1985i]
RG135.W67 1988   618.1′78059   88-6025

ISBN-13:978-1-4612-8288-4   e-ISBN-13:978-1-4613-1005-1
DOI: 10.1007/978-1-4613-1005-1

Based on the Second Annual Workshop on Laboratory Techniques in In Vitro
Fertilization and Embryo Culture, sponsored by University of Wisconsin — Extension,
August 20–24, 1985

© 1988 Plenum Press, New York
Softcover reprint of the hardcover 1st edition 1988

A Division of Plenum Publishing Corporation
233 Spring Street, New York, N.Y. 10013

FOREWORD

    The use of human in vitro fertilization in the management of
infertility is the outgrowth of years of laboratory observations on
in vitro sperm-egg interaction. The editors of this work have
themselves contributed significantly to basic knowledge of the
mammalian fertilization process. The observations of Don Wolf on
sperm penetration, the block to polyspermy and, most recently, sperm
hyperactivation in the monkey and human, Gregory Kopf's elucidation
of the mechanisms of sperm activation during penetration and the
reciprocal dialogue between sperm and egg, and Barry Bavister's
definition of culture conditions and requirements necessary for in
vitro oocyte maturation, fertilization and development in model
mammalian systems including nonhuman primates have contributed
greatly to our understanding of the mammalian fertilization process.
Wolf, Kopf and Gerrity have enjoyed substantial interaction with
clinicians in Departments of Obstetrics and Gynecology and have been
directly involved with successful IVF programs. Both Wolf and Kopf
have served as research scientists in the Division of Reproductive
Biology at the University of Pennsylvania, which, for more than 22
years, has fostered co-mingling of clinically oriented and basic
science faculty. It is through such interaction, which clearly
exists at many institutions including the University of Wisconsin,
that the process of technology transfer is best served.

    Without an exquisitely coordinated laboratory, there can be no
consistent success in human in vitro fertilization. Quality control
is pivotal, but close collaboration between the laboratory and the
clinic is also essential as information is shared and correlated.

    In vitro fertilization programs will continue to proliferate in
response to an ever increasing need. Initially, the technique was
utilized among patients with occluded or damaged fallopian tubes, but
indications have now been expanded to include a number of other
conditions as well as for diagnostic purposes. Continuing surveillance of the success rate in these subcategories is essential if we
are to counsel our patients properly. In this, as in perhaps no
other field in recent memory, there is an opportunity to bring great
happiness. At the same time, there is the possibility that informa-

tion will be misinterpreted and that patients will be misled, sometimes with the best of intentions. The stakes are high and the all-
too-human tendency to tell patients what they would like to hear
rather than what they need to know must be kept in mind and
modulated.

Time has witnessed a gradual and continuous improvement in the
pregnancy rate following in vitro fertilization and embryo transfer.
There is no doubt that this improvement has been significantly
influenced by the availability of additional knowledge of the fertilization process as it is transferred from the research laboratory to
the clinic. Predictably, there will be further improvements in
overall results. Although this volume provides state-of-the-art
information at this given point in time, the principles which are
addressed will certainly be valid into the future. One or another
technique may be modified in time, but the approaches presented
herein will continue to serve as guideposts in the practice of the
art of in vitro fertilization and embryo transfer.

This practical compendium will be useful to the investigator
concerned with the basic cell biology of fertilization and to the
clinician who is called upon to manage infertility. It will be
especially valuable to those scientists whose responsibility it is to
manage the practical aspects of an in vitro fertilization laboratory.

Luigi Mastroianni, Jr.

ACKNOWLEDGMENTS

This endeavor emanated from workshops on in vitro fertilization and embryo culture techniques offered at the University of Wisconsin-Madison in 1984 and 1985. Dr. Bavister assumed primary responsibility for workshop organization and syllabus preparation in 1984 while Dr. Gerrity assumed this responsibility in 1985. The original workshop manual on which this book is partly based was compiled by Dr. Bavister with the help of Dorothy Boatman. As faculty for these workshops, we would like to express appreciation to workshop participants who, by their participation, expressed confidence in our ability to explain the intricacies of human in vitro fertilization and embryo transfer. Presentation of the first workshop in 1984 was assisted by the unstinting efforts of Dorothy Boatman and Patricia Morgan. Additionally, we are indebted to the following for providing facilities, supplies, equipment and support: Dean Barney Easterday, Assistant Dean Susan Hyland and members of the staff at the School of Veterinary Medicine, University of Wisconsin-Madison; Dr. Thomas Meyer and Ann Bailey, Department of Continuing Medical Education, University of Wisconsin-Extension, Madison, Wisconsin; the Nalge Company; Millipore Corporation; Becton Dickinson; Frank E. Fryer Company; Forma Scientific; Nikon Instrument Division, Nikon, Inc.; Bel-Art Products; Wescor, Inc.; and CryoMed.

A special note of thanks is extended to Drs. William Byrd, Frank Kuzan and Pat Quinn who graciously accepted the challenge, on short notice, of treating subjects not adequately covered in our 1985 workshop.

Appreciation is expressed to Gail Alexander for her secretarial support during Dr. Wolf's tenure at the University of Texas, Houston. Upon moving to Oregon, continuation of this effort was catalyzed very substantially by the participation of Patsy Kimzey who provided not only excellent secretarial support but also valuable editing and proofreading skills. Patsy also took on the responsibility of scanning manuscripts prepared with different word processing systems in order to prepare a camera-ready copy from the system available at the Oregon Regional Primate Research Center.

Dr. Kopf wishes to thank Dr. Luigi Mastroianni, Jr., for his support during his tenure in the in vitro fertilization program at the Hospital of the University of Pennsylvania. Appreciation is also expressed to Ms. Barbara Kucyzinski-Brown and Ms. Keely Harris for their excellent technical assistance, and to the other members of the IVF program at the University of Pennsylvania for their support.

Finally a note of appreciation goes to our colleagues, families and friends for their tolerance and understanding during the preparation of this volume.

                                        Don P. Wolf
                                        Barry D. Bavister
                                        Marybeth Gerrity
                                        Gregory S. Kopf
                                        January 1988

CONTENTS

HUMAN IVF-ET: STATE OF THE ART 1987

Don P. Wolf

1.   INTRODUCTION

Since the birth of baby Louise Brown in 1978, human IVF-ET has evolved to become standard medical practice for the treatment of many types of human infertility. As a correlate to the impressive, if not explosive, growth in the application of IVF-ET procedures in human medicine (some 150 active clinics presently accepting patients in the United States alone), a dearth of trained, experienced mammalian embryologists has occurred. This outcome has both positive and negative connotations. On the plus side of the ledger, those of us who have experience in mammalian embryology, and human IVF specifically, find ourselves in the enviable position of being actively recruited and supported by clinical departments. While the provision of a clinical service by research scientists may at first glance appear to detract from research efforts, the advantages in associating with clinical colleagues and infertility patients and the availability of clinical material (sperm, zonae pellucidae, granulosa cells, follicular fluid, etc.), to my mind, far outweighs the sacrifice. On the negative side of the ledger, the rapid development and application of IVF-ET technology has forced many programs to employ cell biologists or embryologists with limited credentials and/or experience. Herein lies the rationale for the present effort, namely, to increase reference material relevant to the techniques

involved in the conduction of human IVF-ET. This effort emanates
from a workshop dealing with these techniques offered by the co-
editors in 1984 and 1985 at the University of Wisconsin in Madison,
Wisconsin. During that workshop, a laboratory situation was simu-
lated for approximately 30 participants who recovered and cultured
mouse embryos and zona-free hamster eggs. Such "hands-on" experiences were augmented with lectures by the faculty as well as by
invited speakers who were asked to treat various related subjects on
a more theoretical level. Recently, similar workshops have become
available through other organizations such as that sponsored by the
American Fertility Society and the Howard and Georgianna Jones
Institute for Reproductive Medicine in Norfolk, Virginia.

In vitro fertilization and embryo transfer as a treatment
modality for the infertile patient has evolved from its original
application to the hopelessly infertile patient, i.e., irreparably
damaged or absent fallopian tubes, to a relatively broad spectrum of
infertility patients including those with endometriosis, cervical
factor, male factor, immunologic, and idiopathic infertility. The
business base for IVF-ET, initially associated with relatively small,
academic programs, has also been expanded and commercialized as
exemplified by Bourn Hall in England and in this country by IVF
Australia and IVF Care Incorporated. Recent successes in the development of the technology per se include the frozen storage of both
eggs and embryos (Chapter 16), exogenous hormone support of pregnancy
in patients without ovarian function, the GIFT (gamete intrafallopian
transfer) procedure (Chapter 14), and oocyte or embryo donation
involving surrogate mothers. Many of these developments raise
ethical issues which have been dealt with in detail by several
authors (Grobstein and Flower, 1985; Andrews, 1986).

2.   THE ART: 1987

One of the objectives of this introductory chapter is to outline
state of the art, established procedures associated with the current
practice of IVF-ET and to mention briefly some of the latest developments.

While initial success in human IVF-ET was achieved with natural
cycles, ovarian stimulation has now become routine (Chapter 19).
This most often takes the form of administering clomiphene citrate
followed by human chorionic gonadotropin (hCG), clomiphene citrate in
combination with human menopausal gonadotropin followed by hCG, or
human menopausal gonadotropin followed by hCG. Stimulation of follicular growth is commonly initiated on day 2 or 3 of the cycle and
results in the recruitment of at least 4 or 5 mature follicles.
Follicular growth itself is monitored by measurements of follicular
function (circulating estradiol levels; Chapter 15) and size/location
by direct ultrasonographic imaging. When characteristic predeter-

mined criterion of follicular growth have been achieved, the patient
receives a bolus injection of hCG and egg pickup is scheduled at a
fixed time interval thereafter. Two keys to the clinical success of
an IVF-ET program are the synchrony with which the cohort of developing follicles are recruited and the timing of hCG administration.
Early hCG administration undoubtedly results in the pickup of
immature, preovulatory oocytes while delayed administration is associated with a high incidence of endogenous LH surges. If an LH surge
is detected early enough, the treatment cycle can be salvaged by
shortening the interval from surge to surgical intervention. While
acceptable follicular recruitment is being achieved by the protocols
outlined here, experimentation is ongoing with purified preparations
of FSH alone or in combination with LH, GnRH agonists and antago-
nists, and with the use of birth control pills to synchronize
menstrual cycles. The latter can be used to program cycles which in
turn supports the efficient use of personnel and resources. To
further simplify the IVF work load, some clinics are utilizing only
ultrasound for monitoring follicular growth. While many of these
experimental protocols are aimed at increasing the size of the cohort
of recruited follicles yielding fertilizable eggs, a maximum in this
regard has not been defined; many believe that this value is relatively small, perhaps on the order of 10-15.

Over the years, the procedures utilized for oocyte pickup have
expanded from laparoscopic aspiration to include ultrasound-guided
techniques (Wikland and Hamberger, 1984). The latter, while originally developed for patients with inaccessible ovaries, also carries
the advantage of avoiding exposure to general anesthesia. Initially,
the ultrasound-guided techniques involved transabdominal, transves-
ical approaches but more recently transurethral and transvaginal
procedures have been utilized successfully. High oocyte recovery and
pregnancy rates have been achieved by these techniques such that,
combined with cost effectiveness and patient preference, ultrasound-
guided pickups on an outpatient basis may prove optimal in the
future.

Fertilization rates for human eggs retrieved by either a laparoscopic or ultrasound-guided pickup are usually high (on the order of
80%) in patients whose infertility is not associated with a diagnosed
male factor. Because fertilization and early preimplantation development in vitro is readily achieved, the conditions presently
employed to support these events, which are now often complex, will
undoubtedly be simplified. As an example, efforts are being made to
utilize simple, relatively well-defined media containing albumin in
place of serum (Menezo et al., 1984; Feichtinger et al., 1986;
Chapter 4). The culturing of preimplantation stage embryos to blastocysts for cryopreservation purposes has allowed further definition
of culture factors that are critical to preimplantation development
in humans as well as in other mammalian species (see Chapter 18).

One of the most important and perhaps least-studied areas of human reproductive physiology relevant to IVF-ET is implantation. Given the relatively high frequency with which embryos are transferred to patients without the establishment of clinical pregnancies, the conclusion that pregnancy failure reflects abnormalities or inadequacies in the implantation process is inescapable. On the one hand, not all transferred embryos are normal; it is well established that pregnancy rates increase with increasing numbers of embryos transferred and that not all fertilizable oocytes are capable of extensive in vitro development. On the other hand, pregnancy rates appear to level off when three or more embryos are transferred suggesting that despite the presence of high quality embryos in the uterus, if the endometrium is asynchronous or abnormal, then pregnancy will not ensue. Ideally, independent measurements of both embryo viability and uterine receptivity are needed. Statistical evaluations have been applied to define the relative contribution of these two components (Speirs et al., 1983; Walters et al., 1985; Rogers et al., 1986). Fortunately, attention is beginning to focus on the identification of noninvasive markers for evaluating preimplantation stage embryo quality (O'Neill, 1985; O'Neill et al., 1985a,b, 1987). Perhaps even more significant, embryo-induced maternal thrombocytopenia may provide a marker for impending implantation (Gidley-Baird et al., 1986). Additionally, an indirect measure of uterine receptivity can be attained by ultrasound measurements of endometrial thickness during the luteal phase. The contribution of endocrine parameters to these critical events has yet to be defined. Supplementation of the luteal phase with either progesterone or hCG, while commonplace, has not been established as efficacious by rigorous scientific experimentation. Efforts to sort out the relative contribution of embryo viability and uterine receptivity to the process of implantation may ultimately focus on nonhuman primates where transfers of single embryos to synchronized recipients can be undertaken. Alternatively, in the context of an IVF-ET program, the relative contribution of these parameters may be defined by the transfer of cryopreserved embryos during natural cycles (Testart et al., 1987).

In the area of cryopreservation, the reader is referred to several recent reviews (Trounson, 1986a,b) as well as to Chapter 16 for a summary of this rapidly changing technology. Extensive databases are available on the cryopreservation of domestic animal embryos which usually involves late preimplantation stage embryos. While substantial success has already been achieved in man, future studies will undoubtedly optimize conditions for the controlled rate freezing of human embryos. For instance, while several different cryoprotectants have been utilized (DMSO, propanediol, glycerol) in the cryopreservation of early as well as late preimplantation stage human embryos, our database is inadequate to allow selection of the optimal conditions. More recently, the introduction of vitrification

and other rapid freezing techniques has catalyzed attempts to cryopreserve the unfertilized egg (Trounson et al., 1987).

In summary, while procedures for the successful conduction of
IVF-ET have been relatively well defined, the state of the art
changes rapidly where a competitive advantage is associated with
improved clinical pregnancy rates. The detailed presentation of
routine IVF-ET procedures as presented in this volume is designed
primarily for the newcomer to the field; however, it may be interesting to the veteran who hopefully can always benefit from a new
insight or two into this exciting and rapidly changing area of
endeavor.

3.   REFERENCES

Andrews, L. B., 1986, Legal and ethical aspects of new reproductive
    technologies, Clin. Obstet. Gynecol. 29:190-204.
Feichtinger, W., Kemeter, P., and Menezo, Y., 1986, The use of syn-
    thetic culture medium and patient serum for human in vitro
    fertilization and embryo replacement, J. In Vitro Fert. Embryo
    Transfer 3:87-92.
Gidley-Baird, A. A., O'Neill, C., Sinosich, M. J., Porter, R. N.,
    Pike, I. L., and Saunders, D. M., 1986, Failure of implantation
    in human in vitro fertilization and embryo transfer patients:
    the effects of altered progesterone/estrogen ratios in humans
    and mice, Fertil. Steril. 45:69-74.
Grobstein, C. and Flower, M., 1985, Current ethical issues in IVF,
    Clin. Obstet. Gynaecol. 12:877-891.
Menezo, Y., Testart, J., and Perrone, D., 1984, Serum is not neces-
    sary in human in vitro fertilization, early embryo culture, and
    transfer, Fertil. Steril. 42:750-755.
O'Neill, C., 1985, Thrombocytopenia is an initial maternal response
    to fertilization in mice, J. Reprod. Fertil. 73:559-566.
O'Neill, C., Gidley-Baird, A. A., Pike, I. L., Porter, R. N.,
    Sinosich, M. J., and Saunders, D. M., 1985a, Maternal blood
    platelet physiology and luteal-phase endocrinology as a means of
    monitoring pre- and postimplantation embryo viability following
    in vitro fertilization, J. In Vitro Fert. Embryo Transfer 2:87-
    93.
O'Neill, C., Pike, I. L., Porter, R. N., Gidley-Baird, A. A.,
    Sinosich, M. J., and Saunders, D. M., 1985b, Maternal recogni-
    tion of pregnancy prior to implantation: Methods for monitoring
    embryonic viability in vitro and in vivo, Ann. N.Y. Acad. Sci.
    442:429-439.
O'Neill, C., Gidley-Baird, A. A., Pike, I. L., and Saunders, D. M.,
    1987, Use of a bioassay for embryo-derived platelet-activating
    factor as a means of assessing quality and pregnancy potential
    of human embryos, Fertil. Steril. 47:969-975.

Rogers, P. A. W., Milne, B. J., and Trounson, A. O., 1986, A model to
    show human uterine receptivity and embryo viability following
    ovarian stimulation for in vitro fertilization, J. In Vitro
    Fert. Embryo Transfer 3:93-98.
Speirs, A. L., Lopata, A., Gronow, M. J., Kellow, G. N., and
    Johnston, W. I. H., 1983, Analysis of the benefits and risks of
    multiple embryo transfer, Fertil. Steril. 39:468-471.
Testart, J., Lassalle, B., Forman, R., Gazengel, A., Belaisch-Allart,
    J., Hazout, A., Rainhorn, J.-D., and Frydman, R., 1987, Factors
    influencing the success rate of human embryo freezing in an in
    vitro fertilization and embryo transfer program, Fertil. Steril.
    48:107-112.
Trounson, A., 1986a, (Guest Editor) Symposium on Embryo Freezing, J.
    In Vitro Fert. Embryo Transfer 3:2-61.
Trounson, A., 1986b, Preservation of human eggs and embryos, Fertil.
    Steril. 46:1-12.
Trounson, A., Peura, A., and Kirby, C., 1987, Ultrarapid freezing: a
    new low-cost and effective method of embryo cryopreservation,
    Fertil. Steril. 48:843-850.
Walters, D. E., Edwards, R. G., and Meistrich, M. L., 1985, A statis-
    tical evaluation of implantation after replacing one or more
    human embryos, J. Reprod. Fertil. 74:557-563.
Wikland, M. and Hamberger, L., 1984, Ultrasound as a diagnostic and
    operative tool for in vitro fertilization and embryo replacement
    (IVF/ER) programs, J. In Vitro Fert. Embryo Transfer 1:213-216.

2

SELECTION AND USE OF EQUIPMENT

Marybeth Gerrity

1.   INTRODUCTION

The choice of equipment for a human IVF laboratory will depend upon the unique situation at each institution. Some equipment items are absolute requirements while others make the process more convenient or reproducible. Finally, there are pieces of required equipment for which shared access may be acceptable. It is probably not an exaggeration to say that equipment selection may influence the outcome of the IVF procedure. This is not to say that one brand of equipment is superior to another and it is not an intention to endorse a particular brand or manufacturer of equipment. The most important variables in equipment selection are dependability, reproducibility, and service. Hence a final decision should be based largely on the technical service and sales representative that services the section of the country in which your laboratory is located. This is particularly true in the selection of microscopes and incubators. This chapter describes different types of available equipment and indicates manufacturer's product numbers. Table I summarizes a list of equipment that is required for human IVF laboratories with an approximate cost and a comment about use. Notice that this table is broken down into sections: equipment that is required and should be purchased; equipment that is required but shared access may be acceptable; and optional equipment that is a matter of convenience or personal preference.

2.   REQUIRED EQUIPMENT

2.1. Carbon Dioxide Incubator

An incubator which provides controlled temperature and humidity is an essential feature of the IVF laboratory. All protocols currently used employ a bicarbonate/CO_2 buffered tissue culture medium requiring a constant CO_2 environment for pH control. Humidity is also a crucial component and should be maintained at about 99%. Control of the O_2 concentration is more controversial with some groups favoring the low (5%) O_2 environment and others making use of ambient (20%) O_2 conditions. There are three main types of CO_2 incubators.

A continuous flow gas incubator (for example, Forma Model 3137) is the most common type of incubator that can be ordered directly from a distributor's catalog. Flow meters control the amount of entering CO_2 and as the name implies there is a continuous inflow of CO_2. Most of these incubators are water jacketed for temperature control. There are two main disadvantages to this incubator type. The continuous gas flow has a desiccating effect on media so in most cases it is necessary to bubble the gas through water. This bubbling of gas is messy and enhances the probability of contamination through production of aerosols. The second disadvantage is the relatively large consumption of CO_2. While CO_2 is not expensive, it is inconvenient to change the gas tank as frequently as every two or three days. Safeguards should be developed to avoid gas tanks running dry when unattended. Several manufacturers produce devices which switch tanks automatically in this situation.

The type of incubator most commonly used in IVF programs is an automatic injection CO_2 incubator (for example, Forma Model 3158). A microprocessor and a CO_2 sensor controls the injection of CO_2 maintaining the carbon dioxide concentration within specified limits. These incubators can only regulate one gas, usually CO_2. The oxygen level will be at ambient concentration, i.e., 21%. Care must be exercised with these types of incubators because the CO_2 sensor is extremely sensitive to temperature and humidity. Large numbers of door openings which cause fluctuations in temperature and humidity also affect the accuracy of the CO_2 sensor. These incubators must be calibrated in the absence of CO_2, i.e., when the injection system is "zeroed." If the incubator is zeroed with, for example, 5% CO_2 in the chamber and then the sensor is set at 5%, a CO_2 concentration of 5% above the baseline or 10% would result. The CO_2 injection incubator is the type most widely used in IVF programs. It is simple, reliable, and a single tank of carbon dioxide lasts approximately two months.

The third type of incubator used in IVF programs is the gas processing or triple gas incubator (for example, Forma Model 3187). With this incubator three gases can be controlled, e.g., 5% CO_2, 5% O_2, 90% N_2. Generally, ambient air serves as the O_2 source and the incubator is attached to external nitrogen and carbon dioxide tanks. Since there does not appear to be a substantial advantage associated with the use of the triple gas mixture, the popularity of this incubator type has declined. A disadvantage to these triple gas incubators is that they do not contain a sensor. The injection of gas is a timed function and anything that interferes with this timing results in an alteration of gas concentration in the chamber.

In general, whatever type of incubator is used, a stacked double incubator is recommended. It permits the isolation of embryos from different patients or allows the embryologist to use one incubator for embryos and the other for functions that require frequent door openings like sperm preparation. It is important to remember in choosing an incubator that technical services should be available since setup and service is crucial to successful operation. It is inadvisable, therefore, to purchase an incubator out of a catalog. Daily calibration of the incubator temperature and CO_2 concentration should be an essential part of a quality control program in an IVF lab. The digital displays on these incubators should not be accepted as accurate. Temperature should be verified with a thermometer kept in the incubator to make sure the digital display actually represents the chamber temperature. Standard laboratory thermometers should be

Table I. List of equipment that is necessary for an in vitro fertili-
          zation laboratory

Item	Approximate cost	Comment
A) Items that are required		
Incubator	$2,700	See incubator comments section.
Gas regulator	$150 ea.	Required on gas tanks attached to incubator. Type purchased depends on type of gas used by the incubator.
Gas cylinders	$10/month	Incubator that regulates at 5% CO_2 in air will use only one type (CO_2).
Microscope, dissecting	$2,000	For egg retrieval and examination of eggs/embryos. Bottom illumination preferred.
Microscope, compound	$3,000 (with phase)	For sperm preparation, semen analysis and hamster tests.
Microscope, inverted with camera attached	$7,000	For detailed examination of eggs/embryos. Photographic records for medical/legal reason.
Water source	$10,000	See text.
Tissue culture laminar flow hood (horizontal flow)	$2,000	See text.
Slide warmer	$250	For use in egg retrieval.
Water bath	$250	For heat inactivating serum.
Magnetic stirrer	$250	For media preparation.
Centrifuge	$500	Table top/nonrefrigerated. For sperm and serum preparation.
Refrigerator/freezer	$500	Must have manual defrost in freezer cycle.
Variable volume pipets	$159.50 ea.	Accurate measurement of small volumes.
Heating block	$350	All types of warming applications.
Aspiration/embryo transfer devices	$300	See appropriate comments section.
Mice (male & female)	$200/month	Needed for media testing.

Table I. List of equipment that is necessary for an in vitro fertili-
          zation laboratory (continued)

B) Equipment required for which shared access may be acceptable

Analytical balance	Required for media preparation.
Osmometer	Required for media preparation.
Drying oven for sterilization	Embryo transfer catheter heat sterilization; glassware preparation.
Autoclave	Instrument sterilization
pH meter (small electrode)	Quality control and media preparation.
Facilities for gas sterilization	Instruments and embryo transfer catheters
Animal care facilities	Housing of mice required for quality control.

C) Optional equipment and miscellaneous

Wraparound incubator for inverted microscope	$3,500	Maintain temperature while observing embryos.
Low temperature freezer	$2,000	Long-term storage of sera, gonadotropins, etc.

D) Miscellaneous items

In addition to those specific suggestions above, the program will also need:

1) Sterilization supplies: sterilization packets, indicator and sealer for packets.

2) Vacuum source or separate vacuum pump with trap; source of filtered air (these can be accommodated in a pump such as the Little Grant Pump Model 0211-V45N-G8CX [Gast Manufacturing Corporation]; filters available [Forma Scientific, Marietta, OH, catalog #770001]).

3) Tissue culture detergent (Flow Laboratories, McLean, VA, catalog #76-670-94).

4) Routine laboratory equipment such as glassware (especially volumetric flasks, funnels, beakers), microscope slides and coverslips, bunsen burner (or alcohol lamp), plastic tubing, weighing paper, spatulas, dish pans, sanitizing agent (Chlorox), gases, chemicals, salts, stains, equipment for routine semen analysis (hemocytometer), pH and osmometer standards, etc. are necessary.

calibrated against a National Bureau of Standards (NBS) thermometer
on a regular basis. Daily CO_2 measurements using a Fyrite should be
part of the laboratory quality control system (see Chapter 3).

2.2. Microscopes

There are three types of microscopes needed in an IVF
program.

The workhorse of the laboratory is the dissection microscope. A
dissection microscope should have illumination from underneath the
microscope stage. The microscope should have an adjustable mirror to
allow alterations in specimen illumination and a large stage so that
a petri dish cannot readily slide off. The magnification range for a
typical dissecting microscope with 10X eyepieces is 8-40X. Some
manufacturers make available an optional 2X magnifying lens (e.g.,
Bausch and Lomb) to further increase magnification; these do,
however, decrease the working distance. Some groups prefer the use
of dark field illumination for evaluating and manipulating embryos.

A compound microscope is necessary for sperm preparations. If
finances permit, phase contrast optics are desirable because they
allow for varied applications in the laboratory, critical evaluation
of sperm preparations, scoring of sperm penetration assays and the
use of differential stains for evaluating morphology.

An inverted microscope permits visualization of embryos in organ
culture dishes during the fertilization and embryo culture steps.
Look for one with a mechanical stage, a 5X or 10X search lens, and
20X and 40X objectivés. The Nikon Diaphot is a typical example. It
comes with an optional wraparound plexiglas incubator that permits
temperature control. A desirable feature in these microscopes is
through-the-lens photography using 35 mm or polaroid cameras or
videomicroscopy. Records can be made of all eggs and embryos for
teaching or data collection as well as medical/legal purposes.

2.3. Laminar Flow Hood

There is a wide division of opinion on the use of laminar
flow hoods in IVF laboratories. Such hoods are widely used in conventional tissue culture laboratories. In IVF where many procedures
are done microscopically, a laminar flow hood may be superfluous
since a microscope or any other apparatus placed in the hood interferes with air flow. Vibration caused by the circulating fan in the
hood may also interfere with microscopic visualization of specimens.
For this reason, many IVF programs shut off the laminar flow hood
(therefore defeating its purpose) during examination of eggs/embryos.
If the only purpose for the laminar flow hood is to provide a clean
work surface, perhaps money can be more wisely spent on a larger
laboratory bench. On the other hand, if the IVF lab is a high

traffic area, this may be the only measure that will prevent contamination problems. In IVF, where manipulations are rapid, hood use may be more harmful than beneficial since the air flow may result in media desiccation or temperature reduction. In at least one IVF laboratory that I visited, the laminar flow hood was placed directly opposite the incubator. Each time the incubator was opened, the air flow from the tissue culture hood removed all the humidified air from the incubator resulting in a dramatic change in osmolarity of the media contained within the incubator. If a laminar flow hood is used, it should not be placed near the incubator. Also be aware that tissue culture hoods vented to the outside may have disastrous implications for temperature control within the laboratory.

Tissue culture hoods often contain ultraviolet (UV) lights to decontaminate the work surface. These lights should not be left on when laboratory personnel are in the area or when embryos are being manipulated. UV light has disastrous consequences for embryos as well as for the eyes of laboratory personnel. Where UV lights are widely used for decontamination purposes, it is important to protect all exposed plastic surfaces since it has been shown that UV causes cracking, yellowing and premature aging of plastics; this is particularly important if your microscope has a wraparound incubator. All plastic equipment should be covered in thick black fabric while the lights are in use.

Some laboratories have made use of a tissue culture enclosure. These are fiberglass boxes for media preparation placed directly on the bench top. They contain an UV light which can be turned off when the box is in use. There is no control of air flow through these boxes and they cannot be used for manipulation of embryos that require a microscope without significant modification.

In summary, it is unfortunate that laminar flow hoods have become a standard of practice in IVF laboratories because their usefulness is limited and they are expensive. In my experience, they are more often used as a prop for inspection purposes or as a clean work surface than for their actual laminar flow properties. The horizontal flow laminar flow hood which is most commonly used in IVF programs prevents contamination of culture media while the flow is in use. It does not provide user protection since the air flow is across the tissue culture media and at the operator. This practice may have infection control implications; see Chapter 3.

2.4. Slide Warmer

There are several other pieces of laboratory equipment which are necessary for IVF (Table I). A slide warmer is used for petri dishes of follicular fluid during the egg retrieval process. These are available in several sizes. It is usually most convenient to buy one without a hinged cover.

2.5. Water Bath

A water bath is necessary for warming media (at times) and also for heat inactivation of serum. The water bath should be adjustable to both 37° and 56°C. Choose one that will not be damaged if the water inadvertently boils away.

2.6. Magnetic Stirrer

A magnetic stirrer is needed for tissue culture media preparation.

2.7. Centrifuge

A centrifuge is needed for sperm preparation. This should be a table top model and need not be refrigerated. The choice of a rotor will depend upon the anticipated volume of the laboratory. A six- or eight-place rotor may be acceptable if you will only be doing one sperm preparation at a time. It is best to choose a centrifuge that has a built-in tachometer. As centrifuges age, they do not run as true to speed. It is much easier to monitor the speed of a centrifuge with a built-in tachometer so that it can be adjusted for each run. Centrifuges with built-in tachometers are more expensive than those without and checking the speed of the centrifuge with an external tachometer periodically may suffice. Most sperm preparations are centrifuged at 300-500x g.

2.8. Refrigerator/Freezer

The selection of a refrigerator and deep freezer is dictated by availability. A simple domestic refrigerator is adequate for storing media and laboratory supplies, etc. The freezer section is sufficiently cold (usually -20°C) to permit storing noncritical items. If you do not have access to a -70°C freezer, select a refrigerator-freezer that has a manual defrost cycle. While regular manual defrosting is a nuisance, the automatic defrost on all refrigerators has a built-in warming cycle. Repeated sample exposure to such freeze-thaw cycles is undesirable. If finances permit, the purchase of an upright or chest deep (-70°C) freezer may be useful. Freezing of serum samples at -70°C allows stockpiling of serum for long periods of time for use as needed.

2.9. Glassware

Glassware used for IVF should be borosilicate glass, not glass made from soda lime. Pieces of glassware that are most frequently used in IVF programs are one liter volumetric flasks, miscellaneous size beakers and graduated cyclinders. Avoid the use of tap water in this glassware since mineral deposits can result in deterioration of water quality. Heat sterilize glassware wherever possible. Avoid autoclaving glassware because it leaves mineral deposits.

2.10. Analytical Balance

Items for which shared access may be acceptable are listed in Table I-B. The majority of these items are required for media preparation. An analytical balance is needed for weighing out the different components of the tissue culture media. If the only purpose of the laboratory is performing IVF and the volume is low, it may be acceptable to share a balance. However, proper judgement should be exercised when sharing with a laboratory that makes use of carcinogens or toxic chemicals.

2.11. Osmometer

There are two main types of osmometers available: freezing point depression and vapor pressure models. Freezing point depression osmometers require a large sample size (0.2-2.0 ml) and are widely used in IVF where there is little restriction on sample size. Vapor pressure osmometers are more widely used in research applications where sample size is limited. If the only purpose of the program is to perform IVF and the program is in a hospital setting, explore the possibility of sharing an existing osmometer. For example, clinical laboratories and dialysis units will probably have an osmometer. Most often these osmometers are calibrated frequently and maintenance programs are in place.

2.12. pH Meter

A pH meter is also required for quality control in media preparation. If you have shared access in a clean laboratory, consider buying a microelectrode. Such an electrode is small in diameter and can be inserted into a tissue culture tube or petri dish so that direct pH measurements can be made. Portable pH meters that can be hand-held are also useful for testing media in dishes while still in the incubator.

2.13. Sterilization Facilities

Facilities for autoclaving, gas and heat sterilization are also needed in many phases of IVF, i.e., sterilization of instruments, preparation of embryo transfer catheters and aspiration devices, and glassware washing.

2.14. Heating Blocks

Heating blocks are useful for a variety of applications in the IVF laboratory. These consist of a heat source into which

predrilled blocks that act as racks are placed. The heating elements
may be adjusted to any temperature. Predrilled holes in the blocks
are available in sizes to accommodate a range of test tube diameters.
Therefore, one can use them to hold the tubes of follicular fluid at
the time of aspiration (sizes are available to accommodate the 17-mm
culture tube or the larger diameter DeLee trap, see Table II). These
blocks can be used for warming fluids, thawing sera and a variety of
other uses.

2.15. Pipets

Reproducible measurement of fluid volumes is an important
function in an IVF laboratory. Large volumes of fluid can be readily
aliquoted using sterile, disposable serological pipets. Mouth
pipetting is not acceptable in the IVF laboratory. Falcon and others
manufacture electric pipetting aids for attaching to the pipet. Bel-
Art makes a simple, inexpensive plastic mechanical PiPump that is
very useful and comes in different sizes for each pipet diameter.
These are especially useful when attached to a 5 3/4-inch Pasteur
pipet for manipulating eggs and embryos. Small volume pipetting is
necessary for insemination of oocytes, sperm preparation and
dilution. Adjustable volume pipets with sterile disposable tips such
as those manufactured by the Gilson Corporation are useful. These
permit accurate measurement of a range of small volumes.

2.16. Aspiration Devices

Aspiration needles for ovum retrieval are available from a
variety of sources (Table II). The choice of a needle depends upon
personal preference and method of oocyte retrieval. Needles used for
laparoscopic egg retrieval are generally larger gauge (14 or 15
gauge) and shorter than those used for ultrasound-guided egg
retrieval. Ultrasound needles are etched at the tip for easy
visualization. Some needles have adapters that permit use of the
needle with a DeLee trap; others are fitted with a silicone rubber
bung that fits into a 17 x 100 (Falcon 2001) culture tube. Aspiration devices must be attached to a vacuum source. Free-standing
aspiration units are available from several manufacturers including
Rocket of London and Forma Scientific. Standard wall suction available in the operating room can be used effectively by attaching a
regulator so that pressure can be adjusted. A foot pedal is
available from Storz instruments (Table II) that can be used with
wall suction.

2.17. Embryo Transfer Catheters

Available from several sources, embryo transfer catheters
are discussed in detail in Chapter 11. A summary is contained in
Table III.

Table II. Aspiration needles

Aspiration needles are available from several sources.

A) Laparoscopic egg retrieval

    1) Resnik Instrument, Inc.
       7308 N. Monticello
       Skokie, IL 60076
       Phone (312) 673-3444

       In vitro fertilization needles catalog #6400 (for use in single puncture) or
       #6640 (for use in double puncture). This needle does not include a trocar which
       should be ordered separately. These needles attach to DeLee trap (Argyle 8888-
       257527; Trap size 20 cc; Catheter size 10 Fr).

    2) Cook-Australia - variety of IVF needles and needles for ultrasound-guided
       retrievals. All of these make use of tubing running through a rubber bung. The
       rubber bung must be inserted into a sterile 17 x 100 culture tube (Falcon 2001).

    3) Horace Barnaby
       Carnegie Building, Rm. 865
       601 N. Broadway
       Baltimore, MD 21205
       Phone (301) 955-3590

       These are 14-gauge, 25-cm needles which can be attached to a DeLee trap (see
       above). Trocars should be ordered separately. Allow three months for delivery
       at minimum.

B) Ultrasound-guided egg retrieval

    1) Cook Ob/Gyn
       925 South Curry Pike
       P.O. Box 489
       Bloomington, IN 47402
       Phone (800) 541-5591

       Product number K-OPS-15-30-U. (This needle is 15-gauge, 30-cm long, but is
       available in other sizes.)

C) Foot pedal for aspiration

    Available from Storz Instruments (Orthopedic Division), Product Number - N 1691-80.

    This should be attached to wall suction that can be regulated to 100 mm Hg. An
    additional standard gauge should be available to check the wall suction at the level
    of the foot pedal (usually this is the same model regulator used at the wall).

2.18. Mice

Mice are required for testing of media as part of the
quality control program; see Chapter 5 for more information on
testing of media. Animals should be housed at a distance from the
IVF lab; both males and females are required.

Table III. Embryo transfer catheters

Monash Embryo Transfer Catheter   METS-2 (side opening)

1)   Cook Australia
     925 South Cury Pike
     P.O. Box 489
     Bloomington, IN 47402
     Phone (812) 329-2235

Also available as end opening catheter (METS-1). These include an
introducer.

2)   Some investigators also like to use Tom Cat catheters. These
     have no introducer, but are disposable.

     Central Veterinary Supply
     One Central Park
     Westminster, MA
     Phone (617) 345-6166
     Catalog #491941 (5 1/2 inch Tom Cat catheter)

3)   Syringes used for embryo transfer can be disposable 1-cc tuber-
     culin syringes. Some prefer to use Hamilton Syringes because of
     their reproducible volume for delivery (Hamilton #80920).

2.19. Optional Equipment

        Some optional equipment is listed in Table I-C. A wrap-
around incubator is available for use with Nikon inverted microscopes
(Diaphot). This adaptation allows maintenance of temperature while
observing the embryos and isolates them (physically) from the
observer. The heat source is a dry one so that media desiccation may
be a problem unless observations are rapid.   pH control is not
provided by this incubator.

        A low temperature (-70°C) freezer for long-term storage of sera
and gonadotropins is useful if finances permit or if shared access is
not possible.

2.20. Miscellaneous Equipment and Supplies

        Miscellaneous equipment and supplies are summarized in
Table I-D. Sterilization supplies for packaging Pasteur pipets and
instruments are necessary. These should be peel packs with appropriate indicators (gas or steam).

        A vacuum source for filtration of media is necessary (see also
Chapter 4 on media preparation). Nalgene Corporation makes a hand-
held pump for use with large volume filtration units. A mechanical

vacuum pump like the one listed in Table I-D provides a source of both vacuum and filtered air. Vacuum can be used for filtration while filtered air can be used for drying Pasteur pipets prior to sterilization.

Tissue culture detergent is necessary for glassware washing. Finally, laboratory equipment such as that listed in Table I-D is required as well as supplies for routine semen analysis and the appropriate standards for pH and osmolarity determinations.

2.21. Plasticware and Disposable Glassware

The disposable plastic products suggested for use in an IVF laboratory are itemized in Table IV. The product number from Falcon Plastics is listed next to the appropriate item; these products, available from many distributors, are all tissue culture grade and are designed specifically to enhance cell growth. However, testing the embryotoxicity of plasticware should be part of the quality control program (see Chapters 3 and 5). Filtration devices are required for media sterilization. Small volume filtration devices are useful for sterilization of sera. Since mouth pipetting is not acceptable in an IVF laboratory, pipetting aids are necessary. Borosilicate glass Pasteur pipets are used for embryo manipulation (5 3/4 inch) and for sperm preparations (9 inch). These should be rinsed with ultrapure water, blown dry with filtered air, packaged in sterilization pouches and sterilized by heat or autoclave. Disposable needles and syringes are needed for animal injections and the mouse embryo bioassay.

2.22. Chemicals

Chemicals and powdered media used in the IVF laboratory are summarized in Table V. See Chapter 4 for a detailed discussion of media preparation.

3.   ALTERNATIVE EQUIPMENT

In some settings, the IVF embryo lab may not be adjacent to the operating room. Gerrity and Shapiro (1985) described a mobile laboratory cart which can be used in this situation. This cart consists of a gassed incubator, dissecting microscope, slide warmer and all the equipment generally used for egg retrieval. The cart can be pushed directly into the operating room and used for egg collection or embryo transfer. It also permits greater flexibility for the site of egg retrieval. For example, as ultrasound-guided egg retrievals become popular, IVF labs adjacent to the operating room may become obsolete. Of course, there is no substitute for rigorous attention to quality control or a fully-equipped IVF laboratory.

Table IV.   Disposable plastic products suggested for IVF laboratory.
             (Listed by manufacturer number; can be ordered through
             local distributor such as American Scientific Products,
             Fisher Scientific, VWR Scientific)

Manufacturer No.	Description	# Start Up	Unit
Falcon 3002	60-mm tissue culture dish	2	case
3001	30-mm tissue culture dish	1	case
2001	17 x 100 culture tubes	1	case
2003	12 x 75 culture tubes	2	case
2095	17 x 100 conical tubes	1	case
3037	Organ culture dish	1	case
3013	50-ml culture flask	1	case
7566	Pipet aid with filter	1	each
Nalgene 450-0020	500-ml filtration flasks	4	case
Millipore SLGV 025LS	Millex GV, small volume filter	4	box
Any manufacturer, but must be in peel pack (no plastic sleeves)	Sterile disposable TD individually wrapped pipets (tissue culture grade)		
	1 ml	1	case
	5 ml	1	case
	10 ml	1	case
Bel-Art	Pipet pump GREEN	5	each
	BLUE	5	each
Any manufacturer	5 3/4 inch and 9 inch Pasteur pipets (borosilicate glass)	1 1	case case
Becton Dickinson	Disposable needles		
	25 gauge x 5/8 or 1 inch	2	box
	30 gauge x 1/2 inch	2	box
	1 cc syringes	2	box/100

     Testart et al. (1982) described an effective modification of a
pediatric isolette (similar to the one used in neonatal intensive
care units) for use in IVF. This approach has been further modified
recently by Chetkowski et al. (1985) permitting observation and
manipulation of oocytes and embryos in a CO_2 gassed and humidified
environment. This gassed isolette is portable and permits more
extensive teaching functions to be accomplished than when embryos are
exposed to room air in a laboratory environment. This group has
emphasized the importance of minimizing embryo exposure to changes in
pH or temperature. Attention to these variables is reflected in an
excellent pregnancy rate in their program.

Table V.   Chemicals and reagents needed for IVF

Manufacturer name and number	Description	Number to start	Units
Gibco (Grand Island, NY)			
430-1200	Nutrient medium Ham's F-10	10	1 liter pkg.
450-1300	Dulbecco's phosphate buffered saline (1 liter packets only; no bottles or larger packets)	10	1 liter pkg.
	Calcium lactate (ACS grade)		
	Sodium bicarbonate (ACS grade)		
Sigma PEN-NA S6501	Penicillin G, sodium salt Streptomycin sulfate	25 5	million units grams

4.   WATER

The water system used in IVF may be the single most important
variable. Water is, of course, the largest component of tissue
culture media. The optimal water system for IVF should include three
separate processes: demineralization and deionization; reverse
osmosis; and finally, organic absorption. Prior to delivery or use,
such water, which is low in organic contaminants and dissolved
pyrogens, should be passed through a 0.22 μ microporous filter to
remove bacteria and particulates. Mather et al. (1986) demonstrated
that water prepared in this manner was significantly better at
supporting growth of cells in serum-free media than standard tap
water or distilled water. Addition of known quantities of organic or
inorganic contaminants (commonly found in tap water) to purified
water inhibited cell growth. Although water purity can best be
evaluated by reversed phase high performance liquid chromatography
(HPLC; see Reust and Meyer, 1982), the most important test of water
quality is its ability to support cellular development. Cell growth
as evaluated above involved somatic cells which grew in sheets on the
surface of the culture vessel requiring diffusion of contaminant into
the cell layers. Since oocytes and sperm are exposed on all sides to
medium, they may be even more sensitive to water impurities.

The method of water storage also appears to be critical. Gabler
et al. (1983) have demonstrated that water quality declines after

even 1 hr of storage. While the concentration of organic contaminants rises most rapidly when water is stored in plastic containers,
the effect is not eliminated by storage in glass. Therefore, since
water quality will decline on storage, it is desirable to have a
system where water can be produced on demand. If storage of water is
necessary, pure borosilicate glass containers with glass stoppers
should be employed. The water should be used within 24 hr and
plastic and rubber stoppers should not be used. Water storage in
large Nalgene plastic jugs defeats the entire purpose of water purification. Since the water purification system makes use of house
water as the starting point, variations in the water source prior to
purification may account for some of the reported regional differences in IVF success rates. Mather et al. (1986) have shown that tap
water quality varies with the season, that is, organic contamination
is highest in warmer months. Some IVF programs have successfully
used HPLC grade water. While this water may be chemically free of
organic contaminants, it is expensive (about $28 a gallon). Since,
as noted above, water quality declines with storage, it is important
to obtain this water from a supplier that has a rapid turnover of
stock. Manufacturers of HPLC grade water have different specifications (including acceptable pH ranges) and each batch may differ in
its ability to support embryo development. Variability in water
batches, therefore, may be just as significant with HPLC grade water.
Furthermore, because of its expense, it is impractical for use in
glassware cleaning and other routine laboratory applications. Is
there really any point in using ultrapure water with glassware that
has been cleaned with tap water?

     In my opinion, the water system represents one of the largest
expenditures for IVF but the one on which most programs skimp.
Simple glass distillation of water from the tap is probably inadequate (see Mather et al., 1986). As described in Chapter 5, one
should not be lulled into a false sense of security by the fact that
the water can support mouse embryo development. At least one group
has demonstrated that culture media prepared with sewage water can
support mouse embryo development (personal communication). This
experiment points up the extraordinary insensitivity of the mouse
embryo system to poor water quality.

     Water quality can be monitored by several methods including
HPLC. Bavister and Andrews (in press) have described a bioassay for
water quality which makes use of hamster sperm in a defined medium.
The College of American Pathologists, Commission on Laboratory
Inspection and Accreditation, has published a useful bulletin
entitled "Reagent Water Specifications" which describes practical
methods for monitoring water quality including tests for bacterial
contamination and pH fluctuations.

     Shared access of a water system may be feasible in certain
settings. Adding an Organex filter to a demineralized, deionized,

reversed-osmosis water system may be satisfactory. Again, as with
all steps in IVF, the quality control measures used are essential to
consistent water quality.

5.   LABORATORY DESIGN

     Since the procedures used in IVF require heavy reliance on
microscopic observation, it is desirable to design the laboratory
with a large number of kneeholes at desk height. Standard laboratory
height benches will be unwieldy for microscope use. Positive air
flow in the area is desirable; however, this must be filtered. There
shouldn't be through-the-wall air-conditioning units in an IVF lab;
this provides a major source of contamination. Heat and cooling
should be maintained within relatively narrow ranges, particularly
since the equipment used generates heat and many incubators will not
function correctly in extremes of heat or cold. The IVF lab should
be physically isolated and secure from other laboratory functions; it
is not acceptable to merely take over a corner of a regular laboratory for IVF use. The lab should have access to emergency power and
all incubators should be plugged into emergency power. Surge protec-
tors should be utilized for all equipment that is microprocessor
controlled (e.g., incubators). In situations where the laboratory is
adjacent to the operating room, a pass-through window from the
operating room to the laboratory is desirable. The IVF laboratory
should be a clean, low traffic area that can be easily secured. It
is probably desirable not to mark it with signs, etc. advertising its
purpose. Glassware washing activity should not be conducted within
the embryo culture lab if at all possible since it represents an
additional source of contamination. Animals should never be housed
or sacrificed in the IVF laboratory. Personnel who have extensive
animal contact should change laboratory coats or clothing and scrub
their hands before entering the IVF suite to prevent inadvertent cell
culture contamination. Check the location of your IVF laboratory to
make sure that the main compressors for the heating and cooling
system of your building are not close to the lab. The constant
vibration from this type of mechanical equipment will make it virtually impossible to conduct microscopic observations, especially
photomicrography. Finally, investigate possible pest control
problems in the area selected for the IVF program since commercial
insecticides cannot be used in an IVF laboratory. All laboratory
surfaces should be washable, i.e., standard laboratory grade tops.
Floors should not be carpeted but should have washable tile or
linoleum. Surfaces should be able to withstand daily cleaning with a
10% bleach solution.

     In conclusion, the establishment of an IVF program is an
expensive venture. In some cases, it is possible to make use of
existing equipment. The ability to share equipment in other labs
depends largely on accessibility, frequency of use, and the types of

laboratory activities taking place in the location where the shared
equipment is used. If the laboratory where the shared equipment is
available makes use of radioisotopes, carcinogens, or toxic chemicals, it is advisable to avoid those facilities. In all cases, the
equipment that is used for IVF should be in a physically isolated
location with limited access. The IVF lab requires emergency power
to run the equipment, appropriate surveillance to guard against
equipment failure, and appropriate security to guard against equipment theft and tampering. While the cost of equipping an IVF lab
varies dramatically and can be scaled up or down depending on the
financial resources of the institution, there are some pieces of
equipment that are absolutely necessary for the minimal operation of
an IVF laboratory. Furthermore, some pieces of equipment may serve
as the "gold standard" for use in the IVF lab. This is especially
true of microscopes and incubators. Equipment used in IVF programs
cannot be plugged in and left to function on its own. Equipment
requires constant monitoring, calibration and checking and this
should be taken into consideration in the selection and purchase of
the equipment and in the establishment of the quality control
program.

6.   REFERENCES

Bavister, B. D. and Andrews, J. C., A rapid sperm motility bioassay
     procedure for quality-control testing of water and culture
     media, J. In Vitro Fert. Embryo Transfer, in press.
Chetkowski, R. J., Nass, T. E., Matt, D. W., Hamilton, F., Steingold,
     K. A., Randle, D., and Meldrum, D. R., 1985, Optimization of
     hydrogen-ion concentration during aspiration of oocytes and
     culture and transfer of embryos, J. In Vitro Fert. Embryo
     Transfer 2:207-212.
College of American Pathologists, Commission on Laboratory Inspection
     and Accreditation, Reagent Water Specifications.
Gabler, R., Hegde, R., and Hughes, D., 1983, Degradation of high
     purity water on storage, Journal of Liquid Chromatography
     6:2565-2570.
Gerrity, M. and Shapiro, S. S., 1985, The use of a mobile laboratory
     cart in a successful university-based human in vitro fertiliza-
     tion program, Fertil. Steril. 43:481-484.
Mather, J., Kaczarowski, F., Gabler, R., and Wilkins, F., 1986,
     Effects of water purity and addition of common water contami-
     nants on the growth of cells in serum-free media, Biotechniques
     4:56-63.
Reust, J. B. and Meyer, V. R., 1982, Determination of organic con-
     taminants in ultrapure water by reversed-phase high performance
     liquid chromatography with ultraviolet detection, Analyst
     107:673-679.
Testart, J., Lassalle, B., and Frydman, R., 1982, Apparatus for the
     in vitro fertilization and culture of human oocytes, Fertil.
     Steril. 38:372-375.

3

QUALITY CONTROL AND LABORATORY MONITORING

Marybeth Gerrity

1. INTRODUCTION

While groups establishing new IVF programs often spend the largest period of time deciding what ovulation induction protocol to use or what types of equipment to purchase, the single most important aspect is the quality control program. More than drug treatment, equipment choice or even personnel, an excellent quality control program put in place will ultimately determine the success of the IVF program. The purpose of this chapter is to describe the components of a comprehensive quality control program and a system for laboratory monitoring in the IVF setting. There are three purposes of a quality control program, the first and most important of which is reproducibility of results. The laboratory aspect of IVF includes literally hundreds of variables that must be controlled.

   The second reason for a quality control program is to comply
with state, federal and institutional requirements. IVF programs
within hospitals generally fall under one of two accrediting
agencies: the Joint Commission on Accreditation of Hospitals (JCAH)
or the College of American Pathologists (CAP). The hospital administration can tell you which of these two bodies has jurisdiction in
your case. Table I summarizes some of the accrediting agencies and
advisory boards (and addresses) that inspect or accredit laboratories. Each individual state also has relevant codes. The example
given in the chart is for the state of Wisconsin. The state of
Illinois has published a detailed code (Illinois Administrative Code,
Chapter I, Section 250.510, Subchapter b, Section J, Quality Control)
concerning laboratory quality control programs. These are useful
illustrations but each group should check with its own state concerning requirements. There are also specific agencies that inspect
and accredit laboratories where Medicare or Medicaid and, in some
cases, other types of insurance coverage are involved. There are

   Table I. Quality control and laboratory monitoring;
             accrediting agencies and advisory boards

CAP - College of American Pathologists
5202 Old Orchard Road
Skokie, IL 60077

JCAH - Joint Commission on Accreditation of Hospitals
875 North Michigan Avenue
Chicago, IL 60611

State Agencies
e.g., State of Wisconsin
Department of Health and Social Services
Section of Laboratory Certification

Federal Agencies - Federal regulations and survey forms for Medicare-
approved laboratories, independent clinical labs and interstate labs
including Health Care Financing Administration Bureau of Health
Insurance

Specific Blood Bank Inspection Checklist and Report
Food and Drug Administration

American Fertility Society
1608 13th Avenue South
Suite 101
Birmingham, AL 35282

American College of Obstetrics and Gynecology
600 Maryland Avenue, S.W.
Washington, D.C. 20024

American Association of Tissue Banks
12111 Parklawn Drive
Rockville, MD 20852

---

also federal agencies which inspect independent clinical labs outside hospital or university settings. Virtually all laboratories that provide services to patients who expect to obtain insurance reimbursement are inspected or accredited. Finally, there are certain advisory boards that offer suggested or voluntary standards for special function laboratories such as IVF or andrology. These include, to name a few, the American Fertility Society, the American College of Obstetrics and Gynecology, and the American Association of Tissue Banks.

The final reason for establishing a quality control program is the issue of "standard of care." In the medical and legal sense, standard of care is defined by what is appropriate procedure and practice in your locale and in your specialty. Since the American Fertility Society and the American College of Obstetrics and Gynecology have defined minimal standards (although voluntary), these will, in many situations, become the standard of care for IVF. Therefore, while one is not required in any legal sense to adhere to these guidelines, if they become the least common denominator of care and one is involved in a law suit, it is probably crucial to document that these standards have been adhered to.

To summarize, the reasons for establishing a quality control program include: wanting to (i.e., striving for the best possible results); having to (i.e., being required by the accreditation of your hospital or institution); and, finally, needing to because someone else feels that you should be adhering to them (i.e., the consumer). Consumerism in the field of IVF has expanded dramatically in recent years. IVF patients are provided with written guidelines by many consumer and/or support groups and they will often ask if your program adheres to any of the standards listed above.

2.   ACCREDITATION

A detailed listing of the standards of the Joint Commission on Accreditation of Hospitals (JCAH) can be found in Table II and for the College of American Pathologists (CAP) in Table III. These standards or general guidelines should be considered in establishing a laboratory. They include general guidelines with regard to personnel, physical configuration, the laboratory, channels of communication and reporting, required records and reports, quality assurance, and quality control. It is important to keep in mind the mentality of the accrediting body and inspection process. The purpose of these standards is to assure the efficient reproducible functioning of the laboratory. While at times specific requirements may seem picky or inappropriate for an IVF subspecialty lab, in reality these standards provide a rational approach to everyday conduct in the lab. If the lab is inspected by JCAH or CAP, it will usually be inspected under the heading "Special Function Laboratory."

Table II.   Joint Commission on Accreditation of
            Hospitals (JCAH) standards

JCAH Standard 1. The pathology and medical laboratory services shall
be directed by a physician who.is qualified to assume professional,
organizational, and administrative responsibility for the facilities
and for the services rendered. There shall be sufficient qualified
personnel with documented training and experience to supervise and
conduct the work of the laboratory.

JCAH Standard 2. There shall be sufficient space, equipment, and
supplies within the pathology and medical laboratory services to
perform the required volume of work with optimal accuracy, precision,
efficiency, timeliness, and safety.

JCAH Standard 3. Channels of communication within the pathology and
medical laboratory services, with other departments/services of the
hospital and the medical staff, and with outside services and
agencies, shall be appropriate for the size and complexity of the
hospital.

JCAH Standard 4. Required records and reports shall be maintained
and, as appropriate, shall be filed in the patient's medical record
and in the pathology and medical laboratory services.

JCAH Standard 5. Quality control systems and measures of the
pathology and medical laboratory services shall be designed to assure
the medical reliability of laboratory data.

JCAH Standard 6. Specific requirements shall be observed when any
anatomic pathology, clinical laboratory, and blood transfusion
services are offered.

There are specific guidelines for this subheading of laboratory which
are summarized in Table IV.

3.   INFECTION CONTROL

    Infection control in the IVF (and andrology) laboratories is
essential to success as well as for medical-legal reasons. Since in
most IVF programs serum is still the main source of protein used in
culture media, due caution must be observed in handling this potentially infectious body fluid (Table V). Semen is a known carrier of
infectious diseases (Mascola and Guinan, 1986). Gloves should be
worn at all times and care taken to prevent aerosol formation (bubble
formation and potential spraying). In view of a recent report of an
IVF patient who was a carrier for hepatitis surface antigen (Axelrod
and Talbot, 1986), care should be taken to avoid contamination of
laboratory personnel as well as equipment which will come into
contact with patients. Table VI summarizes infection control
measures that should be used in screening serum for use as a protein
source in IVF programs. In situations where donor serum is being

Table III.   College of American Pathologists (CAP)
             standards

CAP Standard 1. The pathology and clinical laboratories shall have
sufficient space, equipment and facilities for the performance of the
required volume of work with optimum accuracy, precision, efficiency
and safety.

CAP Standard 2. A hospital department of pathology and clinical
laboratories shall be directed by a physician who is qualified to
assume professional, organizational, and administrative responsibility for the department. The director of an independent laboratory
must be a physician with training and experience in pathology or a
clinical scientist with adequate training and experience in clinical
laboratory work to meet the requirements of a laboratory director
under the Laboratory Improvement Act of 1967. Sufficient personnel
with training and experience adequate to supervise and conduct the
work of the clinical laboratories shall be provided.

CAP Standard 3. Channels of communication within the laboratories as
well as with all other closely affiliated sections and services of
the hospital and the medical staff shall be appropriate for the size
and complexity of the organization.

CAP Standard 4. The quality control systems of the laboratory shall
be designed to assure the medical reliability of laboratory data.

CAP Standard 5. All specimens removed at operation must be clearly
identified and sent to the pathologist accompanied by the pertinent
clinical information. The extent of the examination is to be determined by the pathologist.

CAP Standard 6. Space, equipment and personnel shall be provided for
necropsy service adequate for the needs of the institution.

CAP Standard 7. A blood transfusion must be maintained and directed
by a pathologist or a physician qualified in immunohematology and
blood banking.

CAP Standard 8. Reports of all clinical pathology examinations shall
be a part of the patient's hospital record.

---

used (serum from placentas or from sources other than the recipient),
screening of the serum is essential. The efficacy of heat inactivation at 56°C and filter sterilization in removing potential
infectious contamination remains unproven. Therefore, all donor
serum should be screened for both HTLV-III antibodies and hepatitis
surface antigen. Any serum that is positive for either of these two
should be discarded. These measures will prevent contamination of
laboratory personnel and equipment. A separate but related issue is
contamination of eggs, embryos or embryo recipients. A recent review
by Axelrod and Talbot (1986) described infectious disease screens and
cultures appropriate for IVF patients, i.e., designed to protect the
recipient from ascending infection of the reproductive tract as a
result of embryo transfer (Table VII). The authors also feel that

Table IV. Records that are required to be on
           hand in the laboratory (Special
           Function Laboratories)

A. Inservice records
B. Orientation
C. Safety and infection control policies
D. Specimen collection
E. Arterial puncture authorization (where relevant)
F. Procedures manual
G. Requisitions
H. Accession record
I. Quality control policy and records
J. Proficiency testing (where relevant)
K. Preventive maintenance
    1. Contractual
    2. In house

---

some IVF patients have a history of pelvic inflammatory or tubal
disease that may be associated with specific infectious components.
In view of a recent IVF failure attributed to Chlamydial infection
(Rowland et al., 1985), it seems rational to test IVF patients and
treat them appropriately if the pathogen can be isolated. Since
collection of a sterile sample of semen is impossible, and since this
fluid is a known biohazard, care should be taken in its handling.
There have been scattered reports of contamination of eggs and
embryos by semen pathogens (for example, Barriere et al., 1985). It
is apparent that the small amounts of penicillin and streptomycin in
culture media are inadequate to prevent overgrowth of contaminated
cultures. Finally, Axelrod and Talbot (1986) have suggested that
infectious disease screens of IVF patients provide a relatively
inexpensive and risk-free method of preventing infections of the
neonate should the patient become pregnant. The use of sterile
technique, appropriate disease screens, screening of serum and good
laboratory procedure can prevent contamination of eggs, embryos,
recipients, newborns and laboratory personnel.

4. LABORATORY SAFETY

In addition to infection control measures, laboratory safety
policies are defined by the institution as well as by the relevant
accrediting body for your hospital. Safety procedures include appropriate firefighting equipment, including fire extinguishers and fire
blankets, safe storage of flammable, explosive or corrosive
materials, and electrical circuitry and outlets that are appropriate
for the equipment in use.

Table V. Infection control in the in vitro fertiliza-
          tion and andrology laboratory--laboratory
          procedures

All blood and body fluids (i.e., semen) should be treated as potentially infectious and due care observed.

A. Lab coats should be worn and kept in the lab area.

B. Eating, drinking, smoking, and application of cosmetics should
    be permitted only in designated nonclinical area.

C. Food should be stored in a separate refrigerator in a nonclin-
    ical area.

D. Gloves should be worn for direct contact with all clinical
    specimens (including blood, semen and serum) and for cleaning
    all spills.

E. Wash hands after manipulating biological specimens.

F. No mouth pipetting.

G. Minimize aerosolization of fluids (including use of caps in the
    centrifuge).

H. Red bag all blood/body fluid contaminated specimens and lab
    ware.

I. Needles and disposal sharps should be placed in appropriate
    containers.

J. Disposable plasticware and glassware should be used for all
    laboratory procedures involving blood or semen.

K. Blood/body fluid spills should be cleaned immediately with 1:10
    solution of bleach or phenolic disinfectant/detergent.

L. Personnel with frequent exposure to blood/body fluids and/or
    biological samples should consider hepatitis B vaccine
    (Heptavax).

5.   STAFFING REQUIREMENTS

Both the American Fertility Society and the American College of
Obstetrics and Gynecology have itemized specific (voluntary)
standards for personnel in IVF programs. These include a board-
eligible or a board-certified reproductive endocrinologist to monitor
follicular stimulation and an experienced pelvic reparative surgeon
and/or laparoscopist to undertake the egg retrievals. The laboratory
is to be staffed by a person experienced in embryology/tissue
culture, as well as an andrologist or person experienced in the
handling of semen. These need not be separate people. Since IVF is
dependent on time factors beyond the control of the physician or

Table VI.   Infection control measures to be used on
            serum samples obtained from placentas used
            for IVF

A. All applicable lab procedures described in the infection control
    protocol (Table V) will be used in handling these serum samples.

B. All fetal cord serum will be screened for HTLV-III, hepatitis
    core antibody and surface antigen. Any sample that is positive
    for any of these screens will be discarded.

C. Maternal serum or donor serum will not be used in the IVF
    program except in exceptional situations where the above
    infection control measures are adhered to. If maternal or donor
    serum is used, it will be screened for antisperm antibodies.

---

embryologist (e.g., start of menses; rate at which patients progress
to laparoscopy), it is difficult to provide regular time off. In a
discipline in which burnout comes early, it is therefore very important to provide adequate support personnel to free up the embryologist and physicians from the day-to-day routine aspects of performing
IVF. While media preparation and testing, insemination of eggs,
observing the fertilization and cleavage of embryos, and day-to-day
monitoring of ovulation induction should be under the strict control
of the primary embryologist and physician, the day-to-day support
activity, such as glassware washing, pipet sterilization, and all the
other monumental aspects of supporting an IVF program can best be
assigned to appropriate support personnel. The choice of personnel
for the IVF laboratory may in part be determined by the institution.
Some hospital laboratories require that laboratory assistants be
medical technologists. Such personnel are attuned to quality control
and documentation of laboratory activities since this is an important
component of their training. However, there is nothing in their
training that would prepare them specifically for IVF. In selecting
personnel, it is important to evaluate transferable experience or
skills. Attributes include microbiologic, tissue culture, quality
control and laboratory documentation experience, and manual dexter-
ity. Embryologists experienced in human IVF are in very short
supply. For this reason they command a salary that is higher than
average, a fact that should be kept in mind in preparing an initial
start-up budget for an IVF program. The credentials of the embryologist may depend in large part on the volume of the anticipated
program. Although it may seem like a contradiction, it is probably
best to employ the most experienced embryologist possible if you only
plan to do a few IVF patients. In a low volume program, the novice
embryologist will simply not get enough practice to come on to the
learning curve. A high volume program (greater than 200 patients per
year) is a more appropriate place to train personnel in performing
human IVF.

Table VII. Suggested infectious disease screens for
patients undergoing IVF. (Since these
patients are being treated by private
physicians, we can only make recommenda-
tions about these screens. The care of the
patient is left to the individual physicians within their own practice.)

A. All women scheduled for IVF-ET

1. Serologic test for syphilis
2. Cervical culture for Neisseria gonorrhoeae
3. Rubella titer

B. Participants at high risk for disease

1. Hepatitis B surface antigen
2. Cervical culture for herpes simplex virus
3. Cervical or urethral culture (or immunofluorescent stain)
    for chlamydiae
4. HTLV-III ELISA antibody titer

All patients who are receiving embryo transfer will receive prophy-
lactic Doxycycline, 100 mg, twice a day beginning on the day of their
laparoscopy and continuing for five days, therefore overlapping their
embryo transfer.

In addition to these specific laboratory personnel recommenda-
tions, the JCAH and CAP have requirements concerning documentation of
orientation of new personnel, training in the technical aspects of
each procedure and inservice continuing education activities. The
need for an established training program for IVF embryologists was
the impetus for the workshops at the University of Wisconsin upon
which this book is based. Ongoing efforts to provide training and
certification of IVF embryologists is crucial.

6. SUPPLIERS AND SOURCES

Each lab should keep written records of where and how supplies
for the IVF program have been purchased; batch numbers, shipment
dates and conditions of shipment when received should also be noted.
Chemicals and plasticware can vary and when problems arise, it is
best to be able to go back and verify whether a change in a supplier,
a new batch of chemicals, or inappropriate storage and shipment might
have contributed to a bad outcome. Each lab should establish
standard operating procedures in terms of ordering equipment,
chemicals and supplies. In institutional settings where purchasing
agents may buy the low bid item, it may be best to establish a known
manufacturer from whom all supplies are purchased. This will

eliminate variations in plastics or in purity of different lots of
culture media. Furthermore, reordering will be expedited.

7.   RECORDS AND PROTOCOLS

     A list of required records for all Special Function Laboratories
is summarized in Table IV. Many of these have been mentioned
previously. A procedures manual is one of the most important components of the quality control program. Explicit step-by-step instructions of every protocol in the laboratory from glassware washing to
IVF should be developed. This is not meant as a cookbook to allow an
inexperienced assistant to follow directions and perform IVF. It is
rather a mechanism for keeping track that every step of every
protocol is followed exactly. Table VIII summarizes the components
of a written protocol including the principle of the test, requirements for specimen handling, specific instructions about reagent
preparation and expiration dates, calibration procedures, equipment
operation, the procedure itself and the quality control measures that
should be followed. When a result must be reported, establishment of
normal ranges and a check and balance system to flag absurd
(incorrect) values is also important. This may not be essential for
your IVF laboratory unless you are actually performing radioimmuno-
assays for estradiol. However, these types of normal ranges must be
established in an andrology lab where sperm penetration assays or
semen analyses are performed. Again, it is very important that this
type of a protocol be developed for every single procedure that is
performed in the laboratory. It is not unusual to find that one of
the most seemingly insignificant aspects of laboratory operation is
the point of failure in IVF. Procedures for sanitizing the water
system, glassware washing and sterilization, handling and preparation
of transfer catheters and aspiration needles should be included in
this protocol. Specific listings of where components of the culture
media can be obtained and how they should be tested must also be
included.

8.   MAINTENANCE, CALIBRATION AND EQUIPMENT MONITORING

     The maintenance manual is an essential portion of the record
keeping for accreditation purposes and its timely completion is
crucial to the success of IVF. For reference purposes, see the CAP
(College of American Pathologists) suggestions for maintenance
manuals. CAP provides an incredibly detailed listing of the types of
maintenance that should be performed on equipment from heating blocks
and slide warmers to microscopes and incubators. These guidelines
include maintenance that must be performed daily (such as a humidifi-
cation or documentation of temperature) and weekly, monthly and
yearly preventive measures. For accreditation/inspection purposes,
maintenance must be documented in writing at prescribed time

Table VIII. Components of written protocols adapted
from Illinois Administrative Code

A. Principle of the test including a brief statement of purpose and
    reaction involved
B. Specimen handling and preparation procedure
    1. Method for obtaining sample
    2. Criteria for acceptable sample
    3. Handling conditions
    4. Identification and storage
C. Reagent preparation
    1. Specific methods.
    2. Detailed listing of material needed
    3. Documentation of batch and supplier
    4. Identification and labeling
    5. Expiration date and procedure for discarding
D. Calibration procedures
    1. Standards used where appropriate
    2. Acceptable tolerance limits and procedures to be followed
        if the results fall outside these limits
    3. Expiration dates for standards used
E. Procedure - a detailed description of exact procedure (including
    safety or biological hazards)
F. Calculations - presented in a stepwise fashion with a specific
    example
G. Quality control
    1. State reference material used
    2. Minimum frequency of performing standards and performing
        procedure
    3. State corrective action and troubleshooting activities to
        be followed
    4. State number of replicates, etc., that are necessary
H. Reporting of results
    1. State expected and/or normal ranges
    2. State any information about methodology that may be impor-
        tant to the interpretation
    3. Establish a system for flagging "absurd" values and/or
        critical values
    4. Report the results in an acceptable format and file them in
        a method that permits easy access
I. Procedural notes
    1. List possible sources of error
    2. Plan for alternate means of specimen handling should the
        procedure fail
J. References

intervals. It is not sufficient to merely perform the maintenance or
equipment monitoring; written records must be available (see, for
example, Figures 1-3).

In my opinion, the best maintenance protocols provide a system
of checks and balances to determine when any portion of the equipment
is malfunctioning. In other words, try to imagine anything that
could go wrong and build in a check to be sure that it does not. For
example, consider the incubator, the two main functions of which are

Monthly Checklist

MONTH	Jan	Feb	Mar	Apr	May	June	July	Aug	Sept	Oct	Nov	Dec			
Centrifuge Timer Checked (5 min ± 15 sec)															
Hot Plate Cleaned															
UV Cabinet Cleaned															
Heating Block Calibrated															
Water Bottle Washed															
Mass Spec. Gas Contents															
Door Seal on IVF Incubator Checked and OK															
Door Seal on Lab Incubator Checked and OK															
Shelves Level in IVF Incubator															
Shelves Level in Lab Incubator															

Note: If centrifuge timing check falls outside bracketed range, notify supervisor.

Corrective Actions:

Fig. 1. Example of record keeping form for laboratory quality control items checked on a monthly basis.

Date	1	2	3	4	5	6	7	8	9	10	11	12	13	14	15	16	17	18	19	20	21	22	23	24	25	26	27	28	29	30	31
Still off																															
E$_2$O Baths off																															
Oven off																															
Freezing Machine off																															
Benches Bleached																															
Microscopes Covered																															
Liquid N$_2$ Levels OK																															

Fig. 2. Example of record keeping form for laboratory procedures conducted daily in an IVF laboratory.

WEEK	1st wk Jan	2nd wk Jan	3rd wk Jan	4th wk Jan	1st wk Feb	2nd wk Feb	3rd wk Feb	4th wk Feb	1st wk Mar	2nd wk Mar	3rd wk Mar	4th wk Mar	1st wk Apr	2nd wk Apr	3rd wk Apr	4th wk Apr	1st wk May	2nd wk May	3rd wk May	4th wk May	1st wk June	2nd wk June	3rd wk June	4th wk June
Heating Block Disinfected																								
Interior of IVF Incubator Cleaned																								
Interior of Lab Incubator Cleaned																								
Circulating Fan on IVF Incubator Checked																								
Circulating Fan on Lab Incubator Checked																								
Overtemp Safety Thermostat on IVF Incubator Checked																								
Overtemp Safety Thermostat on Lab Incubator Checked																							·	

NOTE: If any problems are noted with the circulating fans or overtemp thermostats, please notify the supervisor.

Corrective Actions:

Fig. 3. Example of record keeping form for laboratory procedures conducted weekly in an IVF laboratory.

to maintain a constant gas environment and temperature. A daily
written log should be kept of the temperatures of the incubators.
However, one should know that a digital readout is not necessarily
accurate. Therefore, you should keep a thermometer inside your
incubator to verify your digital readout. On a monthly basis, the
digital readout thermometers should be calibrated against a National
Bureau of Standards (NBS) thermometer to insure their accuracy. NBS
thermometers may be borrowed, generally speaking, from the clinical
laboratory of your hospital. The gas content of the incubator is
also registered in a digital form on some models of incubators. The
digital readout is only as good as the calibration. If your incubator was calibrated with CO_2 in the chamber, your digital readout is
incorrect. For example, if your incubator is set to run at 5% CO_2,
but the incubator was zeroed with 2% carbon dioxide in the chamber,
the actual gas content of the chamber is 7%. The best way to verify
the CO_2 content of an incubator is by performing multiple checks.
For example, daily Fyrite (Bachrach) determinations of the CO_2
content are recorded. In addition, there is always a tube of culture
media containing phenol red indicator in the incubator. Observation
of indicator color each morning provides a gross check of whether the
pH control of the indicator is adequate. Blood gas analysis of the
culture media in a sealed tube may be used to determine the CO_2
content of the incubator. Whatever method is used, its limitations
must be recognized. For example, Fyrite measures 5% CO_2 within 1%,
that is, when it reads 5%, the CO_2 content of the incubator can be as
low as 4% or as high as 6%, both of which may have disastrous consequences for the embryos. It is, therefore, important to have a
method to calibrate the Fyrite. Mass spectrometry of the gas content
of the incubator can be used to this end and such equipment is
usually available in the anesthesiology department, respiratory
therapy section or the intensive care unit of a hospital.

Daily determinations of temperature or CO_2 concentration are
useful only if normal ranges are defined. There is little point in
dutifully documenting refrigerator temperature each day if the docu-
mentor does not realize that a temperature of 10°C is abnormal.
Therefore, one must state normal ranges and document corrective
action to be taken in situations where values fall outside this
accepted range. For example, a simple quality control sheet can
provide a method for recording daily temperatures, daily CO_2s,
indicating acceptable ranges and documenting corrective actions
(Figure 4). In addition, a calibration sheet is useful for the
morning of IVF setup (Figure 5). In some situations, it may be
important to attach continuous reading, calibration or recording
devices to equipment, for example, strip chart recorders on low
temperature freezers. A single temperature each morning is clearly
inadequate if the entire freezer thawed and refroze overnight. A
poor man's substitute for a strip chart recorder is to freeze a small
aliquot of water in a test tube, invert the test tube and tape it to
the inside of the freezer. Check the test tube each morning. If the

MONTH _____

Date	1	2	3	4	5	6	7	8	9	10	11	12	13	14	15	16	17	18	19	20	21	22	23	24	25	26	27	28	29	30	31
IVF Refrigerator Temperature (0° – 8°)																															
IVF Freezer Temperature (-6° – -14°)																															
Lab Refrigerator Temperature (0° – 8°)																															
IVF Incubator CO_2 Fyrite (4% – 5%)																															
Lab Incubator CO_2 Fyrite (4% – 5%)																															
IVF Incubator Temperature (36° – 37°)																															
Lab Incubator Temperature (36° – 37°)																															
CO_2 Flow Rate IVF Incubator (1.0 – 2.0)																															
IVF Incubator CO_2 Tank Pressure (above 800)																															
Lab Incubator CO_2 Tank Pressure (above 800)																															
Hamsters Checked for Postestrous Discharge																															

Note: If any of these values fall outside the bracketed ranges, please notify the supervisor.

Corrective Actions:

Fig. 4. Form for monitoring daily incubator temperature and gas composition.

Laboratory Quality Control Sheet

Patient Name _____

Date of Egg Retrieval _____

Preparation of Insemination Medium:

Ham's F-10 Batch _____
  Date Prepared _____

Dulbecco's PBS Batch _____
  Date Prepared _____

Fetal Cord Serum Batch _____
  TM _____
  GM _____
  Transfer _____

Lab Setup   Date _____

  Gas Phase Used ___5% CO$_2$ in air___

  Tank Reading _____

	Digital Readout				
	Temperature	CO$_2$	Fyrite	pH Medium	Phenol Red
Upper Incubator					
Lower Incubator					

Mobile Lab Unit:

  A) Temperatures

     1) Incubator _____
     2) Temp Block _____
     3) Slide Warmer _____

  B) Other:
     1) Tank Reading _____
     2) Microscope bulb _____

Fig. 5.   Form for quality control record keeping on the morning of an IVF procedure.

contents have fallen down into the cap of the test tube, you know that the freezer thawed and refroze overnight. Common sense approaches such as this will help to keep your equipment operating properly and your program successful.

9.   SUMMARY AND TROUBLESHOOTING

The general components of a quality control program as described in this chapter are summarized in Table IX. Strict attention to each aspect of this list coupled with written documentation will assure that laboratory operations continue within specific tolerance limits. As mentioned earlier, the goal of a quality control program is to assure reproducibility of results. It has been a convention in the

Table IX. Components of a quality control program

A.   Written prodecures and policies manual

    1.   Detailed description of all procedures
    2.   Sample handling protocols

B.   Documentation and reporting of results

    1.   Acceptable limit for results
    2.   Documentation of corrective action
    3.   Record keeping

C.   Equipment maintenance

    1.   Daily equipment monitoring
    2.   Monthly/yearly preventive maintenance
    3.   Establishment of acceptable operation standards
    4.   Documentation of corrective actions taken
    5.   Instrument calibration

D.   Safety procedures (including appropriate storage of materials)

E.   Infection control measures

F.   Staffing requirements

    1.   Credentials of staff
    2.   In-service and continuing education
    3.   Evaluation and performance appraisal
    4.   Orientation procedures

G.   Documentation of suppliers and sources of chemicals and
    supplies

---

field of IVF that a start-up program follows the protocol of an
established program until a satisfactory pregnancy rate is reached.
At that point, the group feels free to strike out on their own path
and modify procedures. Too often, however, strict protocols are not
adhered to and when pregnancies don't occur, there is a tendency to
start changing many variables at once. The biggest fear of the
fledgling program is lack of success, and the fear of the established
program is a long "dry spell." When these occur, a rational approach
to correcting problems is a return to the basics as summarized in
Table X.

Finally, the troubleshooting of problems in the IVF laboratory
is possible through the use of the general principles of quality
control described in Table XI. Known test materials may take the
form of media, mice or equipment that is borrowed or purchased from
another program. Adequate controls in media testing have been
described in Chapter 5. Comparison of results should be made to
previous laboratory records as well as literature values (this holds

**Table X. Components of a quality control program for in vitro ferti-
lization**

1. Equipment monitoring
    a.   Daily monitoring of incubator function including temperature and gas content
    b.   Daily monitoring of maintenance and operation of other lab equipment
    c.   Emergency generator back-up to all equipment
    d.   Regular recalibration scheduling including use of internal standards (e.g.,
         NBS thermometers)
2. Disposable supplies
    a.   Documentation of supplier and batch/lab number
    b.   Testing of all disposables which come into contact with culture medium for
         ability to sustain mouse embryo development
    c.   Rinsing of disposable supplies (e.g., transfer catheters, etc.) with culture
         medium before use
3. Nondisposable supplies
    a.   Should be segregated for IVF use only (e.g., glassware should be etched)
    b.   Washing with ultrapure water, no tap water, heat sterilization (170°C for 2
         hr) where possible; avoid autoclaving. Where gas sterilization is used, allow
         equipment to degas for 48 hr before use, rinse vigorously with culture medium.
    c.   Glassware that is scratched, clouded or stained should be discarded
    d.   Nondisposable aspiration needles and transfer catheters should be numbered and
         this number should be recorded for each procedure
    e.   All nondisposable supplies should be replaced at predetermined intervals
4. Culture medium preparation and testing
    a.   Batch and storage and expiration requirements on all culture medium should be
         documented in writing
    b.   ACS grade chemicals should be purchased for medium modifications
    c.   Water used for preparations should be from an unopened bottle or freshly
         obtained (no standing water) from the point of delivery
    d.   All culture medium and serum should be tested in the mouse embryo culture
         bioassay
    e.   Media should be discarded seven days after the completion of testing
    f.   All plasticware should be tested to determine its ability to support mouse
         embryo development
5. Written protocols should be prepared for every step of the procedure
6. Written records should be kept of
    a.   Induction protocol including ultrasound and estradiol monitoring
    b.   Number of follicles aspirated
    c.   Number of oocytes obtained and their stage of maturation
    d.   Detailed analysis of semen parameters
    e.   Number of oocytes fertilized
    f.   Number of oocytes cleaved
    g.   Number of oocytes transferred and fate of extra embryos

true for every step from mouse superovulation to human oocyte
fertilization rates and pregnancy statistics). Human embryos that
develop slowly compared to established growth schemes (e.g., Cummins
et al., 1986) will suggest possible problems. Embryo viability may
be suggested by measuring patient platelet responses after transfer
(O'Neill et al., 1985). Documentation of corrective action prevents
repetition of problems, allows one to trace where problems occurred
and prevents a "shotgun" approach where many variables are changed at
once. After correcting a problem, repeat the quality control
procedures and see how they affected a specific outcome. (For

Table XI.   Summary of quality control criteria

A.   Perform all procedures using known reference material. The
     individual performing the tests must record, date and sign the
     results.

B.   Assay controls often enough to assure the reliability of test
     results.

C.   Compare results of quality control studies with previously
     established values. Identify and document any discrepancies.

D.   Evaluate the components of the procedures for which discrep-
     ancies are identified.

E.   Document the appropriate corrective action after identifying the
     cause of the discrepancy.

F.   Repeat the quality control procedures and document acceptability
     of the repeated results before repeating procedures on patient
     materials.

G.   Document the use of preventive maintenance schedules and equip-
     ment service records.

H.   Perform internal and external testing and control programs.

I.   Perform systematic retrospective review of services to include:

     1.   Technical and professional services.
     2.   Delivery systems.
     3.   Appropriateness of services.
     4.   Adherence to established protocols by all practitioners, as
          appropriate.

example, did recalibration of the incubator result in improved mouse
or human embryo development?) If you have not corrected a problem,
do not proceed to the use of human materials. Documentation of
preventive maintenance can prevent problems that arise as systems are
used over time (e.g., poor human embryo development as water purification systems decay). Internal and external controls of all procedures are useful. Sending your media to another lab for testing and
use may pinpoint a problem (or an absence of a problem) that could
not be detected in your lab. Some labs use GIFT (gamete intrafallopian transfer) as an internal control by handling some oocytes from
the same patient for both GIFT and IVF. Failure of in vitro fertilization while a pregnancy is established by GIFT in the same patient
is strong evidence for a lab problem. Finally, periodic retrospective analysis of the program can only make the good program better.

Success in in vitro fertilization requires tight control of many
variables; the "black magic" is gone from the field. Adherence to a
stringent quality control program insures continued success and helps
pinpoint problems.

10. REFERENCES

American Association of Tissue Banks (12111 Parklawn Drive, Rockville, MD 20852), Guidelines for the Banking of Human Semen.

American College of Obstetrics and Gynecology (600 Maryland Avenue, S.W., Washington, D.C. 20024), Human In Vitro Fertilization and Embryo Placement.

American Fertility Society, 1984, Ethical statement on in vitro fertilization, Fertil. Steril. 41:12.

American Fertility Society, 1984, Minimal standards for programs of in vitro fertilization, Fertil. Steril. 41:13.

Axelrod, P. and Talbot, G. H., 1986, Infection control considerations for in vitro fertilization and embryo transfer programs, Infect. Control 7:373-378.

Barriere, P., Lopes, P., Boiffard, J. P., L'Hermite, A., and Lerat, M. F., 1985, An unusual cause of failure of in vitro fertilization: report of a case, J. In Vitro Fert. Embryo Transfer 2:170-171.

Cummins, J. M., Breen, T. M., Harrison, K. L., Shaw, J. M., Wilson, L. M., and Hennessey, J. F., 1986, A formula for scoring human embryo growth rates in in vitro fertilization: its value in predicting pregnancy and in comparison with visual estimates of embryo quality, J. In Vitro Fert. Embryo Transfer 3:284-295.

Ethics Committee of the American Fertility Society, 1986, Ethical considerations of the new reproductive technologies, Fertil. Steril. 46(Suppl. 1), 94S pp.

Mascola, L. and Guinan, M. E., 1986, Screening to reduce transmission of sexually transmitted diseases in semen used for artificial insemination, N. Engl. J. Med. 314:1354-1359.

O'Neill, C., Pike, I. L., Porter, R. N., Gidley-Baird, A. A., Sinosich, M. J., and Saunders, D. M., 1985, Maternal recognition of pregnancy prior to implantation: methods for monitoring embryonic viability in vitro and in vivo, Ann. N.Y. Acad. Sci. 442:429-439.

Rowland, G. F., Forsey, T., Moss, T. R., Steptoe, P. C., Hewitt, J., and Darougar, S., 1985, Failure of in vitro fertilization and embryo replacement following infection with Chlamydia trachomatis, J. In Vitro Fert. Embryo Transfer 2:151-155.

4

CHOICE, PREPARATION AND USE OF CULTURE MEDIUM

Gregory S. Kopf

1.   INTRODUCTION

When selecting the proper media for human in vitro fertilization and embryo transfer (IVF-ET), a variety of physiological and biochemical parameters must be considered. Preinsemination oocyte (or egg) culturing and incubation of sperm under conditions conducive to motility maintenance, capacitation and the acrosome reaction are essential for successful sperm-egg interaction. Once fertilization has occurred, proper incubation conditions must be met to support early preimplantation development. Although there is much yet to learn about the optimal conditions for oocyte, egg, sperm and embryo culturing in human IVF-ET, a number of criteria have been shown to be essential for success. These criteria will be addressed only briefly here, and the reader is encouraged to examine several excellent reviews (Edwards, 1980; Mastroianni and Biggers, 1981; Wolf and Quigley, 1984) and Chapter 18 in this book dealing with the role of environmental factors on embryonic development.

A variety of media have been successfully used for the culture of human gametes, as well as for fertilization and embryo development. These include simple balanced salt solutions (i.e., modified Earle's, modified Whitten's, Whittingham's T6) as well as more complex media containing amino acids and vitamins (i.e., modified Ham's F-10, Menezo's B_2). The simple salt solutions are extensively used in the IVF-ET programs in Australia and the United Kingdom, whereas Ham's F-10 appears to be the most popular medium in the United States. These media usually contain either heat inactivated maternal serum or fetal cord serum as a primary protein source. However, simplified protein sources such as human serum albumin have been used successfully. A protein requirement has been described in many other cell culture systems, and appears to be important for sperm capacitation (Rogers, 1978; Wolf and Quigley, 1984; Lee et al., 1987). The presence of exogenous serum proteins might also influence oocyte, egg or embryo quality and viability through the ability to chelate ions, and through the potential provision of heretofore unknown growth factors (Leung et al., 1984; Saito et al., 1984; Ball et al., 1985). A variety of energy sources are also present and appear to be important both as oxidizable substrates for sperm (pyruvate, lactate) (Mann and Lutwak-Mann, 1981) and as energy sources for early preimplantation embryos (pyruvate, glucose) (Edwards, 1980). Studies in animal models have suggested that these substrates may play some role in sperm capacitation and the acrosome reaction (Rogers, 1978; Mann and Lutwak-Mann, 1981). Although the optimal osmotic pressure for human IVF and embryo culture has not been determined, pressures of 280–300 mOsmol/kg have been used successfully. Likewise, pH ranges of 7.2–7.9 are compatible with IVF, but an optimum has not been established. A consideration of these variables is addressed in greater detail in Chapter 18. Since all of these media are bicarbonate buffered, it is essential that they be gassed to maintain the proper CO_2 environment and to minimize large pH shifts. This is accomplished using humidified incubators with 5% CO_2:air or 5% CO_2:5% O_2:90% N_2 gas mixtures. Equilibration of the media with the gas mixtures is usually accomplished by placing the media in loosely capped culture flasks in the incubator for a period of 12–16 hr prior to use.

It must be emphasized, however, that it is not known for certain that the culture media currently in use for IVF-ET is critical for the successful culture of eggs and preimplantation embryos, since successful pregnancies have been obtained using both 100% heat inactivated serum (Kemeter and Feichtinger, 1984) and serum-free culture media (Menezo et al., 1984). Synthetic media have also been used successfully (Feichtinger et al., 1986). It remains to be seen whether the introduction of simplified media will be successful; such an approach may circumvent the batch-to-batch variation of these more complex media.

2.   TECHNIQUE 1 - UNIVERSITY OF PENNSYLVANIA

The following procedures are utilized for medium preparation in the IVF-ET program at the Hospital of the University of Pennsylvania.

2.1. Chemicals

a.   Highly purified H_2O (purified through ion exchangers and/or glass distillation). Our program uses H_2O purified through charcoal adsorption and mixed bed ion exchange resins with a 0.2 μm sterilizing filter at the exit port.

b.   Ham's F-10 powder (Gibco #430-1200)

c.   Penicillin G-streptomycin solution (Gibco #600-5140)

d.   Calcium lactate (Calbiochem #4272)

e. $NaHCO_3$

f.   Heparin (5000 USP units/ml saline, sterile)

g.   Dulbecco's phosphate buffered saline (Gibco #450-1300)

2.2. Supplies (all glassware washed for IVF)

a.   Sterile 1-liter volumetric flask (pyrex)

b.   Two 50-ml sterile glass beakers (pyrex)

c.   Sterile spatulas, weighing papers, and labeling tape

d.   10-ml sterile, disposable pipets

e.   500-ml sterilization filter unit (Nalge #450-0020)

f.   115-ml Type S Nalgene filter sterilization units (Nalgene #120-0020) - for filter sterilizing media

g.   115-ml Falcon filter sterilization units (Falcon #7103) - for filter sterilizing blood serum

h.   50-ml tissue culture flasks (Falcon #3031)

i.   100-ml sterile, disposable plastic graduated cylinders (Corning #25500)

j.   Disposable gloves

2.3. Equipment

a.   pH meter

b.   Analytical balance

c.   Magnetic stirrer

d.   Osmometer

e.   Hand-operated vacuum pump

f.   Laminar flow hood

g.   Clinical table top centrifuge

h. H_2O bath adjusted to 56°C

2.4. Procedures

     a.   General tissue culture techniques should be utilized for all of these procedures:

     (1) Use a laminar flow hood at all times.

     (2) Absolutely no mouth pipetting. NOTE: handling of biological fluids (i.e., serum, semen) dictates that proper precautions be taken to insure that the potential for personnel contact with such fluids be minimal, since it is known that these fluids can be contaminated with such pathogens as hepatitis B and AIDS virus. It is advisable that the laboratory personnel be acquainted with all of the necessary precautionary procedures, and that the laboratory maintains a notebook of protocols describing methodology for the clean up of contaminated areas. An appropriate and functioning disease screening program of both patients and personnel should be a part of every IVF program. The design and institution of such a program should be done with the consultation of specialists in infectious diseases and epidemiology (see Chapter 3).

     (3) All glassware, plasticware, and filters should be prerinsed with stock Ham's F-10 medium prior to use.

     b.   Preparation of 1 liter of stock Ham's F-10 medium (once prepared this stock can be stored at 4°C for up to two weeks):

     (1) Add approximately 600 ml of IVF water to a sterile 1 liter volumetric flask.

     (2) Using a powder funnel, add 1 packet of Ham's F-10 powder. Do not rinse foil pouch. Mix by inversion of the flask using parafilm.

     (3) Add 10 ml of the penicillin-streptomycin solution (stock concentration: 10,000 units penicillin/ml, 10,000 mcg streptomycin/ml). (Alternatively, weigh out 0.075 g each of penicillin G and streptomycin sulfate and dissolve in the IVF H_2O.)

     (4) Dissolve 0.2452 g Ca^{2+}-lactate in 50 ml of IVF H_2O, and then add slowly, while swirling, to the Ham's F-10.

     (5) Dissolve 2.1 g of $NaHCO_3$ in 50 ml IVF H_2O and then add slowly, while swirling, to the Ham's F-10. (NOTE: the Ca^{2+}-lactate and $NaHCO_3$ solutions are added as diluted solutions separately because the solubility product of $CaCO_3$ is low. If a white precipitate, which is cloudy or hazy, is formed at this stage, start the procedure over again because insoluble $CaCO_3$ has been formed.)

(6)   Add IVF H_2O to a final volume of 1 liter.

(7)   Determine the osmolarity of an aliquot of this solution and make the proper volume adjustments to insure that the final osmolarity is between 275–280 mOsmol/kg. This adjustment can be made using the following formula:

Vol (ml) stock solution to be removed and replaced with IVF H_2O =

$$\text{final vol (ml)} \times \frac{\text{initial reading (mOsmol/kg)} - 280 \text{ (mOsmol/kg)}}{\text{initial reading (mOsmol/kg)}}$$

NOTE: follow closely the protocols for the operation and maintenance of the osmometer so that accuracy is insured. See Chapter 2 pertaining to the use of this equipment.

(8)   Recheck the osmolarity after the initial adjustment has been made.

(9)   Filter sterilize the medium into two 0.22 µm Nalgene 500 ml sterilization filter units. Label with date, initials of person making the medium, and then store capped at 4°C.

(10)   Equilibrate an aliquot of this medium overnight in CO_2 incubator. The pH of the medium the next day should be 7.3–7.5. If the pH is not in the acceptable range check, 1) the incubator to be certain that the gas composition and temperature in the chamber is correct, and 2) recheck and adjust, if necessary, the pH of the stock medium. If this is done, make certain that the osmolarity is rechecked as described above and that the medium is resterilized.

(11)   This medium is now checked to insure that it supports the growth of mouse embryos by the methods outlined in Chapter 5. Once the stock solution passes this quality control check, it can be used for IVF.

c.   Preparation of patient's serum for use in the Ham's F-10 medium (use gloves when handling blood or serum):

(1)   Blood (8 red-topped tubes) is drawn over a 3-day period during the early luteal phase in one of the patient's cycles previous to follicular induction for IVF-ET (for example, in a 28-day cycle blood would be drawn between days 15 and 17). This is done so that this serum can be screened in the mouse embryo development assay. It must be emphasized that there is no firm scientific support, to date, for the use of serum at this particular stage of the cycle. We have also been successful using serum drawn from the follicular phase of the cycle.

(2) Blood is allowed to clot for 2 hr at room temperature.

(3) Tubes are centrifuged at 900 x g (10 min; room temperature).

(4) Serum is drawn off with a Pasteur pipet, placed in a culture flask, and then heat inactivated at 56°C for 1 hr in a H_2O bath.

(5) Serum is then filter-sterilized (150-ml Falcon filter unit) and divided into aliquots for testing in the mouse embryo development assay and for the preparation of the IVF working medium. The aliquots are then stored frozen at -70°C and serum thawed only once for use. NOTE: Human fetal cord serum (obtained within 4 hr of collection), if used instead of maternal blood, is prepared in the same manner.

d. Preparation of working media for treatment cycles (University of Pennsylvania): Media is usually prepared the day prior to the scheduled oocyte retrieval. At this time the media required for the entire treatment cycle is prepared. (All of these procedures should be carried out under the laminar flow hood using sterile technique.)

(1) Instrument flush: 65.0 ml stock Ham's F-10 in two 115 ml tissue culture flasks.

(2) Insemination medium and sperm washing medium: 64.75 ml stock Ham's F-10 plus 5.25 ml patient's serum (7.5% serum final). Transfer 50 ml to one culture flask (label "Insemination medium") and 20 ml to a second culture flask (label "Sperm washing medium").

(3) Growth medium: 42.5 ml stock Ham's F-10 plus 7.5 ml patient's serum (15% serum final) in a culture flask.

(4) Embryo transfer medium: 1 ml stock Ham's F-10 plus 9 ml patient's serum (90% serum final) in a culture flask.

(5) Follicle flushing medium: 74.0 ml stock Ham's F-10, 6.0 ml patient's serum (7.5% serum final), plus 1.6 ml heparin (8000 USP units). Divide equally into two culture flasks.

(6) Cul-de-sac flushing medium: 50.0 ml stock Ham's F-10 plus 5000 USP units heparin in a culture flask. NOTE: After media are prepared, they are immediately filter sterilized (0.2 μm) using a Nalgene 115 ml type S filter unit, placed in the tissue culture flasks with the caps loosely tightened, and stored overnight at 37°C in the CO_2 incubator. It is strongly recommended that an

extensive coding system be used for labeling the media flasks. This
is especially important when more than one treatment is being carried
out in the same day (or at the same time). Our IVF laboratory labels
the flasks with the patient's name, treatment (or patient) number, a
color code, and the media description.

3.   TECHNIQUE 2 - UNIVERSITY OF WISCONSIN

      The following additional procedures are utilized in the IVF-ET
program at the University of Wisconsin.

      3.1. Chemicals

            a.   Human serum albumin (Sigma, fraction V)

      3.2. Supplies

            a.   50-ml disposable syringes
            b.   Millex GS filters with 0.22 µm pore size
            c.   Sterile 10-ml serological pipets/Pipump
            d.   Beakers

      3.3. Procedures used in the preparation of working media for
            treatment cycles:

            a.   Only use medium that has been prepared and tested for
IVF.

            b.   Prepare 30-40 ml of medium for each patient.

            c.   Preparation of Ham's F-10 containing human serum
albumin (HSA)

                  (1)   Weigh out the appropriate amount of HSA to give a
final concentration of 3 mg/ml (Example: for 30 ml of medium, weigh
out 90 mg of HSA).

                  (2)   Using a sterile pipet and Pipump, place the
appropriate volume of Ham's F-10 into a beaker.

                  (3)   Pour the preweighed HSA on top of the medium;
re-cover the beaker with the aluminum foil that covered it during
storage. Put the beaker aside for a few minutes, stirring occasionally; the HSA will go into solution with continuous stirring.

            d.   Remove the plunger from a 50-ml syringe and place a
Millex filter on the end of the barrel. (NOTE: If you put the
filter on first and then pull out the plunger, you will rupture the
filter.) TAKE CARE to keep the filter sterile.

Table I.   Troubleshooting guide for the preparation and use of
           culture media for human IVF-ET

Outcome	Probable reasons	Response
Failed fertilization and/or embryo development	Problem with media	Check the following parameters: -quality of H_2O -pH -osmolarity -ability to support mouse embryo development -check incubator gas concentrations
	Problem with protein source in media	-Check above parameters
Sperm do not survive	Problem with media and/or serum source	-Check above parameters
Sperm agglutinate	Presence of antisperm antibodies	-Test serum source for the presence of antibodies; use different serum source if necessary
Formation of a white precipitate in media	Precipitation of $CaCO_3$	-Add Ca^{2+}-lactate and $NaHCO_3$ as dilute solutions with stirring
Bacterial growth in media	Improper use of laminar flow hood	-Use proper tissue culture techniques in hood
	Dirty laminar flow hood	-Clean hood and check filters
	Incubator contamination	-Clean incubator
	Caps placed on culture tubes too loosely	-Make sure cap is screwed on but loose
	Ruptured sterilizing filter	-Use a hand operated vacuum pump; check rest of filters for damage

  e.   Pour the F-10 plus HSA into the barrel of the syringe
and hold the filter over a sterile 50-ml tissue culture flask.
Replace the plunger and express all of the medium through the filter
and into the flask. Do not force this as it will rupture the filter.
Discard the syringe and filter.

  f.   Label the flask with Contents, Date Prepared, and the
preparer's initials. Place the flask with cap loose into the incubator to equilibrate overnight.

g.   Medium for egg retrieval: Dulbecco's phosphate buffered saline with 10 IU/ml heparin sulfate.

h.   Insemination medium: Ham's F-10 plus 7.5% human fetal cord serum.

i.   Medium for moats of organ culture dishes: Ham's F-10 plus 3 mg/ml HSA or 7.5% patient's serum.

j.   Growth medium: Ham's F-10 plus 15% fetal cord serum.

k.   Transfer medium: Ham's F-10 plus 75% fetal cord serum.

4.   TROUBLESHOOTING

Table I lists a variety of potential problems in the preparation and use of culture medium for IVF-ET as well as possible solutions to such problems. The reader is also encouraged to examine Chapters 2 and 3 when troubleshooting laboratory problems.

5.   REFERENCES

Ball, G. D., Coulam, C. B., Field, C. S., Harms, R. W., Thie, J. T., and Byers, A. P., 1985, Effects of serum source on human fertilization and embryonic growth parameters in vitro, Fertil. Steril. 44:75-79.

Edwards, R. G., 1980, Conception in the Human Female, Academic Press, New York, 1087 pp.

Feichtinger, W., Kemeter, P., and Menezo, Y., 1986, The use of synthetic culture medium and patient serum for human in vitro fertilization and embryo replacement, J. In Vitro Fert. Embryo Transfer 3:87-92.

Kemeter, P. and Feichtinger, W., 1984, Pregnancy following in vitro fertilization and embryo transfer using pure human serum as culture and transfer medium, Fertil. Steril. 41:936-937.

Lee, M. A., Trucco, G. S., Bechtol, K. B., Wummer, N., Kopf, G. S., Blasco, L., and Storey, B. T., 1987, Capacitation and acrosome reactions in human spermatozoa monitored by a chlortetracycline fluorescence assay, Fertil. Steril. 48:649-658.

Leung, P. C. S., Gronow, M. J., Kellow, G. N., Lopata, A., Speirs, A. L., McBain, J. C., du Plessis, Y. P., and Johnston I., 1984, Serum supplement in human in vitro fertilization and embryo development, Fertil. Steril. 41:36-39.

Mann, T. and Lutwak-Mann, C. (eds.), 1981, Male Reproductive Function and Semen, Springer-Verlag, New York, 495 pp.

Mastroianni, L., Jr., and Biggers, J. D. (eds.), 1981, *Fertilization and Embryonic Development In Vitro*, Plenum Press, New York, 371 pp.

Menezo, Y., Testart, J., and Perrone, D., 1984, Serum is not necessary in human in vitro fertilization, early embryo culture, and transfer, *Fertil. Steril.* 42:750–755.

Rogers, B. J., 1978, Mammalian sperm capacitation and fertilization in vitro: a critique of methodology, *Gamete Res.* 1:165–223.

Saito, H., Berger, T., Mishell, D. R., Jr., and Marrs, R. P., 1984, The effect of serum fractions on embryo growth, *Fertil. Steril.* 41:761–765.

Wolf, D. P. and Quigley, M. M. (eds.), 1984, *Human In Vitro Fertilization and Embryo Transfer*, Plenum Press, New York, 440 pp.

5

MOUSE EMBRYO CULTURE BIOASSAY

Marybeth Gerrity

1. INTRODUCTION

The purpose of the mouse embryo bioassay, in the context of this presentation, is to serve as a quality control measure to assess the culture conditions used in human in vitro fertilization. Investigators using this assay often conclude, optimistically, that those conditions supporting mouse embryo development will also permit human embryo development. However, the converse is probably more accurate;

conditions which halt or retard mouse embryo development will
probably do the same in the human system. At best, the mouse embryo
bioassay is a crude assay that only rates the embryo toxicity of a
treatment or compares the relative merits of two conditions that are
known to support embryo development (for example, human fetal cord
serum versus maternal serum). The assay cannot differentiate between
growth promoting treatments. Furthermore, although mouse embryos
grow well in culture (Biggers et al., 1971; see also Chapter 18),
supporting their use for quality control comparisons, the transfer of
animal embryos cultured in vitro to surrogate wombs seldom results in
viable offspring. Therefore, it can be argued that the mouse embryo
system is artificial at best.

Despite its limitations, mouse embryo culture is the only simple
reproducible method available for testing the embryo toxicity of
culture conditions used in human IVF. The recent minimal standards
for IVF programs presented by the American College of Obstetrics and
Gynecology require use of this assay. Until a more sensitive system
is developed, therefore, this assay is an essential component of a
quality control program in human IVF.

The purpose of this chapter is to describe in detail the step-
by-step methods for performing this test. It is meant to be an
abbreviated format that can be followed as a laboratory manual. The
chapter also points out commonly encountered problems and how to
overcome them and suggests a protocol for appropriate setup and use
of the bioassay.

2.   MATERIALS

2.1. Chemicals/Media

2.1.1. Pregnant mare serum gonadotropin (PMSG) (Diosynth
        Corporation, Chicago, IL; Gestyl)

2.1.2. Human chorionic gonadotropin (Sigma Chemical
        Company, St. Louis, MO; #CG-10)

2.1.3. Dulbecco's phosphate buffered saline (Gibco Labs,
        Grand Island, NY; #450-1300)

2.1.4. Ham's F-10 culture media (Gibco Labs, Grand Island,
        NY; #430-1200)

2.1.5. BWW media (optional)

2.1.6. Human serum albumin (Fraction V, Sigma Chemical
        Company, St. Louis, MO; #A-1653)

2.1.7. Serum, media or plasticware to be tested.

2.2. Supplies Needed

    2.2.1. 35 mm disposable petri dishes, tissue culture grade (Falcon 3001, Becton Dickinson, Oxnard, CA)

    2.2.2. Syringes and needles for injection (usually 1-cc tuberculin syringe with 25-gauge needle)

    2.2.3. Dissecting instruments (coarse scissors, coarse mouse tooth forceps, iris scissors and fine watch-maker forceps)

    2.2.4. 30-gauge x 1-inch hypodermic needle (Yale B-D) and 1-cc syringe for flushing oviduct

    2.2.5. Pipets for handling mouse embryos and Pipump (Bel-Art Corp., Pequannock, NJ) (usually Pasteur pipet pulled to a fine bore over a low flame)

    2.2.6. 1-ml microtiter plates (optional) (Falcon 3047)

    2.2.7. Squeeze bottle of 95% alcohol (optional)

3. EQUIPMENT

3.1. Dissecting microscope

3.2. Slide warmer

3.3. Standard CO_2 incubator (37°C, 5% CO_2 in air, 99% humidity)

4. ANIMALS

    F_1 of C_3B_6 (Charles River, Wilmington, MA): male and female. This strain of mice responds well to superovulation and will undergo embryo development from 2-cell stage to blastocyst in culture in 72 hr.

    In selecting a strain of animals for use in this bioassay, it is essential to determine whether development from 2-cell to blastocyst does occur in the strain selected (i.e., is there a 2-cell block in the strain that is intended for use?). The presence of a 2-cell block can be investigated by harvesting both 2-cell and 4-8 cell embryos from the strain of mouse in question by simply altering the embryo collection times. These embryos are then cultured under the usual conditions. If development of the 4-8 cell embryos proceeds to

the blastocyst stage while the 2-cell embryos have stopped development, a 2-cell block exists. If development is poor in both groups,
a problem with the culture medium should be suspected. The response
of the mouse strain to superovulation regimens is also important; a
minimum of 30-40 embryos from each animal is the goal. Finally,
expense and availability in each geographic area may be important.
It may also be cost effective to use a less expensive strain of male
mice. For a summary of these characteristics in various strains of
mice, see Ackerman et al. (1983, 1984, 1985).

Mice should be housed under a 12 hr light:dark cycle (lights on
at 0600) with standard laboratory chow and water available ad
libitum. Female mice may be group-housed according to appropriate
American Association for Accreditation of Laboratory Animal Care
standards (usually 4-5 to a cage). Male mice must be individually
housed. For best results in mating, always add female mice to the
male cage, not the reverse. Use only sexually mature animals (6-8
weeks of age); avoid using female animals that are aged (over 3
months). Replace all males every 4-5 months. Female mice may be
randomly selected for injection without attention to the stage of the
estrus cycle. All animals must acclimate to their new environment
for one week after shipping.

5.   GONADOTROPINS (PMSG + hCG)

   a.   Prepare stock solutions at a concentration of 100 IU/ml in
         saline or distilled water.

   b.   Freeze in small aliquots convenient for a single use. Do
         not refreeze gonadotropin after thawing.

   c.   Store PMSG and hCG in a freezer at -70°C or in a freezer
         that is not self defrosting or frost-free. The freeze/thaw
         cycle of a standard kitchen frost-free refrigerator will
         inactivate frozen gonadotropins rapidly.

   d.   Never filter sterilize gonadotropins; they bind tightly to
         filters leaving a biologically inactive (though sterile)
         solution.

6.   ANIMAL PREPARATION (see summary in Table I)

   a.   To superovulate female animals, inject intraperitoneally
         with 7 IU of PMSG. For best results, inject animals
         between 3 and 6 p.m.

   b.   48 hr after PMSG injection, inject animals with 7 IU of
         hCG. For best results, inject animals between 3 and 6 p.m.

TABLE I.   Schedule for superstimulation of $F_1C_3B_6$ mice

Procedure:

PMSG injection (3-6 p.m.)	hCG injection (3-6 p.m.)	Recovery: 1-cell embryo (3-6 p.m.)	Recovery: 2-cell embryo (8-10 a.m.)
Monday	Wednesday	Thursday	Friday
Tuesday	Thursday	Friday	Saturday
Wednesday	Friday	Saturday	Sunday
Thursday	Saturday	Sunday	Monday
Friday	Sunday	Monday	Tuesday
Saturday	Monday	Tuesday	Wednesday
Sunday	Tuesday	Wednesday	Thursday

c.   Immediately after giving the hCG injection, place one female mouse into the cage of one male mouse. For best results, do not exceed 2 female mice per male mouse.

d.   Time of Sacrifice
Approximately 24 hr after hCG injection for 1-cell embryos. Approximately 36 hr after hCG injection for 2-cell embryos. (These times will vary somewhat with the strain of mice used.)

e.   For guidelines on how many female mice to use for each assay, see Section 7 entitled "Setup and Preparation."

7.   SETUP AND PREPARATION

a.   Place 1 ml of prewarmed Dulbecco's PBS in each of several 35-mm petri dishes (one dish per mouse). Place on a slide warmer to maintain approximate 37°C before use. Prepare several extra dishes of PBS for filling the flush syringe.

b.   Preequilibrate media and sera to be tested in incubator containing 5% CO_2 in air preferably overnight.

c.   When testing media or sera, the following guidelines are useful. (These guidelines are illustrated in Table II.)

(1) A control medium that has been previously tested must be used in addition to test medium. Some laboratories make use of a standard medium that is known to support

TABLE II. Example of typical setup for a Falcon 3047 microtiter
           plate to test five different treatments with four
           different mice

Sample 1 ml microtiter plate (Falcon 3047) for
mouse embryo culture

Mouse number[b]	Treatment[a]				
	1	2	3	4	5
A 1	O	O	O	O	O
B 2	O	O	O	O	O
C 3	O	O	O	O	O
D 4	O	O	O	O	O

[a]Treatments (sample to be tested)
  1. Control medium – either BWW containing 3 mg/ml of human serum
      albumin (HSA) or Ham's F-10 (previously tested Batch 100) + 3
      mg/ml HSA
  2. Unknown Batch Ham's F-10 (e.g., Batch 101) + 3 mg/ml HSA
  3. Unknown fetal cord serum (FCS) Batch 001 in Batch 100 Ham's
      F-10
  4. Unknown FCS Batch 002 in Batch 100 Ham's F-10
  5. Unknown FCS Batch 003 in Batch 100 Ham's F-10
  Where Batch 100 = previously tested Ham's F-10
        Batch 101 = unknown new batch of Ham's F-10
[b]All of the embryos from each mouse to be evenly split across the
five treatments.

mouse embryo development, such as BWW medium or Tyrode
solution, for this purpose. I make use of a previously tested batch of Ham's F-10 medium that is less
than 3 weeks old.

(2) When testing unknown (untested) sera, always test in a
     culture medium that has been previously tested.

(3) A protein source must be added to the control medium
     as well as to the medium to be tested. The protein
     source must be a previously tested batch of serum or
     human serum albumin (Fraction V); concentration 3
     mg/ml (this source of protein should not be included
     in medium which is used to screen untested sera).

(4) There is a high degree of mouse-to-mouse variability in the rate of fertilization of ova and in the rate and number of embryos to complete development. Therefore, all of the embryos from one mouse should be pooled and divided evenly among treatment conditions.

(5) Do not pool all embryos from all mice and then divide because of the large interanimal variation in development. If all the poor embryos are not evenly distributed across treatments, it may bias interpretation of the test results. Mice that yield greater than 25% fragmented, unfertilized or 1-cell embryos or embryos that have developed beyond the 2-cell stage should be excluded from the testing procedure.

(6) For a sample of assay setup, see Table II. For convenience, 1-ml microtiter plates (Falcon 3047) containing 1 ml of test media in each well may be used. These trays have 4 rows (A-D) and 5 columns (1-5).

(7) The setup of the test wells (the bioassay scheme) may be recorded in Data Sheet 1 (Figure 1). An example of the use of Data Sheet 1 for the scheme in Table II is presented in Figure 2.

d. Guidelines for deciding on number of mice to inject.

(1) Each mouse should yield 30-40 embryos.

(2) It is best to have at least five embryos per treatment; therefore, one mouse could test five treatments easily. Replicates are required so additional mice must be injected.

(3) Not all mice will superovulate.

(4) For the example given in Table II, five mice would have been injected. At least five embryos from Mouse 1 would be placed in Wells A_1 through A_6. This process would be repeated with Mouse 2 in row B and so on.

e. When testing a new batch of plasticware to determine if it is embryo toxic, always be certain to use embryos obtained from the same mouse grown in the same medium but placed in previously tested plasticware as a control.

    f.   Media and sera for IVF should be tested in the incubator used for IVF as a quality control measure to evaluate the entire embryo culture system.

8.   REMOVAL OF OVIDUCTS

    a.   Sacrifice animals by cervical dislocation.

    b.   Soak the animal with 70% alcohol to minimize hair contamination when the abdomen is opened.

    c.   Use coarse instruments to open the abdomen; pneumothorax the animal. To accomplish this, make a midline cut from the pubic symphysis through the diaphragm using coarse scissors and forceps. Extend the cuts laterally to flap the skin back.

Mouse embryo culture: Data Sheet   (I)

Date of PMSG_____   Date of hCG_____
Date of embryo collection (animal sacrifice) _____
Experimenter _____
Experiment # _____

Treatments to be checked:

Dish/Well	Mouse #	Medium	Protein Source	Other

Fig. 1.   Mouse embryo culture bioassay: Data Sheet I.

    d.   Using fine instruments, isolate the reproductive tract and identify the oviducts (refer to Figure 3).

    e.   When removing the oviduct, it is important to avoid handling or squeezing the oviduct itself. For ease of handling, first sever the uterine horn as shown in Figure 3, Cut 1; use this piece of tissue as a kind of handle for grasping the tissue. Next cut the oviduct between the ovary and the oviduct as shown in Figure 3, Cut 2. At this point, the oviduct is free from the mouse and is held by a piece of uterine horn. Hold this piece of tissue over a 30-mm petri dish containing warm PBS. Holding the uterine horn, cut between the uterine horn and the oviduct as in Figure 3, Cut 3, allowing the oviduct to drop into the warm PBS.

    f.   Repeat with the second oviduct.

Mouse embryo culture: Data Sheet   (I)

Date of PMSG_____   Date of hCG_____
Date of embryo collection (animal sacrifice) _____
Experimenter _____
Experiment # _____

Treatments to be checked:

Dish/Well	Mouse #	Medium	Protein Source	Other
A_{12}	1	Control Media Batch 101	3 mg/ml HSA	
A_2	1	Unknown Batch 101	3 mg/ml HSA	
A_3	1	Batch 100	FCS 001	
A_4	1	Batch 100	FCS 002	
A_5	1	Batch 100	FCS 003	
B_{1-5}	2	Batch 100	FCS 003	
C_{1-5}	3	Batch 100	FCS 003	
D_{1-5}	4	Batch 100	Batch 100	

Fig. 2.   Mouse embryo culture bioassay: Example of data sheet use.

g. For best results, flush these two oviducts before proceeding to sacrifice the next mouse. Do not pool oviducts from
several mice in the same dish.

9. FLUSHING THE OVIDUCTS

a. Fill a 1-cc syringe with warm PBS and place a 30-gauge, 1-
inch needle on the end. Sterile disposable needles of this
size are available from Becton Dickinson and are convenient
and inexpensive to use. The sharp end of this needle is a
problem for some investigators but practice can eliminate
puncturing the sides of the oviduct. Some groups prefer to
blunt the end of the needle using an emery board; these
needles must be resterilized before use. Custom made

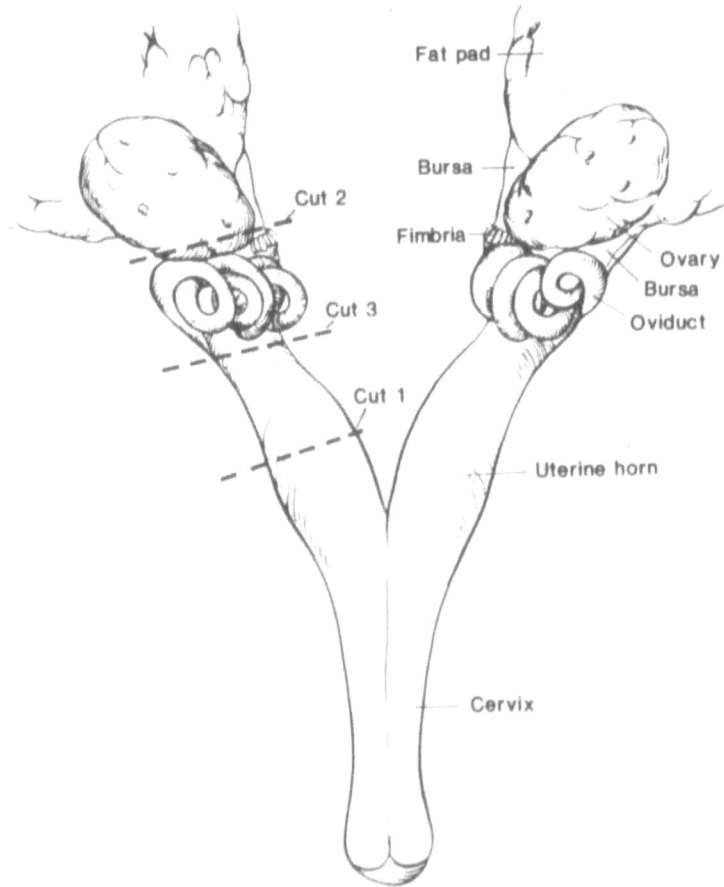

Fig. 3. Schematic depiction of female mouse reproductive tract and
surgical removal of oviducts.

needles that have no bevel or point can also be ordered
from several manufacturers but these needles are expensive
and are not disposable.

b.   Place the 35-mm petri dish on the stage of the dissecting
     microscope and examine the oviduct grossly to identify the
     fimbria (refer to Figure 3) and the uterine end of the
     oviduct.

c.   Carefully work the 30-gauge needle into the fimbriated end
     of the oviduct, using the forceps to grasp the oviduct.
     (Very fine watchmakers' forceps will make handling
     simpler.)

d.   Gently press the plunger on the syringe and watch the
     embryos flush out of the uterine end of the oviduct. If
     the needle is properly placed, the injection of medium
     should partially distend the oviduct. Flushing should
     require no more than 0.1 ml of medium. The smaller the
     volume flushed, the easier will be the recovery of embryos
     once these have been flushed into the dish.

e.   If the needle goes through the wall of the oviduct, with-
     draw it partially and attempt to pass the needle beyond the
     puncture site, grasping the oviduct distally; flush again.

f.   It is also possible to flush from the uterine end of the
     oviduct. This end has a more muscular wall and some will
     prefer to handle the oviduct at this end.

g.   Check the dish to ensure that the embryos have been
     flushed.

h.   If none of the above techniques work, it is also possible
     to use the needle and forceps to break the oviducts
     allowing the embryos to spill out. This method is much
     more time consuming, however, and the number of embryos
     obtained may be reduced.

i.   Repeat the procedure with the second oviduct. Pool the
     embryos from both oviducts. Separate out all embryos that
     are not 2-cell embryos (including fragmented, 1-cell
     embryos and embryos at 4-cell or beyond). When collecting
     2-cell embryos, if more than 25% of the embryos recovered
     are unfertilized, fragmented, 1-cell or have developed
     beyond the 2-cell stage, that mouse should not be used in
     testing and all of the embryos should be discarded. Place
     equal numbers of 2-cell embryos in each treatment to be
     evaluated as discussed in Section 7.

Mouse embryo culture: Data Sheet (II)

Date of Embryo Collection (Animal Sacrifice) _____
Experimenter _____
Experiment # _____

Time after flushing:

Dish/ Well	Total # Embryos	1-Cell or Fragments	Stages of Embryo Development:					
			2-Cell	3-4	6-8	16-32	Morula	Blastocyst

Fig. 4.   Mouse embryo culture bioassay: Data Sheet II.

j.   It is most convenient to handle the embryos using a Pasteur
      pipet pulled to a fine bore over a low flame and attached
      to a PiPump.

k.   Place the microtiter plate containing the embryos in the
      incubator and repeat the procedures above with each mouse
      in sequence.

l.   Record the total number of embryos in each well on Data
      Sheet II (Figure 4).

10. MONITORING AND ANALYZING RESULTS

     a.   Mouse embryos should be observed at 24-hr intervals after
           placing in culture (i.e., 24, 48 and 72 hr after collec-
           tion). Embryos may be examined directly under the
           dissecting microscope in the test wells of the microtiter
           plate or in petri dishes. The method used is a matter of
           personal preference. Petri dishes have a large surface
           area to scan in order to find and examine all of the
           embryos. While microtiter plates confine the embryos to a

small area for observation, the optical properties of these dishes often make it difficult to see embryos which settle around the edges of the well. Flat-sided culture tubes have also been used for these purposes but necessitate a large volume of media and attention to adequate equilibration of the media. Finally, some investigators prefer to place the embryos on a slide with coverslip; this method only permits observation of the embryos at the final observation time point.

b.   The total number of embryos in each well should be verified and the stage of development noted each day that the embryos are observed. (See Data Sheet II, Figure 4, for suggested format. If this Data Sheet is used, a new one will be filled out at each observation time.) Care must be taken to make these observations rapidly since the pH and temperature of the media will change while the embryos are outside of the incubator. Such changes may retard embryonic development. For research applications, it is particularly useful to see if the rate of development is altered and at what stage development stopped. For the strain of mice used here ($F_1C_3B_6$), embryos should progress from 2-cell to hatched blastocyst in 72 hr. Some labs prefer to make a single observation at 72 hr and record the percentage of total embryos that have progressed from 2-cell to hatched blastocyst. If 75% of the embryos do not progress from 2-cell to hatched blastocyst in 72 hr, the serum, media or other test substance should be considered less than optimal and, therefore, discarded. Since mouse embryos are quite hardy when grown in culture, conditions which retard development should not be taken lightly. Media that "comes close to passing" should be retested before use. This would be reflected on Data Sheet 2 at 72 hr after embryo collection. Inspection of results at 24 and 48 hr will indicate when development stopped. Figure 5 shows examples of several different stages of mouse embryo development.

c.   Timing of Development

Stage of development	Time (hours)
1 cell	-12
2 cell	0 (time placed in culture)
4-8 cell	24
Morula	36
Early blastocyst	48
Hatched blastocyst	72

In general, mouse embryos that are placed in culture at the
1-cell stage develop more slowly but should still reach the
hatched blastocyst stage in 84 hr.

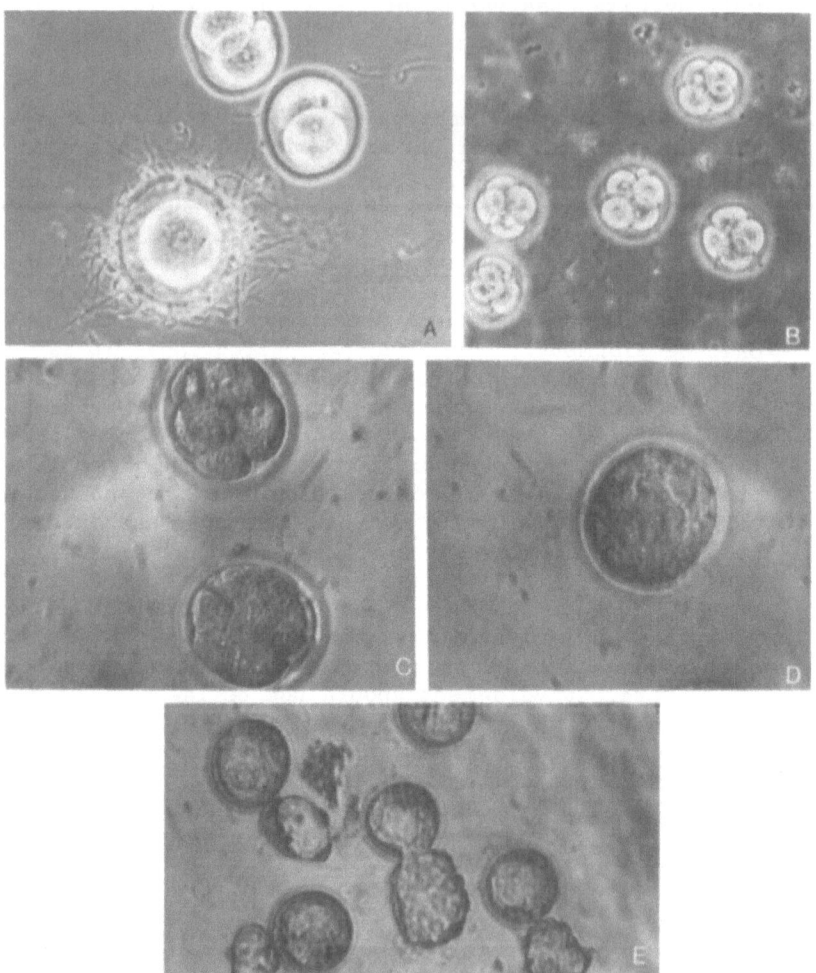

Fig. 5.   Examples of the sequence of development of mouse embryos in
           culture

           A) 1-cell fertilized and 2-cell (beginning point of assay)
           B) 4-cell
           C) morula
           D) early cavitating blastocyst
           E) hatching blastocyst (end point of assay)

d.   Returning to the example presented in Table II, comparisons
      made are within rows (Control A_1 is compared to Treatments
      A_2-A_6). Comparison between rows are an indication of the
      interanimal variation. From a statistical viewpoint, this
      interanimal variation must be taken into account before
      valid statistical comparisons using pooled data between
      mice can be made. If examination of results from Mouse 1
      indicate that in the control treatment (A_1) only 50% of the
      embryos reached hatched blastocyst, this mouse should be
      dropped from the data set. (Please note that the purpose
      of this assay is to screen for embryo toxicity so a good
      control group is essential. Use of this assay for a
      research application would not permit dropping of data.)
      If all controls A_1, B_1, C_1, and D_1 showed slowed develop-
      ment, however, the assay should be repeated.

e.   The use of 1-cell mouse embryos in addition to or even in
      place of 2-cell embryos has been advocated (Quinn et al.,
      1984). These embryos are apparently more sensitive to
      adverse culture conditions than are 2-cell embryos. There
      are also different requirements for embryo development from
      the 1-cell to the 2-cell stage than for 2-cell and beyond,
      making their use a more rigorous quality control measure.

11.   USES OF THE BIOASSAY

All new batches of media should be tested using the mouse embryo
bioassay. Experience indicates that Ham's F-10 medium will support
mouse embryo development for up to a month after preparation but no
conclusions can be drawn concerning effects on human embryo development. Furthermore, the purity of the water (the largest single
component of the medium) will decline with time. In this laboratory,
the medium is discarded two weeks after preparation.

All sera used in the IVF program should be tested in the mouse
embryo bioassay after heat inactivation and filter sterilization.
Experience indicates that far more sera than media will be rejected
due to bioassay results. Although results vary from center to center
(Condon-Mahoney et al., 1985), experience in our program has demonstrated that less than 5% of media tested will be discarded based on
results of mouse embryo culture. However, careful screening of sera
leads to rejection of 25-30% of sera. Programs that do not test sera
might inadvertently expose human embryos to embryo toxic serum and
compromise success rates. The mouse embryo bioassay is not an all or
none test; a test sera may permit growth to a 4-cell, 8-cell or even
unhatched blastocyst stage. Obviously, the same sera used in the
human IVF setting may yield seemingly normal embryonic development
which actually halts after embryo transfer. The fact that human ova
undergo fertilization and cleavage up to the point of transfer should

not be taken as a sign that the sera are safe for embryos unless verified in the mouse system.

Testing serum drawn during the IVF stimulation cycle is often not possible because of time constraints. In addition, if clomid is used during ovulation induction, it has a direct embryo toxic effect (Laufer et al., 1983). If maternal serum is to be used as a protein source, therefore, it must be drawn in a cycle prior to the IVF cycle to permit adequate testing. All sera used for IVF that are not actual matched patient sera should be screened for HTLV-III, hepatitis surface antigen and core antibody as well. In cases where IVF is used as a treatment for immunologic infertility or for infertility of unknown etiology, the patient's own serum should not be used. Alternatives include the use of fetal cord serum or appropriately screened donor serum.

The use of serum may not be necessary in human IVF. Caro and Trounson (1984) compared media without added protein or supplemented with albumin or fetal cord serum for the ability to sustain mouse embryo development in vitro. The resultant embryos, when transferred to pseudopregnant recipients, produced pregnancy rates significantly higher in the group where no protein supplementation was used. The same authors recently did a similar study in a human IVF program (Caro and Trounson, 1986). In this report, 10% maternal serum in T6 culture medium was compared with T6 medium containing no protein. There was no significant difference in pregnancy rate in the group where no protein was used to supplement the media. A similar result has been described by European investigators (Menezo et al., 1984; Feichtinger et al., 1986). Feichtinger's group made use of commercially available Menezo B-2 medium. Thus far, these studies have not been done using Ham's F-10, the culture medium most widely used in the United States.

All new batches of plasticware, tubing, etc. or any material that comes into contact with human eggs or embryos should be tested for embryo toxicity. This is easily accomplished by placing known embryo-safe pretested medium in contact with the test surface (either by washing with medium or actually culturing embryos in it) and performing the assay as described. This is a useful method for testing the effectiveness of glassware washing techniques, cleansing of surgical instruments (aspiration needle), transfer catheters and any other item that comes in contact with the human eggs, sperm or embryos. In our program, only 1 of 15 batches of plasticware tested did not support development; this was observed with all of the plasticware in that particular case. Adverse effects of plasticware on embryo development have also been described by Schiewe et al. (1984).

In summary, when using this strain of mice, any condition that does not result in mouse embryo development from the 2-cell to hatched blastocyst in 72 hr should be discarded or reevaluated.

The limitations of this assay are often overlooked. This assay
can only detect conditions which are grossly and harshly embryo
toxic. It cannot detect or differentiate growth promoting factors.
Perhaps pregnancies in IVF programs are associated with the coinci-
dental use of serum containing a growth promoting factor that cannot
be discerned by the mouse embryo assay. The addition of protein to
the medium may absorb toxic substances as may the amino acids present
in Ham's F-10 medium. Therefore, the medium may quench the system,
damping the sensitivity of the assay. It is possible, therefore,
that mouse embryos will develop in conditions that might slow or stop
human development. In the human IVF-ET procedure, embryos are transferred at an early stage of development. It is possible that these
embryos stagger through one or two cleavage divisions after which
they cease development. This same phenomenon often occurs in the
mouse embryo system in the presence of embryo toxic substances but
only longer periods of observation allow detection of the problem.
Many labs have had experience with mouse embryos growing in a variety
of conditions that are less than optimal; these embryos appear to be
relatively indestructible when grown in Ham's F-10.

Some laboratories formerly made use of sperm viability assays
for testing culture media, sera, etc. However, human spermatozoa
from normal men are even more resistant to adverse conditions than
are mouse embryos. In addition, the end point of sperm viability is
usually a subjective parameter (for example, motility). This method,
therefore, has been largely abandoned. Perhaps this method of media
evaluation will be resurrected when computer assisted, objective
methods of measurement of sperm motility, velocity and even lateral
head displacement are more widely available. Finally, a hamster
sperm viability assay has been recently devised (Bavister and
Andrews, in press) which purports even greater sensitivity to embryo
toxic factors and is simpler, more chemically defined and more
economical than the mouse bioassay. A bioassay system for embryo
derived platelet activating factor has been described for differen-
tiating viable embryos. This assay may be useful for predicting
success of embryo transfer (O'Neill et al., 1985). Frozen mouse
embryos are now available commercially for use in IVF quality
control. Unfortunately, these embryos are frozen at the 8-cell stage
and have therefore bypassed the more sensitive stages of development.
Furthermore, to carry out adequate quality control testing weekly
would be prohibitively expensive.

In conclusion, there is a need for objective, sensitive and
reproducible methods for testing materials used in human IVF for both
embryo toxic and growth promoting factors. The limitations of the
mouse embryo bioassay are obvious; however, it is the standard of
practice in the field and should be required in all settings.

Table III. Troubleshooting in the mouse embryo culture bioassay

Problem	Probable causes	Suggested course of action
1. Animals do not superovulate	Hormones used for super-ovulation are inactive	Examine ovaries of sacrificed mice a) Ovaries smooth in appearance – PMSG inactive b) Ovaries contain many cysts – hCG inactive Verify conditions of hormone preparation and storage a) Date of preparation b) NO filter sterilization c) NO storage of hormones in frost free freezer d) Verify lot with manufacturer Check lighting in animal facility (12 hr light; dark)
2. Eggs unfertilized	Injection and mating schedule not coordinated Problem with male animals	Female animals should be placed in male cage immediately after hCG injection. No more than one male animal per cage. Male animals should be between 8 and 20 weeks old.
3. Abnormal or irregular cleavage (fragmentation)	Aged eggs	hCG should be given 48 hr after PMSG
4. Embryo development stops at 2-cell stage	2-cell block exists in mouse strain selected	Culture both 2-cell and 4-cell embryos to the blastocyst stage. If 4-cell development > 2-cell development, a 2-cell block exists. Change mouse strain used.
5. Embryo development stops at points beyond 2-cell stage including the control groups	Problem with assay conditions	Verify correct operation of incubator. Verify adequate equilibration of media. Check media pH, osmolarity, and preparation. Check batch of plasticware used. Prepare new media. Borrow known media from another lab.
6. Embryo development stops at points beyond 2-cell stage in test treatment	Culture media or condition problem exists	Use appropriate control groups (see this chapter). If the control group development is normal, the test conditions are the problem. Differentiation of this group is the purpose of the assay.

12. TROUBLESHOOTING

There are a number of steps involved in this bioassay and therefore troubleshooting difficulties that arise require attention to each component. A troubleshooting scheme is described in Table III.

13. REFERENCES

Ackerman, S. B., Swanson, R. J., Adams, P. J., and Wortham, J. W. E., Jr., 1983, Comparison of strains and culture media used for mouse in vitro fertilization, Gamete Res. 7:103–109.

Ackerman, S. B., Swanson, R. J., Stokes, G. K., and Veeck, L. L., 1984, Culture of mouse preimplantation embryos as a quality control assay for human in vitro fertilization, Gamete Res. 9:145–152.

Ackerman, S. B., Stokes, G. L., Swanson, R. J., Taylor, S. P., and Fenwick, L., 1985, Toxicity testing for human in vitro fertilization programs, J. In Vitro Fert. Embryo Transfer 2:132–137.

American College of Obstetrics and Gynecology (600 Maryland Avenue, S.W., Washington, D.C. 20024), Human In Vitro Fertilization and Embryo Placement.

Bavister, B. D. and Andrews, J. C., A rapid sperm motility bioassay procedure for quality–control testing of water and culture media, J. In Vitro Fert. Embryo Transfer, in press.

Biggers, J. D., Whitten, W. K., and Whittingham, D. G., 1971, The culture of mouse embryos in vitro, in: Methods in Mammalian Embryology (J. C. Daniel, Jr., ed.), W. H. Freeman and Company, San Francisco, pp. 86–116.

Caro, C. M. and Trounson, A., 1984, The effect of protein on preimplantation mouse embryo development in vitro, J. In Vitro Fert. Embryo Transfer 1:183–187.

Caro, C. M. and Trounson, A., 1986, Successful fertilization, embryo development, and pregnancy in human in vitro fertilization (IVF) using a chemically defined culture medium containing no protein, J. In Vitro Fert. Embryo Transfer 3:215–217.

Condon–Mahoney, M., Wortham, J. W. E., Jr., Bundren, J. C., Witmyer, J., and Shirley, B., 1985, Evaluation of human fetal cord sera, Ham's F-10 medium, and in vitro culture materials with a mouse in vivo fertilization system, Fertil. Steril. 44:521–525.

Feichtinger, W., Kemeter, P., and Menezo, Y., 1986, The use of synthetic culture medium and patient serum for human in vitro fertilization and embryo replacement, J. In Vitro Fert. Embryo Transfer 3:87–92.

Laufer, N., Pratt, B. M., DeCherney, A. H., Naftolin, F., Merino, M., and Markert, C. L., 1983, The in vivo and in vitro effects of clomiphene citrate on ovulation, fertilization, and development of cultured mouse oocytes, Am. J. Obstet. Gynecol. 147:633–639.

Menezo, Y., Testart, J., and Perrone, D., 1984, Serum is not neces-
    sary in human in vitro fertilization, early embryo culture, and
    transfer, Fertil. Steril. 42:750–755.
O'Neill, C., Pike, I. L., Porter, R. N., Gidley-Baird, A. A.,
    Sinosich, M. J., and Saunders, D. M., 1985, Maternal recognition
    of pregnancy prior to implantation: methods for monitoring
    embryonic viability in vitro and in vivo, Ann. N.Y. Acad. Sci.
    442:429–439.
Quinn, P., Warnes, G. M., Kerin, J. F., and Kirby, C., 1984, Culture
    factors in relation to the success of human in vitro fertiliza-
    tion and embryo transfer, Fertil. Steril. 41:202–209.
Schiewe, M. C., Schmidt, P. M., Bush, M., and Wildt, D. E., 1984,
    Effect of absorbed/retained ethylene oxide in plastic culture
    dishes on embryo development in vitro, Theriogenology 21:260
    (abstract).

6

PREPARATION AND ANALYSIS OF SEMEN SAMPLES

Gregory S. Kopf

1.   PREPARATION OF SEMEN SAMPLES

    1.1. Introduction

        Washing of spermatozoa to ensure the removal of seminal
plasma is an important step in the preparation of the semen sample
for in vitro fertilization (IVF). A variety of seminal plasma-
associated factors have been shown to influence sperm function and
fertilization, in both positive and detrimental ways (Van der Ven et
al., 1982, 1983; Rogers et al., 1983; Sokoloski and Wolf, 1983).
Since the seminal plasma constituents of the semen are usually not
present (or are tremendously diluted) at the site of fertilization in
vivo, it is important that the sperm be washed free from these components prior to insemination in vitro. Since the effects of seminal
plasma on sperm may be variable and dependent on the semen source,
the semen sample is routinely washed as soon as possible after liquefaction has occurred.

        There are two basic methods for washing the semen sample to
remove the seminal plasma, and each method has both advantages and
disadvantages. One of the most common methods employed is washing by
repeated centrifugation and resuspension in culture medium (Rogers et
al., 1979). Although this method has been successfully utilized for
the preparation of sperm for IVF, the repeated centrifugations and
pipet manipulations used in this procedure may damage the cells. Its
advantage lies in its effectiveness in separating seminal plasma
components from sperm.

        An alternative method that is employed by many IVF laboratories
is a swim-up procedure in which centrifuged sperm are either overlaid
with culture media or undiluted semen is layered under the media
(Overstreet et al., 1980). These techniques rely on the ability of
the most motile cells to swim up into the culture medium. This
procedure selects for the most motile cell population and results in
a sperm preparation relatively free of dead sperm, immature sperm
(spermatogenic cells), noncellular debris, and nonsperm cells. A

disadvantage to this procedure is the lower recovery of motile sperm as compared to centrifugation-resuspension techniques. This recovery problem may be particularly important to consider when working with patients having extremely low sperm counts and motilities, although reductions in the swim-up volumes may circumvent such problems. The washing of compromised semen samples will be considered in Section 1.5.2, as well as in Section 2, the techniques of semen analysis.

The IVF laboratory at the Hospital of the University of Pennsylvania uses a swim-up washing technique for the preparation of sperm from patients with normal semen parameters. The technique is simple, yields a vigorously motile sperm population, and has given excellent results with regard to fertilization in vitro.

1.2. Chemicals

Sperm washing medium: modified Ham's F-10 medium containing 7.5% heat-inactivated patient's serum, pH 7.35-7.45. This medium (20 ml) is stored in a 115-ml tissue culture flask (Falcon 3013) and has been filter sterilized.

1.3. Supplies

a. Sterile plastic specimen jars
b. Sterile Falcon 2098 15-ml conical centrifuge tubes
c. Sterile, disposable Pasteur pipets (borosilicate glass)
d. Sterile, disposable micropipet tips
e. Microscope slides and coverslips
f. Hemacytometer
g. Sterile, disposable 1-ml and 10-ml serological pipets
h. Pi-pump or pro-pipet
i. Plastic disposable gloves

1.4. Equipment

a. Clinical table-top centrifuge
b. Light microscope
c. CO_2 in air incubator (or triple gas incubator)
d. Micropipets

1.5. Procedures

NOTE: The handling of biological fluids (i.e., serum, semen) dictates that proper precautions be taken to insure that the potential for personnel contact with such fluids be minimal, since it is known that these fluids can be contaminated with such pathogens as hepatitis B and AIDS virus. It is advisable that the laboratory personnel be acquainted with all the necessary precautionary procedures, and that the laboratory maintains a notebook of protocols

describing methodology for the clean up of contaminated areas. An
appropriate and functioning disease-screening program of both
patients and personnel should be a part of every IVF program. The
design and institution of such a program should be done with the
consultation of specialists in infectious diseases and epidemiology.
For example, all husbands entered in the in vitro fertilization
program at the Hospital of the University of Pennsylvania are
required to be screened for hepatitis B prior to acceptance into the
program.

1.5.1.   Normal Semen Specimens

        a.   Husband should abstain sexually for a period
of 48-72 hr prior to producing an ejaculate to be used in IVF.

        b.   Since we generally culture morphologically
mature eggs for a period of 4-5 hr prior to insemination, we request
a semen sample from the husband approximately 2 hr prior to anticipated insemination. This same time schedule would also apply to
morphologically immature oocytes that would be cultured for 30-36 hr
prior to insemination. We routinely have frozen sperm from the
husband as a back-up in the event that he cannot provide us with a
semen sample on the day of the egg retrieval. To date, we have only
had to resort to such a back-up system four or five times. In order
to avert some of the stress factors associated with producing a semen
sample after the oocyte retrieval has been completed, some IVF laboratories (in particular, the University of Texas, Houston) request
that the semen sample be provided prior to the egg retrieval.

        c.   The husband's name and time of semen collec-
tion is labeled on the sterile disposable specimen container upon
receipt in the laboratory.

        d.   The specimen is then allowed to liquefy for
a period of 30-45 min at room temperature prior to processing.

        e.   General techniques for routine semen
analysis should be followed (see Section 2.5.1.).

        f.   The volume of the ejaculate is determined
and the following parameters assessed on the unwashed ejaculate:

            (1)   sperm concentration
            (2)   percent motility
            (3)   sperm progression
            (4)   semen viscosity
            (5)   bacterial contamination and presence of
                  acellular debris.

The techniques used to make these measurements are found in Section 2.5.

       g. The remainder of the semen sample is then transferred to a 15-ml sterile, disposable conical centrifuge tube, and twice the semen volume of sperm washing medium is added (see Chapter 4 on Choice, Preparation and Use of Culture Medium for the preparation of this medium). The tube is then mixed by inversion and centrifuged at 200 x g for 5 min. If a problem is encountered in mixing the semen sample with the washing medium, a sterile Pasteur pipet could be used to mix the sample.

       h. The resultant supernatant is removed with a sterile Pasteur pipet and placed in another tube. Since there will still be sperm in this supernatant fraction, save the fraction in the event that problems are encountered with the preparation of the pellet fraction. The pellet is then carefully overlaid with 0.2-1.0 ml sperm washing medium using a micropipet with sterile pipet tips (the volume of medium to be added is dependent on both the initial sperm count and resultant sperm pellet). The medium should be added very slowly down the side of the tube to avoid disturbing the sperm pellet.

       i. The tube is loosely capped and then carefully placed in the incubator (37°C; humidified 5% CO_2 in air) at a 45° angle for a period of 1 hr to allow the motile sperm to swim up into the upper layer of sperm washing medium (in some cases of low sperm counts, the sperm may be allowed to swim up for 90 min).

       j. Following this incubation, the uppermost layer of the supernatant is drawn off with a sterile Pasteur pipet and placed in a separate loosely-capped sterile tube. Sperm concentration, percent motility and progression is assessed in the sample prior to insemination. The volume of sperm required for insemination is then calculated. We routinely use 100,000 motile sperm/oocyte (in 1 ml) but are also currently inseminating with 50,000 and 25,000 motile sperm under certain circumstances.

       k. All of the sperm fractions obtained during the washing procedure should be retained in the incubator until it is confirmed that the eggs are fertilized. These samples may be useful if some (or all) of the eggs have to be reinseminated (see Section 1.5.3).

       l. Insemination is done in the laminar flow hood under the dissecting microscope, using a Pipetteman® with a sterile micropipet tip. The insemination volume is kept between 4 and 15 µl. Be sure to change pipet tips between insemination of the individual eggs to minimize potential cross-contamination of the in vitro cultures.

    m. The inseminated oocyte is then placed back
into the incubator for an additional 16–21 hr prior to observation
for signs of fertilization.

   1.5.2. Abnormal Semen Specimens

    a. The above procedures are appropriate for
semen specimens displaying normal counts and motilities. When a
semen sample of low sperm count (oligospermic) and/or motility is
encountered, the sample is prepared for IVF in a different manner.
The semen is diluted with the same proportions of sperm washing
medium but is centrifuged at higher speeds (400 x g; 5 min) in order
to recover as many cells as possible. The pellets are resuspended to
the original volume with sperm washing medium and the spin is
repeated. The pellet is resuspended to a small volume (50–200 µl)
with sperm washing medium and the sample is allowed to sit upright in
the incubator for 30–60 min. Dead cells will fall to the bottom of
the tube and the supernatant containing the motile cells is withdrawn
with a sterile micropipet. The sample is then counted and insemination carried out as described above. As described in Chapter 8,
inseminations are usually carried out with higher sperm concentrations (if possible) when an oligospermic patient is encountered.

    b. An additional consideration when working
with a sample displaying abnormal semen parameters might be to
perform a limited sperm survival test using sperm washed in the sperm
washing medium. This test should be carried out prior to the IVF
treatment cycle, and may be done in conjunction with the routine
semen analysis as a part of the male work-up for IVF. The sample is
washed as described above and incubated in the sperm washing medium
overnight in the CO_2 incubator at the concentration that would
normally be used in IVF inseminations. The percentage motility and
the quality of motility is then analyzed after the incubation (see
Section 2.5). Such a test might prove invaluable in assessing the
ability of sperm in these compromised samples to survive the insemination period.

    c. If samples displaying high degrees of vis-
cosity are encountered, they are first treated to reduce viscosity
before washing. Pipetting the sample up and down using a sterile
Pasteur pipet is usually very successful (be careful not to agitate
the sample so that foaming occurs). Alternatively the sample can be
carefully and slowly drawn up into a 5-ml sterile, disposable plastic
syringe with a 23-gauge needle attached and then carefully and slowly
expelled back into the tube. This procedure must be done carefully
so as not to break up the cells; it is only used as a last resort
when the viscosity problem is so severe that the pipetting technique
does not work.

          d.   Once the semen sample is washed, the insemi-
nation is performed in the same manner as that described above for
normal semen samples. If the husband is severely oligospermic and
attempts to optimize the sperm concentration of motile cells meets
with limited success, an alternative insemination technique is to
inseminate all of the eggs in a single culture dish. This is advisable only as a last resort since such a procedure could be potentially
risky (for example, all of the eggs could be lost if the culture dish
was dropped or contaminated). See Table I for a troubleshooting
guide.

1.5.3.   Reinsemination

          In the event that there is no evidence of
fertilization (pronuclear formation; polar body emission) of one or
all of the eggs, consideration should be made to reinseminate the
eggs. The lack of fertilization could be due to a number of reasons
(improper grading of maturity of the egg at the time of the initial
insemination, lack of sperm fertilizing capacity) which are not
usually readily discernible. If the husband has good semen parameters, it is advisable that a fresh semen sample be obtained and
washed for reinsemination. If this is not practical for a particular
reason (husband out of town or a long distance from the hospital),
the washed semen sample from the previous day could be used if the
motility was still adequate. It is for this reason that the entire
washed semen sample is saved after the initial insemination. In the
event of abnormal semen parameters, keep in mind that the fresh
sample might be worse than the original sample. The original sample
should be examined and used if it is better than the fresh sample.
In many cases this is a judgment call on the part of the laboratory.
Nevertheless, reinsemination is a simple procedure with no known
attributed risk. Studies in our laboratory have confirmed the value
of this approach (Ben-Rafael et al., 1986).

1.6. Discussion

          Although centrifugation-resuspension washing techniques are
not routinely used in our IVF laboratory for sperm preparation, one
such technique is described in detail by Wolf and co-workers (Wolf
and Quigley, 1984). We do not routinely utilize long capacitation
times (> 3 hr) prior to insemination and have opted for capacitation
of the highly motile sperm sample in the culture dish containing the
cumulus-enclosed oocyte. Our reasons for doing this are two-fold.
In the first instance, assessment of capacitation times using artificial systems (zona-free hamster eggs) have demonstrated that there
may be a wide interdonor variation in the time course of capacitation
(Perreault and Rogers, 1982; Wolf and Sokoloski, 1982; Rogers et al.,
1983). Secondly, there is evidence in man (Soupart and Morgenstern,
1973; Tarlatzis et al., 1984), as well as in other mammalian systems
(Bavister, 1982; Lenz et al., 1982; Bradley and Garbers, 1983), that

Table I. Troubleshooting guide for the preparation of semen samples
for human in vitro fertilization

Outcome	Probable reasons	Response
Sperm count very low	Oligospermia; inappropriate abstinence time	—Wash sperm by centrifugation and resuspension or reduce overlay volume for swim-up wash —Inseminate multiple eggs in a single culture dish —Inseminate eggs at a higher sperm concentration
Ejaculate displaying high viscosity	Abnormal semen characteristics	—Use a Pasteur pipet to break up sample —Draw sample through a syringe with a 23-gauge needle
Ejaculate displaying slow liquefaction times	Abnormal semen characteristics	—Use a Pasteur pipet to break up the sample
Husband unable to provide a semen specimen	Stress-related factor(s)	—Require that semen sample be provided before oocyte retrieval —Tell husband to relax and assure him that you can wait for the specimen
All sperm dead in the ejaculate	Some factor(s) involved in the method of semen collection	—Use frozen sperm back-up if it is available —If unavailable, determine potential collection problem and request another sample later in the day

the oocyte and/or its surrounding investments (e.g., cumulus
oophorus, corona radiata) may play some role in capacitation. Our
percentage fertilization using the above methods are similar to those
of other successful IVF-ET programs which utilize longer capacitation
times prior to in vitro insemination.

2.   TECHNIQUES OF SEMEN ANALYSIS

    2.1. Introduction

        Semen analysis has become an integral part of the clinical
work-up for male infertility, and is also routinely used in the work-

up for many IVF-ET programs. Assessment of sperm characteristics (e.g., concentration, motility, viability, morphology) as well as other semen components (e.g., contamination by bacteria, white and red blood cells, immature germ cells, epithelial cells) and characteristics (e.g., viscosity, color) may aid in the determination of semen quality. Despite the development of newer clinical laboratory techniques for assessing sperm function (e.g., sperm antibody screening, sperm penetration assay), the basic semen analysis remains the principle diagnostic tool in the evaluation of the male-factor in human IVF-ET programs. Other in vitro tests for assessing fertilizing ability of human spermatozoa should only be used as a corollary to, and not a replacement for, the conventional semen analysis. The IVF-ET program at the Hospital of the University of Pennsylvania requires that two semen analyses of the husband's ejaculate be performed prior to acceptance into the program, the results of which would enable an assessment of whether the couple can proceed into the treatment cycle. Additional analyses are also performed if deemed necessary. The most accurate evaluation of seminal fluid is made by considering the results of two or more semen analyses over a time period of several weeks.

There has been a growing need over the years for the standardization of techniques used in semen examination in order to maintain accuracy and reproducibility of results. This need has recently been addressed (Belsey et al., 1980; Cannon, 1984) and a more uniform basic format for the semen analysis has evolved.

2.2. Chemicals

   a.   Eosin B solution (0.5% in saline); filter sterilized
        (0.22 μm filter)
   b.   Methanol (analytical grade)
   c.   Harleco Diff-Quik® differential staining set

2.3. Supplies

   a.   Sterile plastic specimen jars
   b.   Disposable 5.75- and 9-inch Pasteur pipets
   c.   Microscope slides
   d.   Coverslips
   e.   Wooden applicator sticks
   f.   pH paper (pH range 6-9.5)
   g.   Coplin jars
   h.   Micropipets (20 ul to 200 μl volumes)
   i.   Disposable micropipet tips
   j.   Hemacytometer
   k.   Disposable vinyl gloves
   l.   15-ml sterile, graduated plastic conical test tubes
   m.   Five-place laboratory tally counter

n.   Slide warming plate (37°C)
o.   Disposable, sterile 0.2-μm media filters

2.4. Equipment

a.   Light microscope (10X, 20X, 40X objectives; bright-
     field and phase contrast)
b. CO_2 incubator (or triple gas)

2.5. Procedures

2.5.1.   General Guidelines

The following procedures for semen collection and
handling should be adhered to when preparing a specimen for analysis.
These guidelines will help to ensure the accuracy of semen analysis
results.

a.   The patient or sperm donor is instructed to
refrain from sexual activity (i.e., ejaculation) for 48-72 hr prior
to submitting a semen sample to the laboratory for analysis. A
period of sexual abstinence helps to ensure that the sperm count,
motility and viability are maximized in the analyzed specimen.
Excessive periods of abstinence are not advised as this has been
reported to result in suboptimal semen parameters (Cannon, 1984).

b.   The acceptable method of specimen collection
for semen analyses is masturbation. The specimen should be collected
in a sterile, plastic specimen container. Glass containers should be
avoided since they are not sterile and there is danger of breakage.
Specimens collected following coitus interruptus (i.e., "withdrawal")
should not be analyzed since this method does not provide a reliable,
representative sample. The specimen may become contaminated and/or
there may be a loss of the first portion of the ejaculate, which may
contain the highest concentration of spermatozoa. However, if it is
necessary that the sample be collected following intromission, a
special condom is available for this particular use. It is marketed
as the Seminal Collection Device® (HDC Corporation, 2551 Casey
Avenue, Mountain View, CA) and does not contain any lubricants or
spermicides.

c.   The patient must be instructed to collect
the entire specimen--not just a portion of it. It is important that
the time of collection be noted on the specimen container; if part of
the ejaculate is spilled, this should also be noted.

d.   It is preferable that the patient collect
the semen specimen on the hospital (office) premises, so as to ensure
that the sample is as fresh as possible when analyzed. When specimen
collection cannot be accomplished at the hospital/office (e.g., psy-

chological or physical reasons), it is advised that the sample be
delivered to the lab as soon as possible after collection. If this
situation occurs during cold weather, the specimen must be carried in
an insulated thermos bottle during transit in order to avoid temperature extremes (semen temperature should not fall below 20°C or rise
above 40°C during the interval between ejaculation and analysis).
Patients should be warned against storing semen in a home refrigerator.

            e.   Semen must be placed in a 37°C CO_2 incubator
immediately after collection, and stored there until analyzed. The
specimen container must be labeled appropriately with the patient's
full name, date and time of collection, and number of days of sexual
abstinence. The laboratory technician must record both the time that
the sample was received in the lab and the time of analysis on the
results form.

            f.   Semen must be allowed to liquefy at 37°C for
15-45 min following ejaculation. Before complete liquefaction, the
sample should not be pipetted or mixed and, therefore, cannot be
properly analyzed. Liquefaction of semen is initiated by enzymes
emanating from the prostatic secretions (Mann and Lutwak-Mann, 1981).
The prostate contributes approximately 20% of the volume of the
semen. In a normal sample, liquefaction should be complete within 30
min. Somewhat delayed liquefaction should not be confused with
persistently increased viscosity. For accurate and reliable results,
semen should be evaluated within one hour after collection. If
samples displaying high degrees of viscosity are encountered, they
must be first treated to reduce this viscosity problem before the
semen analysis is continued (see Section 2.5.2). When this problem
is encountered, pipetting the sample up and down using a Pasteur
pipet to break up the sample is usually very successful (be careful
not to agitate the sample so that foaming is observed). Alternatively, the sample can be carefully and slowly expelled back into a
tube. It must be emphasized that this must be done carefully so as
not to break up the cells. This method is only used as a last resort
when the viscosity problem is so severe that the pipetting technique
does not work. Attempting to prepare the semen specimen so that its
components are evenly distributed will aid in the proper evaluation
of the other semen parameters. It must be remembered, however, that
in breaking up the sample one may compromise the accurate measurement
of some of these parameters.

            g.   The semen should next be examined and any
abnormal physical characteristics should be noted in the comments
sections of the semen analysis report. Freshly ejaculated normal
semen is a viscid, opaque, white or gray-white coagulum with a
distinctly acrid odor.

h.   Following complete liquefaction, the sample
appears translucent and may be pipetted or poured into a 15-ml
sterile graduated test tube (prior to this the sample is too thick to
be analyzed).   At this time the total volume is measured and
recorded.   Semen is a composite suspension consisting of spermatozoa
suspended in seminal plasma, which is contributed by the accessory
reproductive organs.   Spermatozoa account for less than 5% of the
semen volume, with approximately 60% derived from the seminal
vesicles, 20% derived from the prostate, and the remaining 10-15%
from the epididymides, vasa deferentia, bulbourethral glands and
urethral glands (Mann and Lutwak-Mann, 1981).   The sample is then
well mixed by gentle pipetting (several times up and down) or by
using a vortex mixer set at low speed.   This must be done very gently
to avoid foaming.   It is essential that the sample be completely
mixed before the microscopic examination is begun.   Accurate results
can only be achieved with a uniform semen suspension.

i.   Personnel performing the analysis should
always wear disposable gloves and follow all procedures required when
handling any potentially pathological biological specimens.   Please
note, however, that the talc normally found in disposable gloves is
spermicidal.

j.   Lay out three microscope slides (cleaned and
prewarmed to 37°C).   Draw up a small amount of semen into a Pasteur
pipet and then place enough on the end of one slide to make a wet
mount for a motility determination (see Section 2.5.3.).   Place a
second, smaller drop on the other end of the slide and mix with
approximately the same volume of eosin B solution using a wooden
applicator for viability testing (see Section 2.5.5.).   Place a third
drop of semen on the second slide and make a smear using the third
slide.   Let this smear air dry overnight.   It will be stained for
morphology determinations (see Section 2.5.8.).   With the remaining
semen in the pipet, check the pH with the indicator paper.   The pH
should be in the 7.3 to 8.3 range in normal samples.

2.5.2.   Viscosity

When the semen is drawn up with the pipet, a
quick assessment of viscosity can be made qualitatively by observing
the ease with which the semen is pipetted.   Normal semen is neither
watery nor gelatinous, but instead should be easily pipettable.
Viscosity is ranked subjectively as follows:

Thin: watery

Normal: sample is readily drawn up into the pipet.

+(thin):   small strands that break easily between the tip of the
pipet and the specimen container.

++(thick): strand formation occurs between the tip of the pipet
and semen sample when pipetting.

+++(very thick): extensive strand formation; the bulk of the
semen comes back out of the pipet and falls back into the
specimen container.

   2.5.3.   Motility Testing

            a.   On the slide with the first drop of semen
(approximately 2-3 mm in diameter), prepare a wet mount and place
under the microscope using the high dry objective (400X magnification).

            b.   Several fields are scanned until a "typical"
field is located to count. This field should not be near the edges
of the coverslip, since these are not representative areas. At least
100 spermatozoa, both motile and immotile, are counted in this and
other representative 400X fields (tallied separately on a tally
counter). Any sperm cell which exhibits active movement is classified as "motile." Motility is then recorded as a direct percentage.

            c.   The speed of the sperm is also assessed and
defined as forward progression. This parameter is a classification
of the quality of sperm movement. It must be evaluated while viewing
the wet mount prepared for the motility determination, and several
microscopic fields should be viewed. This is a qualitative estimate,
and should represent the progression of the majority of motile cells.
Progression is ranked on a 1-4 scale as follows, with pluses (+) and
minuses (-) allowable for each ranking:

   0: no movement.

   1: extremely weak movement that generally results in no forward
   motion. Sperm moving in place.

   2: poor to moderate movement that is often seen as spermatozoa
   simply "fish-tailing" in place. There may be only slight
   forward gain. Erratic movement; movement in circles.

   3: normal or average progression; these cells display determined, purposeful forward movement.

   4: excellent progression; these cells are actively swimming in
   what appears to be almost a frenetic fashion. Tail movements
   are blurred due to increased velocity.

NOTE: While more than one type of progression may be seen in
the same sample, the examiner must make an estimation of the average
progression by viewing several fields.

d.   A normal semen sample will display 60% or greater motile sperm with a majority of the cells exhibiting good (3) forward progression.

2.5.4.   Sperm Agglutination Determination

a.   After a motility determination is made, the same wet mount can be used to determine the extent of sperm aggluti- nation. This is a qualitative determination and is expressed in the following manner:

None: all cells swimming freely.

+: presence of 2 or 3 cells stuck together.

++: presence of large clumps of sperm consisting of many cells, or a lot of smaller clumps.

+++: very large clumps of cells throughout the specimen.

b.   A rough percentage of the type of agglutina- tion can also be determined if it is desired.

c.   The types of agglutination are usually head- head, tail-tail, or head-tail.

d.   It is important to remember that agglutina- tion refers to sperm-sperm attachment, and not sperm attachment to cellular debris or glass surfaces.

e.   Also note that agglutination determinations should be made soon after the wet mount is prepared to avoid drying of the sample under the coverslip.

f.   Severe agglutination of spermatozoa would warrant a repeat semen analysis. This condition is suggestive of an infection in the genital tract or possible immunological problems.

2.5.5.   Viability Testing

The number of live (motile and immotile) and dead sperm is determined by staining the cells with a supravital dye such as eosin B (0.5% in saline). The second drop (2-3 mm in diameter) of semen is used for this purpose. A drop of eosin B stain is added to the second drop and the two are mixed using a wooden applicator stick. It is essential to use the proper amount of stain; the drop should be no more than 1/3 - 1/2 the size of the semen drop. The slide must be examined immediately (keep on a 37°C warming plate for short delays) so that the sperm still remain motile. After the sperm droplet is mixed with the dye, a wet mount is prepared and observed

at 400X under the light microscope. Dead sperm will display a
reddish head region, whereas live sperm do not take up the stain and
the head region will remain colorless. One hundred (100) sperm are
counted, and the following percentages classified as described below:

Alive-motile sperm: any motility by the sperm classifies it as
alive-motile.

Alive-immotile sperm: all nonmotile, unstained sperm.

Dead sperm: staining of the head region will be orange to pink
to dark red in appearance.

2.5.6.   Counts of Non-germ Cells

a.   Using the wet mount made for the motility
determination, examine four to five (400X magnification fields) for
the presence of while blood cells (WBC), red blood cells (RBC) and
epithelial cells (EPITH).

b.   Note in your determination if there are any
averages greater than five (5)/high power field (HPF) of the particular cell type. If below 5, list count as normal.

Example: WBC: normal
         RBC: 8/HPF
         EPITH: normal

c.   NOTE: It is often difficult to distinguish
WBC from germ cells in wet mounts under phase contrast.

d.   The presence of an abnormally high concen-
tration of white blood cells and bacteria in the semen may be indica-
tive of a localized infection. If this is encountered in the semen
analysis prior to the IVF treatment cycle, the patient should be put
on an antibiotic regimen under the advice of the primary care
physician or in consultation with a urologist (for example, a regimen
of tetracycline or vibramycin). It is advisable that the infection
be cleared up before starting the IVF treatment cycle.

2.5.7.   Assessment of Sperm Concentration

a.   The most common method utilized for doing
routine sperm counts involves the use of a Neubauer-type hemacytometer. This method is simple, quick and accurate.

b.   The undiluted semen specimen should be well
mixed prior to making dilutions for counting. The semen is diluted
1:4, 1:9, or 1:19 (part semen:part diluent), depending on the
estimated count (this rough estimate can be quickly made by counting

the number of sperm in a 400X field and multiplying by 10^6/ml; it must be emphasized that this is only a rough approximation and should not be substituted for an accurate count). If sperm concentrations are extremely high (> 150 x 10^6/ml), greater dilutions can be made. The diluent that we use is distilled H_2O (this will immobilize the sperm). Dilutions are made with micropipets (e.g., Eppendorf, Pipetteman, etc.) to insure accuracy, and these dilutions are usually made in a disposable glass test-tube. NOTE: Although commonly used, a white blood cell pipet is not precise enough for making this dilution.

c. The drop of diluted semen is then added to the hemacytometer (both wells) by means of a micropipet. The sample should be applied slowly, letting capillary action draw the sample under the coverslip. Care must be taken to avoid flooding the chambers since this may influence the accuracy of the results. The hemacytometer is allowed to sit for a period of 5 min while the cells settle out into a single plane. For reporting sperm concentration, both sides of the hemacytometer must be counted and the results averaged. Sperm density is then expressed directly in 10^6 cells/ml. Only morphologically mature sperm cells should be included in the count (e.g., whole, intact sperm cells with head, midpiece and tail). The hemacytometer is examined under a phase contrast microscope at a magnification of 200X or 400X.

d. As shown in Figure 1A, the standard Neubauer hemacytometer is constructed of a series of grids. Using the standard hemacytometer coverslip, the volume encompassed by each of the large grids (i.e., No. 1) is 0.1 μl. Three of these grids are each subdivided into 25 equal parts (i.e., No. 1a, 1b, etc.). When counting sperm, 5 of these subdivisions (1a-1e) are counted. Cells crossing the edges of these subdivisions are counted on only two sides of each subdivision (e.g., only upper and left edges) (Figure 1B).

e. NOTE: When dilutions of the semen sample are made, the dilution should be such that an average of 20-40 cells can be counted in each of the 5 boxes (1a-1e) so that a total number of 100-200 cells are counted on each grid. This minimizes counting error. Increased numbers of cells make counting more difficult and fewer cells compromise accuracy.

f. Calculations:

(1) to determine sperm concentration, (x 10^6 cells/ml):

n = a x 5 x b x 10,000, where ...

a = total number of sperm counted in the 5 subdivisions (1a-1e).

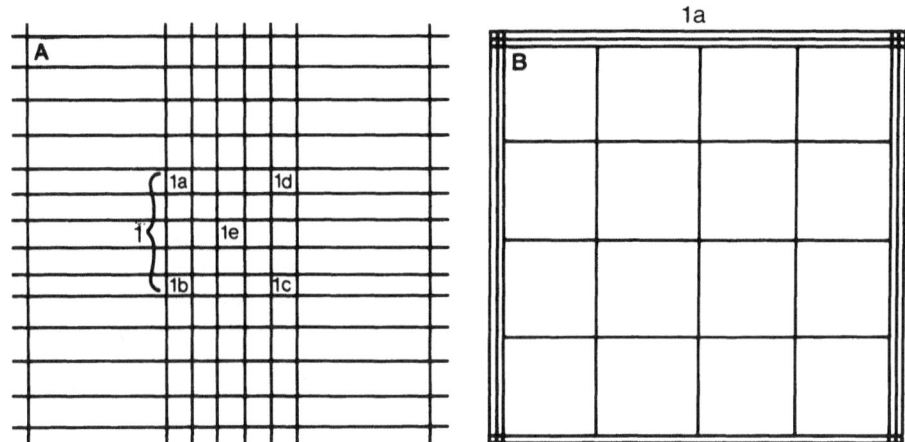

Fig. 1.   Diagrammatic representation of a standard Neubauer-type
           hemacytometer. The volume encompassed by the major cross-
           hatched grid (No. 1 in panel A) is 0.1 µl, using the
           standard hemacytometer coverslip. Panel B represents the
           subdivisions comprising each of the 25 boxes (i.e., 1a) in
           the major crosshatched grid.

b = total dilution factor of sperm.

10,000 = value obtained by dividing 1 ml (1000 µl) by 0.1 µl
(the volume of box No. 1 - see Figure 1A).

5 = correction value to obtain the total number of sperm in the
large grid (No. 1 in Figure 1A).

EXAMPLE: The following sperm numbers were obtained after
counting each of the 5 subdivisions (1a-1e, as in Figure 1A) in
a sample that was diluted 1:19 (20 µl semen into 380 µl of
distilled H_2O):

        20, 24, 22, 30, 29 = 125 = a

        dilution factor, 1:19 = 20 = b, then

        n = 125 x 5 x 20 x 10,000 = 125 x 10^6 sperm/ml.

        Therefore, at a semen dilution of 1:19, the total number of
        sperm in the 5 subdivisions (a) equals the sperm count (x
        10^6 cells/ml).

                        (2)   to determine the total sperm count (z):
z = n x volume of semen (ml)

2.5.8.   Assessment of Morphology

a.   The morphological characterization of semen
is as important to the analysis of the seminal fluid as are sperm
motility and concentration. To assess sperm morphology, stained
smears are prepared and differential counts made of morphologically
normal and abnormal sperm types.

b.   A single smear is made on a precleaned
microscope slide in the same manner in which a blood smear is
prepared. This technique is as follows: Using a Pasteur pipet,
apply a single fine streak of semen across the end of a slide. It is
important to use the proper amount of semen; if too little or too
much semen is applied, the smear will be either too thin or too thick
to read once it is stained. Place the short edge of another clean
slide against this streak of semen at a 45° angle and, using uniform
pressure, carefully draw the semen out so that it is evenly spread
across the length of the slide. Label the slide with the patient's
name and date, and allow to air dry for about 24 hr.

c.   The dry semen smear is placed in a Coplin
jar and stained using the Harleco® Diff-Quik differential staining
set. This is a modification of the Wright Stain technique which
takes only 15 sec to complete. The staining solutions and general
instructions are described below:

(1)   Fixative Solution: 1.8 mg/ml triaryl-
methane dye in methyl alcohol (Note: methanol alone may be used
instead). Dip the slide 5 times, 1 sec/dip. Drain excess.

(2)   Diff-Quik Solution I: 1 g/liter
xanthene dye, buffer and 0.01% NaN_3 (as a preservative). Dip slide 5
times, 1 sec/dip. Drain excess.

(3)   Diff-Quik Solution II: 1.25 g/liter
thiazine dye mixture (Azure A and methylene blue) and buffer. Dip
slide 5 times, 1 sec/dip. Drain excess.

(4)   10 dips in distilled H_2O and rinse.

(5)   Allow slide to air dry overnight (or
until dry to the touch) and examine under an oil immersion lens.

d.   This count includes various abnormal forms
and immature germinal cells (six categories in all) which are counted
on separate tally counters. All morphologies are, therefore,
expressed as direct percentages. At least two different fields
should be counted and the results averaged. If the sperm density is
low and less than 100 cells are encountered in one field, the
examiner must scan other fields until 100 cells are counted. The

reader is referred to Belsey et al. (1980) for excellent color photographic plates to aid in the identification of the following sperm morphologies:

   Oval: This grouping includes both small and large oval heads. It constitutes a majority of cells in the ejaculate which are considered as normal forms.

   Tapered: Cigar-shaped heads.

   Amorphous: Very irregular-shaped, formless, or pitted shaped heads.

   Duplicate: Sperm with two heads and a single tail, or one head and two tails.

   Immature: Immature germinal cells (e.g., spermatogonia, primary and secondary spermatocytes, spermatids). It may be difficult to distinguish between some of these cell types and WBCs. The presence of WBCs, RBCs and epithelial cells is also noted along with the sperm morphologies.

   2.5.9.   Particle Count (10^6/ml)

          A separate particle count should be made simultaneously with the sperm count, using a separate tally counter key. Any particulate debris encountered during the sperm count should be included, such as bacteria, epithelial cells, red blood cells, leukocytes, or immature germ cells. Particles are considered to be at least as large as a sperm head in size. The number of particles is calculated as described above for the sperm count and is expressed in 10^6/ml.

   2.6. Normal Ranges

          It should be stressed that standards of semen quality are only relative, not absolute, indicators of fertility or sterility (Nelson and Bunge, 1974). Shown in Figure 2 is a sample of the semen analysis report used at the Hospital of the University of Pennsylvania.

   2.6.1.   Volume

          Normal semen volume averages 3 to 4 ml. A range of 0.7 to 6.5 ml has been reported. Males treated for infertility tend to have an increased rather than decreased semen volume, which is associated with a depressed sperm count. On the other hand, a volume of less than 0.1 ml may be inadequate for conveying sperm to the endocervical mucus.

SEMEN ANALYSIS

Hospital of the University of Pennsylvania
Department of Obstetrics and Gynecology
Andrology Laboratory
Room 572, Dulles Bldg.
215-662-2988

Luis Blasco, M.D., Medical Director
Gregory S. Kopf, Ph.D., Scientific Director

☐ HUP Patient   ☐ Non-HUP Patient
Record No.:
Name:
(Last, First)

SS No.:
DOB (mo./day/yr.):
Address:

Home Telephone:
(Imprint Patient Registration Plate in Space Provided)

Requesting Physician: _____
☐ HUP   HUP Address: _____

_____

Beeper or Ext: _____

Name of Spouse/Partner: _____

Date/Time of Request: _____
Date/Time of Collection: _____
Date/Time of Arrival in Lab: _____

☐ Other   Hospital/Clinic Address: _____

_____

_____

Telephone No.: _____

Sperm Sample: ☐ Fresh   ☐ Frozen

SEMEN ANALYSIS

Days Abstinence: _____

Volume (ml): _____

Viscosity (+, ++, +++) _____

pH: _____

Agglutination (+, ++, +++): _____

Sperm Count (x10^6/ml): _____

Particle Count (x10^6/ml): _____

MOTILITY

Percent motile: _____

Forward progression (1-4): _____

VIABILITY (%)

Alive - Motile: _____

Alive - Immotile: _____

Dead: _____

MORPHOLOGY (%)

Oval: _____

Tapering: _____

Amorphous: _____

Duplicate heads: _____

Duplicate tails: _____

Immature: _____

CELLS (number/40 x field)

WBC: _____

RBC: _____

Epithelial: _____

COMMENTS: _____

_____

_____

Assayed by: _____

Fig. 2.   Sample semen analysis form used in the Andrology Laboratory
           at the Hospital of the University of Pennsylvania.

2.6.2.   Viscosity

Semen of normal viscosity pipets easily (pours drop by drop) or is slightly thick (+). Semen of watery or thick consistency (++ or +++) is abnormal.

2.6.3.   pH

This varies somewhat, but is most commonly 7.3–7.5. Normal semen falls into the pH range of 7.3 to 8.3. Seminal fluid pH can be altered after a few hours by bacterial growth and sperm metabolic activity.

2.6.4.   Motility

A normal semen sample will display 60% or greater motile sperm that exhibit a good to excellent forward progression.

2.6.5.   Forward Progression

A forward progression ranking of 3 or better is considered normal.

2.6.6.   Agglutination

An agglutination ranking of "none" or "+" is considered normal.

2.6.7.   Viability

While not absolute, the following ranges for viability are considered representative of normal semen samples:

% alive-motile = 50%-60%, or greater. This result should be within 10% of the motility result on the wet mount.

% alive-immotile = not greater than 30%.

% dead = not greater than 20%.

2.6.8.   Non-germinal Cells

An average of 5 or less of any of the three cell types per high power field is considered normal.

2.6.9.   Sperm Concentration

Sperm counts of less than 20×10^6/ml are abnormal and these samples are termed oligospermic. In one large study of sperm counts for over 2,500 men, an overall range of 1 x

10^5/ml to 3.75 x 10^8/ml was reported, with a mean sperm concentration of 70 x 10^6/ml (Smith et al., 1978). Sperm concentration increases with sexual abstinence for at least the first 10 days.

2.6.10.   Morphology

A semen sample containing a minimum of 60-70% morphologically mature and normal spermatozoa (e.g., oval heads) is considered to be normal.

2.6.11.   Particle Count

The particle count should not be greater than 25% of the sperm count. Human semen normally contains a number of granules and globules which probably result from secretions from glandular cells or autolysis of epithelial cells which line the accessory reproductive structures. The particulate debris counted on the hemacytometer should be compared with the count of non-germinal cells on the wet mount.

2.7. Video Imaging Systems for Semen Analyses

With the advent of computer-assisted video imaging analysis (VIA) systems and their use in many facets of the physical and biological sciences, attempts have been made to completely automate the semen analysis. The software for such computer based motion analysis systems as CellSoft® (Cryo Resources, Ltd., New York, NY), CellTrak/S® (Motion Analysis Corp., Santa Rosa, CA) and Motion Analyzer (Hamilton-Thorn, Danvers, MA) have been designed for the quantitative analysis of motility, and attempts are now being made to automate morphology determinations. While VIA systems are extremely useful for the determination of a variety of motility measurements (i.e., curvilinear velocity, progressiveness), attempts to automate morphology determinations have been, to date, far less successful owing to the extreme complexity of software design. Although VIA systems may become an invaluable aid in the quantitation of certain parameters of the conventional semen analysis, it is felt that a fully automated system for making all of these measurements is still unavailable. The prices of VIA systems are also prohibitive to many laboratories. One cannot, however, overlook the potential large-scale use of VIA systems once more sophisticated software designs are available and the costs have reached a more competitive level.

2.8. Quality Control

a.   Always treat a semen sample as a potential pathogen when handling. Be sure to wear gloves, never mouth pipet, and wipe up spills immediately. See Table II for a troubleshooting guide.

Table II. Troubleshooting guide for semen analysis

Outcome	Probable reasons	Response
Abnormal semen analysis	May be indicative of generally abnormal semen parameters	Repeat analysis within a 2-week period
	Patient did not follow a 48-hr abstinence period prior to semen collection	Repeat analysis within a 2-week period
	First portions of ejaculate lost	Repeat analysis within a 2-week period and collect the entire ejaculate
	Too long a time period between collection and analysis	Analysis should be performed within 1 hr of semen collection
	Collection container contaminated	Use sterile, disposable specimen jars
	Temperature extremes during transport of semen specimen	Avoid temperature shifts (insulate container)
Ejaculate displaying slow liquefaction times	Abnormal semen characteristics	-Use a Pasteur pipet to break up sample
Ejaculate displaying high viscosity	Abnormal semen characteristics	-Use a Pasteur pipet to break up sample -Draw sample through a syringe with a 23-gauge needle
Husband unable to provide a semen specimen	Stress-related factor(s)	-Have husband collect sample in more familiar surroundings (i.e., home) -Reschedule analysis to avoid stress
Motility of sample is low	Abnormal motility characteristics	-Repeat motility analysis within 2 weeks
	Microscope slide used for motility analysis has dried out	-Observe motility as soon as sample is placed on slide
Massive sperm agglutination	Presence of antibodies	-Analyze sample for antibodies
Low percentage viability although sperm are motile	Problem with performing viability assay	-Sample under coverslip has dried out -Too much dye has been added
Presence of high concentrations of bacteria or white blood cells	Localized infection	-Treatment by a physician
Poor duplication when doing sperm counts	Improper use of hemacytometer	-Proper loading or use of chamber -Allow cells time to settle

b.   To achieve the most accurate semen analysis results, samples must be analyzed within 1 hr after collection.

c.   Thorough mixing of the specimen to achieve a uniform distribution of the sperm is very important. Avoid vigorous mixing of the sample, as this could damage the cells.

d.   Motility and viability are the most time-dependent parameters in the semen analysis; these must always be evaluated first, since they deteriorate progressively the longer the sperm remain in the seminal plasma.

e.   Microscope slides for wet mounts should be prewarmed to 37 °C.

f.   Be careful not to touch the end of the strip of pH paper where the semen drop will be placed, as this may alter the pH reading.

g.   Adding the proper amount of eosin B stain to the drop of semen is crucial to the viability results. If an insufficient volume is added, it will be difficult to distinguish dead cells; on the other hand, if too much stain is added, it may be toxic to living, motile cells.

h.   Be careful not to contaminate the viability stain needle; the stain drop should be added adjacent to (not directly into) the semen drop.

i.   Agglutination determinations should be made as soon as possible following preparation of the wet mount to avoid misreading a dried sample area under the coverslip.

j.   When assessing motility on the wet mount, it is essential to focus through the entire depth of a given field so as to include nonmotile sperm that have settled down into a different focal plane.

k.   Proper technique in filling hemacytometer chambers (e.g., not flooding) is critical to achieving an accurate sperm and particle count.

l.   When making a judgement of semen quality, all of the parameters of the semen analysis should be considered together.

m.   If any abnormal semen analysis result is found, an additional semen analysis should be performed.

3. REFERENCES

Bavister, B. D., 1982, Evidence for a role of post-ovulatory cumulus
    components in supporting fertilizing ability of hamster sperma-
    tozoa, J. Androl. 3:365-372.
Belsey, M. A., Eliasson, R., Gallegos, A. J., Moghissi, K. S.,
    Paulsen, C. A., and Prasad, M. R. N., 1980, Laboratory Manual
    for the Examination of Human Semen and Semen-Cervical Mucus
    Interaction, Press Concern, Singapore, 43 pp.
Ben-Rafael, Z., Kopf, G. S., Blasco, L., Tureck, R. W., and
    Mastroianni, L., Jr., 1986, Fertilization and cleavage after
    reinsemination of human oocytes in vitro, Fertil. Steril. 45:58-
    62.
Bradley, M. P. and Garbers, D. L., 1983, The stimulation of bovine
    caudal epididymal sperm forward motility by bovine cumulus-egg
    complexes in vitro, Biochem. Biophys. Res. Commun. 115:777-787.
Cannon, D. C., 1984, Seminal Fluid: Clinical Diagnosis and Manage-
    ment by Laboratory Methods, 17th ed. (J. B. Henry, ed.), Todd,
    Sanford, Davidson Pub., Chapter 22, pp. 516-522.
Lenz, R. W., Ax, R. L., Grimek, H. J., and First, N. L., 1982,
    Proteoglycan from bovine follicular fluid enhances an acrosome
    reaction in bovine spermatozoa, Biochem. Biophys. Res. Commun.
    106:1092-1098.
Mann, T. and Lutwak-Mann, C., 1981, Male Reproductive Function and
    Semen, Springer-Verlag, Berlin, 495 pp.
Nelson, C. M. K. and Bunge, R. G., 1974, Semen analysis: evidence
    for changing parameters of male fertility potential, Fertil.
    Steril. 25:503-507.
Overstreet, J. W., Yanagimachi, R., Katz, D. F., Hayashi, K., and
    Hanson, F. W., 1980, Penetration of human spermatozoa into the
    human zona pellucida and the zona-free hamster egg: A study of
    fertile donors and infertile patients, Fertil. Steril. 33:534-
    542.
Perreault, S. D. and Rogers, B. J., 1982, Capacitation pattern of
    human spermatozoa, Fertil. Steril. 38:258-260.
Rogers, B. J., Van Campen, H., Ueno, M., Lambert, H., Bronson, R.,
    and Hale, R., 1979, Analysis of human spermatozoal fertilizing
    ability using zona-free ova, Fertil. Steril. 32:664-670.
Rogers, B. J., Perreault, S. D., Bentwood, B. J., McCarville, C.,
    Hale, R. W., and Soderdahl, D. W., 1983, Variability in the
    human-hamster in vitro assay for fertility evaluation, Fertil.
    Steril. 39:204-211.
Smith, K. D., Stultz, D. R., Jackson, J. R., and Steinberger, E.,
    1978, Evaluation of sperm counts and total sperm counts in 2543
    men requesting vasectomy, Andrologia 10:362-368.
Sokoloski, J. E. and Wolf, D. P., 1983, Effect of seminal plasma on
    human sperm fertility assessment, J. Androl. 4:40 (abstract).
Soupart, P. and Morgenstern, L. L., 1973, Human sperm capacitation
    and in vitro fertilization, Fertil. Steril. 24:462-478.

Tarlatzis, B. C., Laufer, N., Murillo, O., Makler, A., DeCherney, A., and Naftolin, F., 1984, The effect of human oocyte-corona-cumulus complex (OCCC) on sperm motility and acrosome reaction, Fertil. Steril. 41:102S (abstract).

Van der Ven, H., Bhattacharyya, A. K., Binor, Z., Leto, S., and Zaneveld, L. J. D., 1982, Inhibition of human sperm capacitation by a high-molecular-weight factor from human seminal plasma, Fertil. Steril. 38:753–755.

Van der Ven, H. H., Binor, Z., and Zaneveld, L. J. D., 1983, Effect of heterologous seminal plasma on the fertilizing capacity of human spermatozoa as assessed by the zona-free hamster egg test, Fertil. Steril. 40:512–520.

Wolf, D. P. and Quigley, M. M. (eds.), 1984, Human In Vitro Fertilization and Embryo Transfer, Plenum Press, New York, 440 pp.

Wolf, D. P. and Sokoloski, J. E., 1982, Characterization of the sperm penetration bioassay, J. Androl. 3:445–451.

7

ASSESSMENT OF HUMAN SPERM FERTILITY POTENTIAL

Don P. Wolf

1.   INTRODUCTION

     The objective of semen or washed sperm assessment is to deter-
mine the probability of reproductive success, i.e., the fertility
potential of the sample. Diagnostic procedures available for evaluating human sperm fertility potential are limited in number and
effectiveness. Semen analysis is the most frequently employed proce-

dure which includes measurements of semen volume, pH and viscosity,
assessments of sperm density, percentage of motile cells, the quality
of motility, and sperm morphology. A detailed discussion of the
semen analysis is included in Chapter 6. Unfortunately, parameters
of the semen analysis taken either individually or collectively are
poor predictors of fertility status except in the extreme cases of
azoospermia or asthenospermia, i.e., the patient without sperm or
with only nonmotile sperm.

In addition to the conventional semen analysis, the andrology
laboratory has available several assays which purport to monitor
sperm function in vitro. The first deals with the ability of sperm
to penetrate cervical mucus (Overstreet, 1986). The clinician, of
course, uses a postcoital test to evaluate whether or not a cervical
factor exists in the couples' infertility. Basically, this is an
examination of compatibility of husband's sperm and wife's mucus, and
the test, if abnormal, may reflect the presence of hostile mucus,
antisperm antibodies, or a sperm deficiency. The diagnostic procedure of choice to augment the postcoital test is a cross-match test
involving measurement of husband and donor sperm penetration in wife
and donor cervical mucus (Jonsson et al., 1986). Additionally,
capillary tube penetration test kits are available commercially which
employ lyophilized bovine cervical mucus as a substitute for human
cervical mucus (Penetrak®, Serono Diagnostics, Braintree, MA). The
migration of the vanguard sperm into this column of mucus has been
correlated with postcoital test results (Alexander, 1981) and hamster
egg penetration (Takemoto et al., 1985) and may be related to the
fertility potential of the donor. These in vitro assays provide a
mechanism for defining the relative contribution of the male versus
the female in abnormal sperm-mucus interaction.

Diagnostic tests or assays for male fertility under development
include hypo-osmotic swelling (Jeyendran et al., 1984) and acrosomal
status quantitation (Wolf et al., 1985; Byrd and Wolf, 1986).

Over the past decade, two bioassays of sperm function have been
introduced as a means of evaluating fertility potential. One of
these tests involves sperm penetration of the zona pellucida
surrounding nonviable, human ovarian oocytes as developed by Jim
Overstreet and his colleagues at the University of California, Davis.
The second involves human sperm fusion with the zona-free hamster egg
and is based upon pioneering work conducted by R. Yanagimachi and his
colleagues at the University of Hawaii. Since it is the latter
bioassay that has enjoyed widespread application in clinical
andrology laboratories, the remainder of this chapter will be devoted
to a detailed discussion of this bioassay.

The ability of human sperm to fuse with zona-free hamster eggs
has been used for several years as a predictor of human male
fertility potential. For a comprehensive review on how this test

became popular, the reader is referred to review articles by
Yanagimachi (1984), Rogers (1985) and the October 1986 supplement
(#6) of the International Journal of Andrology. While this bioassay
is a welcome and much needed addition to traditional semen analysis,
uniformity as to what score differentiates between a fertile, subfertile, or an infertile response has come only slowly, and the conditions employed are far from standardized. Here the experimental and
physiological basis for the test will be reviewed as well as at least
some of the parameters that contribute to the reproducibility of test
results.

2.   WHAT DOES THE BIOASSAY MEASURE?

The hamster egg bioassay measures the sperm's ability to
complete a series of biochemical and biophysical changes collectively
referred to as capacitation. The assay does this by indirectly
monitoring the presence of capacitated, acrosome-reacted sperm, as
the latter are uniquely capable of fusing with the plasma membrane of
the naked egg.

In the normal fertilization process, capacitation occurs during
sperm transit in the female genital tract, and is completed by the
time the sperm reaches the site of fertilization, the ampulla or
distal end of the fallopian tube. Capacitated sperm bind to the
egg's zona pellucida and, in some species at least, this interaction
between sperm and the zona pellucida's sperm receptor culminates in
the induction of the acrosome reaction, the exposure of acrosomal
enzymes, and subsequent sperm penetration of the zona pellucida
(Bleil and Wassarman 1983). In others, the acrosome reaction may
occur prior to sperm contact with the zona pellucida in the cumulus
(Huang et al., 1981). After the reacted sperm traverses the zona and
the perivitelline space, the fluid-filled area between the zona and
the oocyte, fusion of the limiting membranes of sperm and egg occurs
and the sperm is incorporated into the egg. For a recent review of
fertilization, see Tesarik (1986).

In vitro, when human ejaculated sperm are washed and exposed to
zona-free hamster eggs, it is the capacitated, acrosome-reacted sperm
which binds to and fuses with the egg. Under these conditions, the
rate limiting parameter in hamster egg penetration is probably
capacitation and the occurrence of a spontaneous acrosome reaction.
The reaction is spontaneous in that it occurs in the absence of any
known inducing agent.

Capacitation of human sperm is supported by several types of
culture medium which can be grouped, for convenience, into two major
categories: simple and complex. Simple media, such as BWW (Biggers
et al., 1971) and Tyrode's solution are, in general, bicarbonate
buffered, balanced salt solutions with ionic compositions similar to

human serum (Bavister, 1981). Complex media including Ham's F-10 and
Menezos's B2, commonly employed for human in vitro fertilization, are
supplemented with vitamins, amino acids, and monosaccharides.
Usually, all media also contain a protein source, specifically
albumin. The mechanism by which albumin influences capacitation is
not clearly understood. This protein is notorious for its ability to
bind or chelate ions, fatty acids, or other physiologically active
substances and one possibility is that albumin, acting as a sterol
acceptor, directly induces alterations in lipid constituents of the
sperm plasma membrane, indirectly altering membrane stability (Go and
Wolf, 1983). The requirement for albumin in sperm capacitation may
not be absolute for several studies suggest that at least limited
capacitation occurs in the complete absence of sperm exposure to
albumin (Johnson et al., 1984; Sokoloski and Wolf, 1984). Energy
sources such as pyruvate, lactate or glucose are usually present in
culture media which support sperm capacitation; however, their
presence is presumably more important to the maintenance of sperm
motility. Bicarbonate anion has been implicated in sperm capacitation in at least one species, the mouse (Lee and Storey, 1986).
Other conditions that may contribute include osmolarity, pH, temperature, and the composition of the gas atmosphere. The molecular
correlates of capacitation are poorly defined. We know that cell
surface alterations occur, for instance, in the addition and subsequent alteration of seminal plasma components during ejaculation and
postejaculatory maturation (Yanagimachi, 1982; Boldt and Wolf, 1984;
Wolf, 1985). Presumably, capacitation concerns the cells ability to
control intracellular free calcium ion concentrations in the presence
of substantial calcium gradients. A seminal plasma component which
is involved in the regulation of calcium transport has been described
in the bull (Rufo et al., 1982).

3.   CONDUCTING THE TEST

     The following procedures provided in laboratory manual format
detail conduction of the bioassay. See Figure 1 for a schematic
outline. For the WHO protocol for conducting the bioassay, see
International Journal of Andrology (Supplement 6), October 1986,
p. 197.

     3.1. Materials

          a.   Chemical/media

               (1)   Pregnant mare serum gonadotropin (PMSG) (Diosynth
                      Corporation, Chicago, IL)

               (2)   Human chorionic gonadotropin (Sigma Chemical
                      Company, St. Louis, MO)

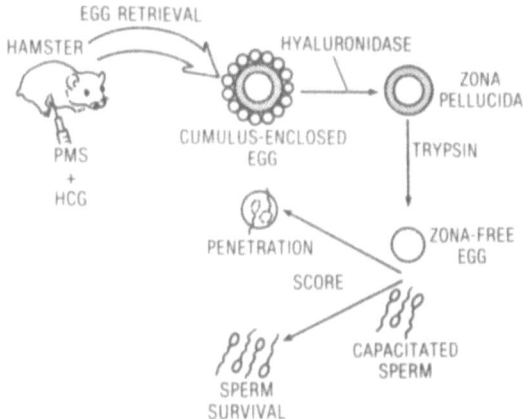

Fig. 1.   Schematic outline of the steps involved in the zona-free
          hamster egg bioassay.

          (3) BWW medium (see below and Table I)

          (4) Human serum albumin (Fraction V, Sigma Chemical
               Company, St. Louis, MO)

          (5) Hyaluronidase (from bovine testes, 200–400 NF
               units/mg solid, Sigma Chemical Company, St.
               Louis, MO)

          (6) Trypsin (from bovine pancreas, Type III, Sigma
               Chemical Company, St. Louis, MO)

          (7) Silicone oil (Dow Corning 200 dielectric fluid,
               Dow Corning Corp., Midland, MI)

          (8) Dulbecco's phosphate buffered saline (Gibco Labs,
               Grand Island, NY)

          (9) Glutaraldehyde; formalin, aceto-lacmoid; ethanol
               (optional)

          (10) Vaseline:paraffin

          (11) Human sperm from control donor and patients

     b.   Supplies

          (1) 35-mm disposable petri dishes, tissue culture
               grade (Falcon 3001, Becton Dickinson, Oxnard, CA)

Table I. Composition and preparation of BWW medium modified from
          Biggers et al. (1971)

Solution A: Concentration

Reagent	g/liter	mM
NaCl	5.540	94.78
KCl	0.356	4.77
$CaCl_2 \cdot 2H_2O$	0.189	1.29
$MgSO_4 \cdot 7H_2O$	0.294	1.19
KH_2PO_4	0.162	1.19

Add to 100 ml of Solution A for working medium:

	g/100 ml	Final concentration (mM)
$NaHCO_3$	0.21 g	25.06
Glucose	0.1 g	5.55
Na Pyruvate	2.8 mg	0.25
Na Lactate (60% syrup)	0.37 (ml)	19.65
HSA		3-35 mg/ml
Penicillin G	1 ml of stock	100 units/ml
Streptomycin sulfate	(100x) solution	100 mcg/ml

(2) Syringes and needles for hamster injection and
     microdrop preparation (usually 1-cc tuberculin
     syringe with 25-gauge needle)

(3) Dissecting instruments (course and fine watch-
     makers' forceps, course and fine scissors

(4) Pipets for handling eggs (Pasteur pipet 5 3/4"
     pulled to a fine bore over low flame), PiPump,
     rubber tubes for mouth pipetting

(5) Glass slides and coverslips

(6) Hemacytometer

(7) Coplin jars (optional)

(8) Disposable sterilizing filters

(9)   Conical centrifuge tubes, 15 ml tissue culture
      grade plastic (Falcon 2095, Becton Dickinson,
      Oxnard, CA)

3.2. Equipment

a.   Dissecting microscope
b.   Slide warmer
c.   Standard CO_2 incubator (37°C, 5% CO_2 in air, 99%
     humidity)
d.   Phase contrast microscope with 10, 20, and 40X objec-
     tives
e.   Analytical balance

3.3.   Animals

Juvenile or sexually mature golden (Syrian) hamsters
(Charles River-Breeding Labs, Wilmington, MA, Engle Lab Animals,
Inc., Farmersburg, IN, or Simonson Labs, Gilroy, CA).

Follicular maturation and ovulation induction are controlled by
the administration of PMSG and hCG. Animals should be housed (5-10
per cage) under a 10/14 or 12/12 light/dark cycle with standard
laboratory chow and water available ad libitum. Allow animals to
acclimate for a week or so before use.

3.4. Superovulation

a.   For preparation of stock solutions of gonadotropins,
see Chapter 5.

b.   To superovulate female hamsters selected at random
with regard to the estrous cycle, inject intraperitoneally with 30 IU
of PMSG. It is very convenient to inject in the morning. Animals
should yield on average 30 eggs. Inject 1-2 animals for each sperm
sample to be tested.

c.   Approximately 56-58 hr later, in the late afternoon or
evening of day 2, inject 30 1U of hCG. The PMSG to hCG time interval
is rather flexible from as short as 50 hr to as long as 72 hr. The
selection of 56-58 hr is based primarily on convenience.

d.   Time of sacrifice. Approximately 16 hr after hCG
administration, animals are sacrificed by asphyxiation in 100% CO_2 or
by cervical dislocation.

e.   For a timetable on assay setup, see Table II.

Table II. Schedule for the sperm penetration assay

Procedure:

PMSG injection 9-10 a.m.	hCG injection 4-6 p.m.	Medium preparation 9 a.m.-12 noon	Sperm preparation 2-5 p.m.	Egg recovery 8-10 a.m.
Monday	Wednesday	Wednesday	Wednesday	Thursday
Tuesday	Thursday	Thursday	Thursday	Friday
Wednesday	Friday	Friday	Friday	Saturday
Thursday	Saturday	Saturday	Saturday	Sunday
Friday	Sunday	Sunday	Sunday	Monday
Saturday	Monday	Monday	Monday	Tuesday
Sunday	Tuesday	Tuesday	Tuesday	Wednesday

3.5. Preparation for the Assay

The stock salt ingredients for BWW medium (Solution A,
Table I) are weighed out and added to 1.0 liter of distilled-
deionized water. The stock can be kept stored at 4°C for up to one
month. The additional components, sufficient for 100 ml of medium,
are weighed on the day of use and added to 100 ml of the stock salt
solution. The pH is adjusted to 7.4-7.6 and the medium is passed
through a 0.22 μm filter prior to use. The addition of Hepes (20 mM)
is optional. This medium contains human serum albumin--0.3% for
sperm washing and 3.5% for sperm preincubation and sperm-egg interaction. Since different lots of albumin may contribute to bioassay
outcome, it is advisable to screen for a satisfactory lot that
supports penetration and then minimize lot changes. Once prepared,
the medium if free of Hepes is preequilibrated in 5% CO_2 in air for
several hours before use.

3.6. Sperm Preparation

The preparation of human sperm for insemination of hamster
eggs is similar to that used for human IVF as described in Chapter 6
and can involve BWW or the medium employed for human IVF. In our
laboratory, 1-ml samples of washed sperm in 15-ml conical centrifuge
tubes (Falcon 2095) are used. Sperm samples are reconstituted to
approximate their original concentrations and incubated overnight,
loosely capped, at a 45° angle in the incubator. We employ BWW
containing 3.5% HSA for routine capacitation.

3.7. Setup for the Assay

    a.   Prepare BWW working medium (Table I) containing 0.3% HSA and hyaluronidase (1 mg/ml) or trypsin (1 mg/ml).

    b.   Prepare five 35-mm petri dishes as outlined in Figure 2 for the preparation of zona-free eggs. A, 100 µl drop of BWW-hyaluronidase; B, 50 µl drops of BWW; C, 200 µl drop of BWW-trypsin; and D, 50 µl drops of BWW. It is convenient to first define the drop diameter in the petri dish by the addition of a small amount of medium. This drop is then covered with silicone oil (the dish is filled to approximately 75% of capacity) and finally the drop is made to volume/size by medium addition.

    c.   Recover sperm samples and evaluate motility and density of populations to be tested (see Chapter 6 for methods).

    d.   Set up 35-mm petri dishes containing 200 µl drops of sperm with silicone oil at 5×10^6 motile cells/ml; overlay with silicone oil. Usually two dishes per sample are used. In the event that sperm concentrations are low, supernatant preparations can be removed, collected by centrifugation at 200 x g and resuspended in smaller volumes to effect an increase in concentration. Once these dishes are set up, they can be held in the incubator until egg collection is finished.

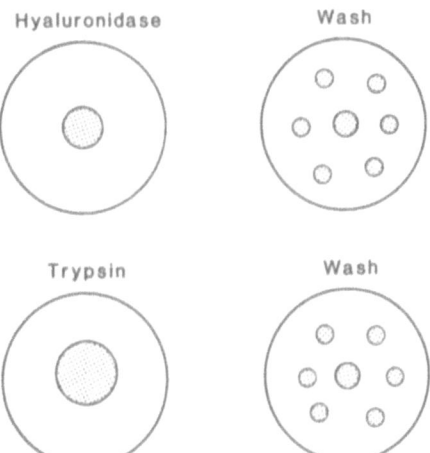

Fig. 2.   Solutions and dish setup for processing eggs in the hamster egg bioassay.

3.8. Preparation of Zona-Free Hamster Eggs

a.   On the morning of the assay, animals are sacrificed. After rinsing with 95% ethanol, the ventral body wall is opened and the reproductive tract is exposed. Oviducts are removed and placed in silicone oil in a disposable plastic petri dish (Falcon #3001) at ambient temperature.

b.   The dish is transferred to the stage of a dissecting microscope at ambient temperature and the cumulus masses are released from the oviducts. Release is accomplished by puncturing the oviduct at the site of the cumulus mass with a 25-gauge needle attached to a 1-cc disposable tuberculin syringe. The cumulus mass is then teased from the oviducts with two needles, one to manipulate the oviduct and the other the cumulus mass.

c.   Cumulus masses from all oviducts are accumulated and transferred together to a drop of culture medium under oil containing hyaluronidase (Figure 2). This transfer can be made with watch-makers' forceps without microscopic observation.

d.   Cumulus dispersal requires approximately 10 min exposure to 0.1% hyaluronidase. Using hand drawn micropipets, eggs are removed from hyaluronidase in batches of 25 to 50 and washed repeatedly in BWW before exposure to 0.1% bovine pancreatic trypsin for several minutes. In processing relatively large numbers of eggs (200-300), it is not always possible to control trypsin exposure times precisely. However, this appears to be of little consequence as long as exposure times are restricted to 10 min. Incomplete zona removal can prevent sperm-egg interaction and often impacts on egg processing as partially digested zona material is gelatinous and can occlude micropipets. If this occurs, the use of a larger drop or a reduction in the albumin concentration may be advantageous. Distortion of the egg commonly occurs during zona removal. The zona-free state can be assessed by observing whether treated eggs are touching one another.

e.   Zona-free eggs are removed from the trypsin suspension, 25-50 at a time, and washed through five drops of BWW under oil and held for insemination. The estimated viable lifetime of the zona-free egg at 37°C in vitro is limited and eggs should be inseminated within approximately 1 hr of harvesting from the animal (Binor et al., 1982b); eggs can be held for longer time periods at reduced temperature (Syms et al., 1984; Barros et al., 1986).

3.9. Insemination

a.   To initiate the insemination reaction, zona-free eggs are added to petri dishes containing the sperm suspensions. Two dishes each containing 10–15 eggs are superior to one dish of 20–30, since the intersample variability can be assessed.

   b.   The insemination reaction is allowed to proceed at 37°C in 5% CO_2 in air for 3 hr.

   c.   The reaction is terminated by simply removing eggs from the insemination drop and washing them through several drops of culture medium. An estimate of the number of attached sperm should be made since it usually reflects the penetration level. During this washing procedure, which does not involve an oil overlay, eggs are aspirated in and out of finely drawn micropipets to remove loosely adherent sperm on the egg surface.

   3.10. Mounting and Staining Eggs; Scoring Penetration

   a.   Upon completion of washing, eggs are transferred in a small drop to a glass slide (Figure 3).

   b.   A coverslip is prepared with a small amount of vaseline:paraffin (20:1) mixture added to each corner.

   c.   The coverslip is centered over the drop of medium containing the eggs and depressed onto the vaseline:paraffin supports until the eggs are compressed. This is a delicate procedure, for too little compression of the egg precludes an accurate assessment of the mean number of sperm/egg and will cause the egg to be washed off the slide while too much compression results in egg lysis.

   d.   A drop of 1% glutaraldehyde is added at the edge of the coverslip and coaxed to fill the area under the coverslip by gentle compression of the coverslip.

   e.   The slide is transferred to a Coplin jar containing 10% formalin in Dulbecco's PBS where it can be held indefinitely.

   f.   When penetration is to be scored, slides are removed from the 10% formalin solution, rinsed briefly in a stream of distilled water and transferred to a Coplin jar containing 95% ethanol.

   g.   Following dehydration, the slide is removed and stained with a 0.25% lacmoid and 45% acetic acid solution. The stain can conveniently be drawn under the coverslip by using a paper tissue as a wick.

   h.   The stained eggs are located with the aid of a dissecting microscope, marked on the bottom of the slide with a felt-tip pen, and scored with a phase contrast microscope at 200–400X. An alternative procedure involves egg exposure to calcium-free BWW (substitute 0.1% polyvinyl alcohol for HSA). Wash eggs through 5

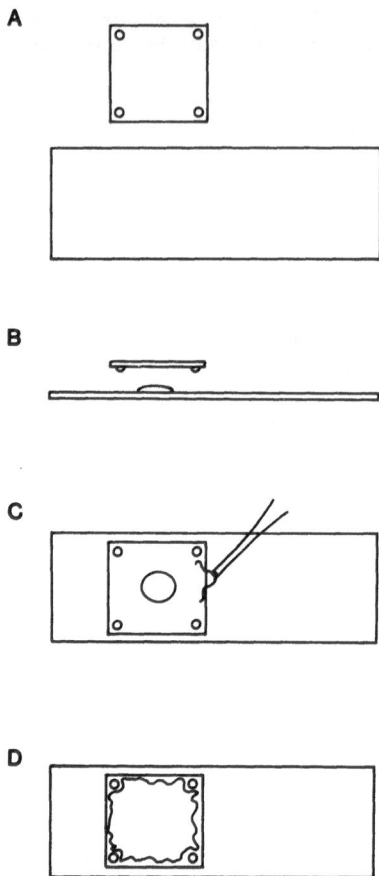

Fig. 3.   Schematic representation of the processing of zona-free
           hamster eggs for penetration scoring. A) Glass microscope
           slide and coverslip with vaseline-paraffin column at each
           corner. B) Placing the coverslip over the drop of medium
           containing the eggs.   C) The coverslip is depressed onto
           the vaseline-paraffin supports until the drop is contacted
           and flattened slightly. Glutaraldehyde is then added at
           the edge of the coverslip. D) The fixative spreads under
           the coverslip fixing and compressing the eggs.

drops of this medium to remove sperm, and then rinse through calcium-
free BWW containing 1% glutaraldehyde in 0.1 M cacodylate buffer at
pH 7.0 (1:1). These eggs can be held in the refrigerator in drops of
BWW and glutaraldehyde under oil for up to 1 week before mounting and
scoring. Mount and score directly or stain with aceto-lacmoid.

i.   Eggs are considered penetrated when a swollen sperm head (or heads) or male pronucleus (or pronuclei) and a corresponding sperm tail(s) are found within the ooplasm (Figure 4, 5). Results are recorded as the number of attached sperm before extensive washing, the percentage of eggs penetrated and the mean number of penetrating sperm per inseminated egg. Yet another option is to utilize unfixed preparations at 400X. The advantage associated with scoring penetration on live preparations is that it is faster. The disadvantages are that since one sequentially mounts and scores eggs recovered from dishes, the sperm/egg contact times will be different for each dish. Secondly, it is not possible to store slides for later scoring and this may be inconvenient when large numbers of eggs are processed. A third disadvantage is that some sperm may be missed when penetration is scored on living cells.

4.   BIOASSAY MODIFICATIONS

A number of alterations of the bioassay have been described which produce higher sperm/egg fusion levels. These include the use of the ionophore A23187 (Aitken et al., 1984; Tesarik, 1985), TES-Tris egg yolk buffer (Bolanos et al., 1983; Syms et al., 1984), and exposure to cold shock at 4°C (Cohen et al., 1985). These techniques may improve the discriminating power and the reliability of the zona-free hamster egg bioassay. The mechanism by which they operate, although not confirmed, probably involves the production of increased populations of motile, acrosome-reacted sperm. The disadvantages to their use are that sperm motility is often compromised and interpretation requires a different, sometimes unavailable or limited, data base. Because insemination with sperm pretreated by one of these augmented protocols results in high levels of egg penetration (usually 100% and up to 50 sperm/egg), increased discriminating power results and the possibility that the eggs may contribute to bioassay failure is largely eliminated. Contrast, for example, a standard assay where 50% of the 30 eggs analyzed are penetrated by 1 sperm (15 fusions total) versus an augmented protocol where 15 eggs are penetrated by an average of 25 sperm (375 fusions total).• The details of several augmented techniques have been included here.

The first involves prolonged exposure of sperm to TES-Tris egg yolk buffer (Johnson et al., 1984). The procedure is as follows: After liquefaction, whole semen samples are mixed one-to-one with a buffer consisting of 211 mM TES, 96 mM Tris, 11 mM dextrose, and 1% penicillin-streptomycin solution to which 20% fresh hen egg yolk has been added. Fresh hen egg yolk (approximately 20 ml/egg) should be recovered from eggs collected within 1-2 hr of laying. One- to two-hundred-ml batches of TES-Tris egg yolk buffer can be prepared and stored frozen in 10-ml fractions. Alternatively, these preparations are available commercially (Irvine Scientific, Irvine, CA). Semen samples, after mixing one-to-one with the buffer system, are slowly

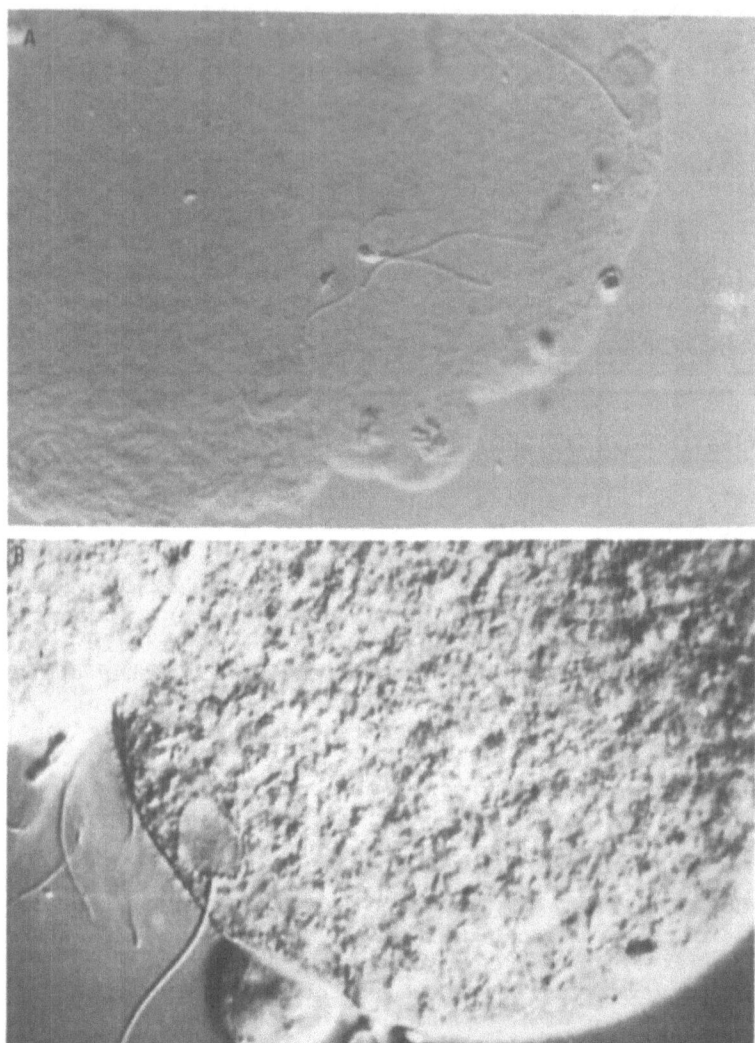

Fig. 4.   Scoring penetration of human sperm into zona-free hamster
           eggs. A) Nomarski differential interference micrograph of
           living eggs as a whole mount. Note two enlarged decon-
           densing sperm heads with corresponding tails and one intact
           sperm. The maternal chromosomal compliments can be visual-
           ized in the top of the micrograph, one set presumably
           within the second polar body and one set in the oocyte.
           B) Phase contrast micrograph of a penetrating human sperm.
           Note the presence of an intact second polar body. Courtesy
           of Marybeth Gerrity.

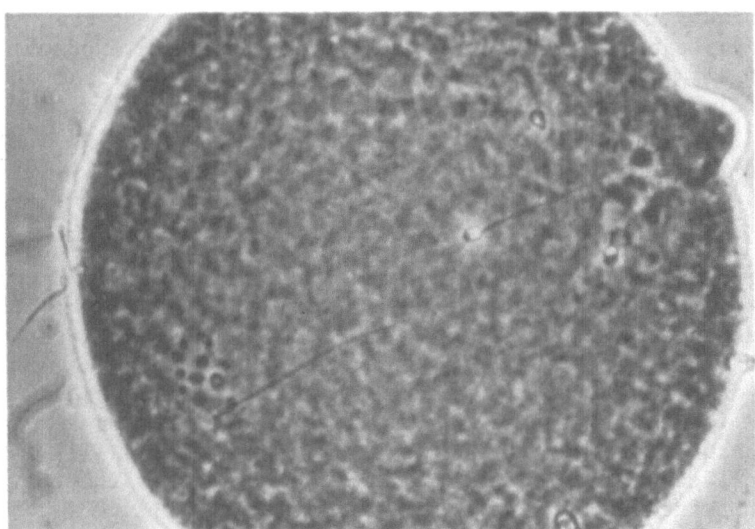

Fig. 5.   Human sperm penetration of zona-free hamster egg. Phase
            contrast micrograph of an aceto-lacmoid stained whole mount
            preparation. A male pronucleus is visible along with a
            sperm tail at the lower left. Several intact sperm are
            also visible.

cooled to 4°C and incubated for up to 66 hr. At the end of the
incubation, sperm are washed in BWW containing 0.3% HSA and recovered
in a swim-up into BWW supplemented with 1% HSA. For results with
this protocol, the reader is referred to Johnson et al., 1984; Syms
et al., 1984; Hirsch et al., 1986. In our hands, donor levels
average 98% of eggs penetrated with 30 sperm/egg after overnight
exposure of sperm to TES-Tris egg yolk buffer. We use a value of 2
sperm/egg as a cutoff. Below this level, a male factor is probable.

     Another modification of the zona-free hamster egg test has
appeared recently (Aitken et al., 1984). These authors have incorporated 50-100 μM A23187 into the culture medium used for sperm capacitation in an attempt to "circumvent problems relating to the rate and
efficiency of sperm capacitation, thereby making test results more
reflective of sperm functional capacity." Washed sperm are mixed
with BWW containing ionophore to give 50-100 μM ionophore and 10 x
10^6 sperm/ml. Note the effective ionophore concentration must be
much lower than 50-100 μM if any motility is retained. In our hands,
10 μM A23187 is sufficient to completely inhibit motility (Byrd and
Wolf, 1986). Incubations are conducted for 0.5 to 3 hr. Sperm are
then collected by centrifugation and resuspended in ionophore-free
BWW at 10 x 10^6/ml before eggs are added to begin the bioassay.

A third modification of the zona-free hamster egg test which leads to increased penetration levels involves sperm exposure to 4°C (Cohen et al., 1985). The procedure that we have utilized involves incubation of washed sperm in BWW containing 3.5% human serum albumin for 3 hr at 37°C followed by overnight incubation in a refrigerator. Refrigerator temperatures vary from 4–8°C. Incubations of washed sperm at 37°C are conducted in 15-ml conical tubes maintained at a 45° angle in a gassed (5% CO_2 in air) incubator. The low temperature exposure is conducted by transferring a sample from the incubator to a 100-ml beaker containing approximately 30 ml of room temperature water and then to a refrigerator. Following completion of the overnight preincubation, samples are removed from the refrigerator and incubated for 30–60 min at 37°C in an atmosphere of 5% CO_2 in air. Aliquots of the supernatant sperm suspension are removed and evaluated for sperm density and motility. Some loss of motility is usually associated with the augmented protocol (Table III). Inseminations are conducted as described elsewhere in this chapter. The results obtained for 6 donors are summarized in Table III. Spontaneous capacitation involved 18 hr in BWW at 37°C. Acrosomal status as determined by an end point immunofluorescence assay is also included and supports the contention that augmented protocols result in larger populations of acrosome reacted cells (see Wolf et al., 1985, and Byrd and Wolf, 1986, for acrosomal status quantitation).

Table III.   SPA and acrosomal status; correlation for donors of proven fertility (matched data from 6 donors and 37 ejaculates)

	Results (Mean ± SEM)	
	Spontaneous capacitation	Augmented capacitation
Motility, %	79.7 ± 1.5	56 ± 2.9
Bound sperm (Max = 200)	57.5 ± 8.2	176.6 ± 7.7
SPA		
% penetrated	30.9 ± 5.4	92.8 ± 2.9
mean sperm/egg	0.6 ± 0.2	5.3 ± 0.9
Acrosomal status		
% reacted	12.5 ± 0.9	19.9 ± 1.4

5.   FACTORS INFLUENCING THE BIOASSAY

The necessity of applying quality control criteria to this bioassay should be obvious. In selecting protocols for the bioassay, practical considerations are also important. For instance, it may be more convenient to begin sperm capacitation late in the day with the bioassay itself conducted on the following day. Alternatively, an early morning start with a 3-hr sperm capacitation time followed by a 3-hr insemination allows conduction of the assay by an experienced individual in the course of one working day. Because of donor-dependent differences in the kinetics of capacitation, bioassays conducted with two or more time points are desirable. Once the conditions have been selected and standardized for the laboratory, accuracy, sensitivity and specificity of the test must be evaluated. These issues have been addressed extensively in a recent symposium published in International Journal of Andrology (Supplement 6), October 1986. Troubleshooting the bioassay can be difficult since a large number of parameters must be controlled simultaneously; however, Table IV summarizes several of the more obvious approaches.

6.   CORRELATION OF ASSAY RESULTS WITH DONOR STATUS

A significant body of information has accumulated over the past decade correlating results of the zona-free hamster egg bioassay with donor status. These results pertain to standard assay conditions, not the augmented protocols described above. Males of proven fertility with normal semen parameters penetrate on average 50% of the eggs. Bioassay results for infertile males with normal or abnormal semen parameters summarized by Yanagimachi (1984) are 24% and 10%, respectively.

There is not much controversy over whether or not the mean scores for the different donor groups differ. Rather, the major controversy concerns whether or not there is overlap in groups. In other words, is it appropriate to define an arbitrary but absolute cutoff point that differentiates between the score of a proven fertile donor and that of a subfertile or clinically infertile male? The relatively high incidence of false negatives when a 10-15% cutoff is utilized indicates that fertility prognosis at the present time cannot be made solely on the basis of penetration scores. Thus, males of proven fertility do occasionally score poorly in repeated testing (Yanagimachi, 1984) and a subset of infertility patients who score poorly in the bioassay father children (Binor et al., 1982a). Perhaps the middle ground in this controversy over arbitrary cutoffs would simply be to take any penetration as a sign of fertility and a penetration score of less than 15% as a suggestion of subfertility which must be confirmed by repeat testing on a second ejaculate under the same conditions as well as under altered conditions of sperm capacitation.

Table IV. Troubleshooting the sperm penetration assay

Problem	Probable causes	Suggested course of action
Animals do not superovulate	Hormones used for superovulation are inactive	Examine ovaries a) Ovaries smooth in appearance – PMSG inactive b) Ovaries contain many cysts – hCG inactive Verify conditions of hormone preparation and storage a) Date of preparation b) NO filter sterilization c) NO storage of hormones in frost-free freezer d) Verify lot with manufacturer; check lighting in animal facility (12; 12 hr light; dark)
Sperm do not survive preincubation	Medium or incubator failure if occurs in fertility – proven donor sperm preparations	Check a) Medium pH b) Incubation temperature and gas composition c) If occurs exclusively with patient sample, it may reflect unique sperm dysfunction
No sperm–egg fusion	Inadequate capacitation and/or spontaneous acrosome reactions	a) Confirm presence of vigorously motile sperm to eliminate gross deficiencies in medium b) Repeat assay with different albumin lot if unproven lot in use c) Repeat assay with augmented capacitation protocol, i.e., pretreat sperm in ionophore, TES-Tris egg yolk buffer or low temperature
Highly variable penetration levels between eggs in the same assay	Incomplete zona removal	a) Aspirate eggs during trypsin exposure in and out of micropipets b) Use large drops of trypsin-containing medium

7.   CORRELATION WITH HUMAN IN VITRO FERTILIZATION

The "acid-test" for evaluating the accuracy of the hamster egg assay involves comparison of test results with the in vitro fertilization of human eggs. The availability of human in vitro fertilization-embryo transfer programs provides a unique opportunity to assess this relationship by comparing results directly with observations of human sperm-human egg interaction in a controlled setting. The predictive capability of the bioassay has been examined by a number of investigators (Margalioth et al., 1983, 1986; Foreman et al., 1984; Wolf and Sokoloski, 1984; Ausmanas et al., 1985). When applied to males with normal semen parameters, approximately 85% showed correspondence between the ability to penetrate wives' eggs in vitro and the ability to penetrate zona-free hamster eggs. All the errors occurred as false-positive results, that is, sperm penetrated zona-free hamster eggs but did not penetrate wives' eggs. These conclusions are based upon the arbitrary decision to use a generous interpretation of fertility indication as any sperm penetration of the eggs.

Although male infertility in patients with normal semen may occasionally be diagnosed on the basis of bioassay results, men with abnormal semen parameters represent the patients most often referred for hamster egg testing. The expectation in this case is that the penetration score will distinguish between fertility, subfertility associated with a decreased quantity of normal sperm and infertility where both quantity and quality are altered. Again, the "acid-test" for evaluating the accuracy of the bioassay involves comparisons with the in vitro fertilization of human eggs (Foreman et al., 1984; Wolf and Sokoloski, 1984; Margalioth et al., 1986). The overall efficacy of the hamster egg bioassay in predicting the results of IVF when the semen analysis is abnormal is approximately 50%. Interestingly, however, higher correlations have been observed when an augmented bioassay is employed (Hirsch et al., 1986) which needs to be confirmed by a larger data base and/or other investigators. On considering augmented protocols where sperm are preincubated under highly nonphysiologic conditions, the evaluation of significance must rely on correlations with the fertility status of the sperm donor. Whether or not these protocols will be useful in developing new treatment modalities for the infertile male remains speculative.

Perhaps the last question that needs to be addressed is whether or not zona-free hamster egg testing should be employed for screening potential participants in a human IVF-ET program. While I do not advocate excluding patients from IVF based upon failed zona-free hamster egg testing, the day may come when this expensive, time-consuming, and occasionally capricious testing may be used to determine clinical treatment. At the very least, a failed test at present should suggest the appropriateness of donor backup during an IVF treatment cycle.

8. REFERENCES

Aitken, R. J., Ross, A., Hargreave, T., Richardson, D., and Best, F.,
    1984, Analysis of human sperm function following exposure to the
    ionophore A23187. Comparison of normospermic and oligozoo-
    spermic men, J. Androl. 5:321–329.
Alexander, N. J., 1981, Evaluation of male infertility with an in
    vitro cervical mucus penetration test, Fertil. Steril. 36:201–
    208.
Ausmanas, M., Tureck, R. W., Blasco, L., Kopf, G. S., Ribas, J., and
    Mastroianni, L., Jr., 1985, The zona-free hamster egg penetra-
    tion assay as a prognostic indicator in a human in vitro fer-
    tilization program, Fertil. Steril. 43:433–437.
Barros, C., Herrera, E., Fuenzalida, I., and Argüello, B., 1986,
    Hamster oocyte fertilizability after 4°C storage, Gamete Res.
    14:149–157.
Bavister, B. D., 1981, Analysis of culture media for in vitro fer-
    tilization and criteria for success, in: Fertilization and
    Embryonic Development In Vitro (L. Mastroianni, Jr., and J. D.
    Biggers, eds.), Plenum Press, New York, pp. 41–60.
Biggers, J. D., Whitten, W. K., and Whittingham, D. G., 1971, The
    culture of mouse embryos in vitro, in: Methods in Mammalian
    Embryology (J. C. Daniel, Jr., ed.), W. H. Freeman and Company,
    San Francisco, pp. 86–116.
Binor, Z., Rao, R., and Scommegna, A., 1982a, Further studies on
    dysfunctional human spermatozoa using the zona-free hamster
    oocyte penetration assay, Fertil. Steril. 37:300 (abstract).
Binor, Z., Sokoloski, J. E., and Wolf, D. P., 1982b, Sperm interac-
    tion with the zona-free hamster egg, J. Exp. Zool. 222:187–193.
Bleil, J. D. and Wassarman, P. M., 1983, Sperm-egg interactions in
    the mouse: sequence of events and induction of the acrosome
    reaction by a zona pellucida glycoprotein, Dev. Biol. 95:317–
    324.
Bolanos, J. R., Overstreet, J. W., and Katz, D. F., 1983, Human sperm
    penetration of zona-free hamster eggs after storage of the semen
    for 48 hours at 2°C to 5°C, Fertil. Steril. 39:536–541.
Boldt, J. and Wolf, D. P., 1984, Sperm capacitation, in: Human In
    Vitro Fertilization and Embryo Transfer (D. P. Wolf and M. M.
    Quigley, eds.), Plenum Press, New York, pp. 171–211.
Byrd, W. and Wolf, D. P., 1986, Acrosomal status in fresh and capaci-
    tated human ejaculated sperm, Biol. Reprod. 34:859–869.
Cohen, J., Fehilly, C. B., and Walters, D. E., 1985, Prolonged
    storage of human spermatozoa at room temperature or in a
    refrigerator, Fertil. Steril. 44:254–262.
Foreman, R., Cohen, J., Fehilly, C. B., Fishel, S. B., and Edwards,
    R. G., 1984, The application of the zona-free hamster egg test
    for the prognosis of human in vitro fertilization, J. In Vitro
    Fert. Embryo Transfer 1:166–171.
Go, K. J. and Wolf, D. P., 1983, The role of sterols in sperm
    capacitation, Adv. Lipid Res. 20:317–330.

Hirsch, I., Gibbons, W. E., Lipshultz, L. I., Rossavik, K. K., Young,
    R. L., Poindexter, A. N., Dodson, M. G., and Findley, W. E.,
    1986, In vitro fertilization in couples with male factor
    infertility, Fertil. Steril. 45:659-664.
Huang, T. T. F., Fleming, A. D., and Yanagimachi, R., 1981, Only
    acrosome-reacted spermatozoa can bind to and penetrate zona
    pellucida: a study using the guinea pig, J. Exp. Zool. 217:287-
    290.
Jeyendran, R. S., Van der Ven, H. H., Perez-Pelaez, M., Crabo, B. G.,
    and Zaneveld, L. J. D., 1984, Development of an assay to assess
    the functional integrity of the human sperm membrane and its
    relationship to other semen characteristics, J. Reprod. Fertil.
    70:219-228.
Johnson, A. R., Syms, A. J., Lipshultz, L. I., and Smith, R. G.,
    1984, Conditions influencing human sperm capacitation and pene-
    tration of zona-free hamster ova, Fertil. Steril. 41:603-608.
Jonsson, B., Eneroth, P., Landgren, B.-M., and Wikborn, C., 1986,
    Evaluation of in vitro sperm penetration testing of 176
    infertile couples with the use of ejaculates and cervical mucus
    from donors, Fertil. Steril. 45:353-356.
Lee, M. A. and Storey, B. T., 1986, Bicarbonate is essential for
    fertilization of mouse eggs: mouse sperm require it to undergo
    the acrosome reaction, Biol. Reprod. 34:349-356.
Margalioth, E. J., Navot, D., Laufer, N., Yosef, S. M., Rabinowitz,
    R., Yarkoni, S., and Schenker, J. G., 1983, Zona-free hamster
    ovum penetration assay as a screening procedure for in vitro
    fertilization, Fertil. Steril. 40:386-388.
Margalioth, E. J., Navot, D., Laufer, N., Lewin, A., Rabinowitz, R.,
    and Schenker, J. G., 1986, Correlation between the zona-free
    hamster egg sperm penetration assay and human in vitro fertili-
    zation, Fertil. Steril. 45:665-670.
Overstreet, J. W., 1986, Evaluation of sperm-cervical mucus interac-
    tion, Fertil. Steril. 45:324-326.
Rogers, B. J., 1985, The sperm penetration assay: its usefulness
    reevaluated, Fertil. Steril. 43:821-840.
Rufo, G. A., Jr., Singh, J. P., Babcock, D. F., and Lardy, H. A.,
    1982, Purification and characterization of a calcium transport
    inhibitor protein from bovine seminal plasma, J. Biol. Chem.
    257:4627-4632.
Sokoloski, J. E. and Wolf, D. P., 1984, Laboratory details in an in
    vitro fertilization and embryo transfer program, in: Human In
    Vitro Fertilization and Embryo Transfer (D. P. Wolf and M. M.
    Quigley, eds.), Plenum Press, New York, pp. 275-296.
Syms, A. J., Johnson, A., Lipshultz, L. I., and Smith, R. G., 1984,
    Reduced ability of motile human spermatozoa obtained from
    oligospermic males to penetrate zona-free hamster eggs, Fertil.
    Steril. 41:105S (abstract).
Takemoto, F. S., Rogers, B. J., Wiltbank, M. C., Soderdahl, D. W.,
    Vaughn, W. K., and Hale, R. W., 1985, Comparison of the penetra-

tion ability of human spermatozoa into bovine cervical mucus and zona-free hamster eggs, J. Androl. 6:162–170.

Tesarik, J., 1985, Comparison of acrosome reaction–inducing activities of human cumulus oophorus, follicular fluid and ionophore A23187 in human sperm populations of proven fertilizing ability in vitro, J. Reprod. Fertil. 74:383–388.

Tesarik, J., 1986, From the cellular to the molecular dimension: the actual challenge for human fertilization research, Gamete Res. 13:47–89.

Wolf, D. P., 1985, How accurate is the hamster zona-free ovum test? Contemporary Ob/Gyn 25:199–208.

Wolf, D. P. and Sokoloski, J. E., 1984, Fertility potential evaluation with the zona-free hamster egg bioassay, in: Human In Vitro Fertilization and Embryo Transfer (D. P. Wolf and M. M. Quigley, eds.), Plenum Press, New York, pp. 297–326.

Wolf, D. P., Boldt, J., Byrd, W., and Bechtol, K. B., 1985, Acrosomal status evaluation in human ejaculated sperm with monoclonal antibodies, Biol. Reprod. 32:1157–1162.

Yanagimachi, R., 1982, Requirement of extracellular calcium ions for various stages of fertilization and fertilization-related phenomena in the hamster, Gamete Res. 5:323–344.

Yanagimachi, R., 1984, Zona-free hamster eggs: Their use in assessing fertilizing capacity and examining chromosomes of human spermatozoa, Gamete Res. 10:187–232.

ANALYSIS OF OOCYTE QUALITY AND FERTILIZATION

Don P. Wolf

1.   OOCYTE RECOVERY

     Before discussing quality evaluation of human oocytes, it is
appropriate to comment on oocyte recovery techniques. In general,
when searching for oocytes in follicular fluid, the operator must be
aware of the possible detrimental effects of oocyte exposure to room
air, undesirable pH in CO_2-bicarbonate buffered solutions, temperature changes, and to ambient lighting. While it is difficult to
quantitate adverse effects of these parameters when carefully controlled, it is well established that extremes of pH, temperature and
light exposure are detrimental to mammalian embryos. Several options
are available for controlling in vitro conditions including the use
of buffers that are stable in room air such as phosphate or Hepes,
working with an oil overlay and the use of heated stages or enclosed
temperature controlled environments under reduced lighting conditions. Regardless of the procedures employed, a skilled, experienced
operator should make every effort to minimize the exposure of eggs in
follicular fluid or in culture medium to adverse environmental conditions.

The quality of the follicular aspirate is critical to laboratory
success, if not to the clinical outcome of the treatment cycle.
Since recovery of a clear, straw-colored fluid is ideal, the surgeon
should be encouraged to stop aspirating at the first appearance of
blood in the aspiration fluid. Moreover, aspiration can be conducted
into a tube containing heparinized phosphate buffered saline to
control pH. Re-aspiration and irrigation, if necessary, can be
accomplished after microscopic examination of the initial aspirate.
In order to minimize the time required to recover oocytes, especially
relevant when follicular fluid samples are contaminated with blood,
only enough follicular fluid should be decanted to cover the bottom
of the petri dish to a depth of approximately 0.5 cm. With this
level of liquid, a rapid scan should reveal all cumulus masses which
can quickly be removed to fresh culture medium for further examination. Often the cumulus mass will appear to float in decanted fluid
and is, therefore, very easy to visualize. If the sample does not
contain a cumulus mass or other particulate material that requires
further observation, the decanted fluid can be set aside or discarded
and additional sample aliquots can be added to the petri dish. Note
also that if a cumulus mass containing an egg is found, you should
not discard the remainder of the sample. More than one follicle may
have been aspirated and biovular follicles have been observed (Figure
1). Oocytes, when identified, should be transferred through a wash
drop of insemination medium and into the center well of individual
organ culture dishes for rapid transfer to the incubator. In the
absence of an obvious cumulus mass, the presence of granulosa cell
plaques confirms that the fluid is of follicular origin. Such granu-

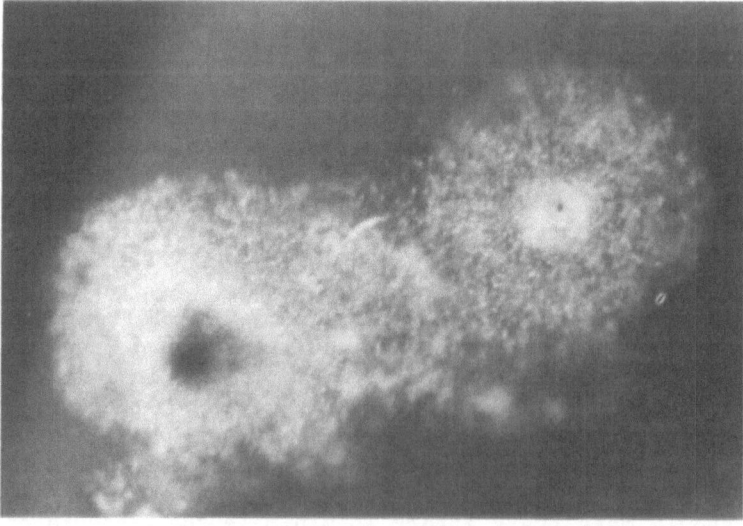

Fig. 1.   Example of two cumulus enclosed oocytes recovered from one
           aspirant. This may have been a biovular follicle.

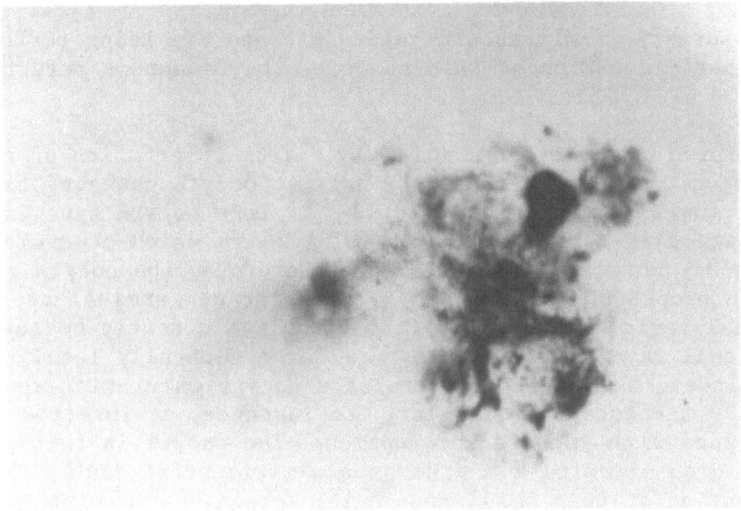

Fig. 2.   Granulosa cell plaques associated with expanded cumulus
           mass.

losa cell plaques can readily be confused with a cumulus mass when
the sample under analysis is contaminated with blood. However,
granulosa cell plaques are readily distinguishable from the cumulus
cell masses in culture medium at relatively low magnifications
(Figure 2).

        During culturing, in addition to controlling the parameters
mentioned above, another parameter of importance is humidity. This
can be controlled by several techniques; the most common is to
incubate oocytes in moated organ culture dishes in humidified incubators. The medium is changed every 24 hr. Alternatively, individual
dishes can be placed in gas-tight desiccators which are gassed and
placed in incubators and, finally, individual cultures can be overlaid with oil.

2.   OOCYTE QUALITY

        A determination of oocyte quality or maturity is important since
it dictates the length of time required for in vitro maturation prior
to sperm/egg mixing. Obviously, the more immature the oocyte, the
longer the maturation period required before sperm addition is appropriate. Several parameters should be recorded and evaluated in an
effort to judge oocyte maturity. First is the size of the follicle
from which the oocyte is collected, estimates of which can be made by
direct observations at laparoscopy, the volume of the follicular
fluid recovered and ultrasonography on the day the patient receives

hCG or at surgery if ultrasound-guided pickups are being performed. For a further discussion of this subject, the reader is referred to Quigley et al. (1982).

In determining oocyte maturity by direct observation of the cumulus-corona complex, perhaps the easiest oocyte category to define is that of the immature oocyte (Table I). This oocyte type is characteristically recovered from a follicle in which preovulatory maturation has not been initiated and, therefore, the oocyte remains arrested in prophase I of meiosis with an intact germinal vesicle. The germinal vesicle is difficult to visualize directly because of obscuring cell layers but appears as a large centrally localized nucleus. Indeed the ability to evaluate oocyte maturation requires high quality microscopic equipment, for instance, an inverted microscope equipped with interference optics. The oocyte is then placed in a small drop of culture medium in a plastic petri dish. The surface tension of this small drop helps spread the cumulus matrix thereby facilitating observation of the oocyte (Veeck, 1985). Beware, however, of the rapid desiccation that can occur. This procedure is best perfected or tested with nonviable human oocytes or nonhuman material. Immature oocytes are normally confined to or found in follicles of less than 10 mm mean diameter in which less than 0.5 ml of follicular fluid has been recovered. They will be surrounded by several layers of tightly adhering granulosa cells but are usually not associated with a discernible cumulus mass (see Veeck, 1985, for further details). Germinal vesicle intact oocytes are actually rather difficult to find in normal follicular aspirates since they are not embedded in a large, highly visible cumulus mass. When immature oocytes are recovered, however, they should be placed in culture and observed for prolonged periods of time until the appearance of the first polar body signals the completion of preovulatory maturation. The oocyte can then be transferred to fresh insemination medium and mixed with sperm. In our experience, premature addition of sperm to maturing oocytes interferes with preovulatory maturation and, therefore, should be avoided. Clinical pregnancies have been reported following the transfer of only in vitro matured germinal vesicle intact oocytes (Veeck et al., 1983). However, the low success associated with the transfer of immature oocytes does not justify efforts to recover such oocytes unless for research protocols.

With the follicular recruitment protocols now in use for patients undergoing IVF-ET, oocytes are usually recovered from follicles of mean diameter exceeding 15 mm (for further information on follicular recruitment, see Chapter 19). These follicles have grown to competency and will, therefore, invariably contain oocytes which have resumed meiosis from prophase I progressing to metaphase II, the stage normally ovulated in most mammalian species. In the human, the LH peak which triggers meiosis resumption should occur approximately 24 hr prior to ovulation, i.e., preovulatory maturation

Table I. Classification of viable human ovarian oocytes

Classification	Numerical grade	Follicular size	Meiosis	First polar body	Cumulus	Corona	Recommended in vitro incubation time prior to sperm addition
IMMATURE							
GV Intact	1	5–9 mm	Prophase I	Absent	Absent	1–3 layers tightly adherent to oocyte	Until 1st PB 18–36 hr
Maturing (No GV)	1	5–9 mm	Between Prophase I and Met II	Absent	Sparse, if present	As above	As above
PREOVULATORY							
Immature	2	10–14 mm	Between Prophase I and Met II	Absent	Dense	Tightly packed	Until 1st PB 6–12 hr
Mature	3	$\geq$ 15 mm	Met II	Present	Expanded, fluffy, large	Expanded	Within 6 hr
Postmature	4	$\geq$ 20 mm	Met II	Present	Expanded, often scanty	Expanded, often partially lost	Immediately
ATRETIC/DEGENERATIVE	5	Any	Any	Sometimes	Variable	Variable	Discard

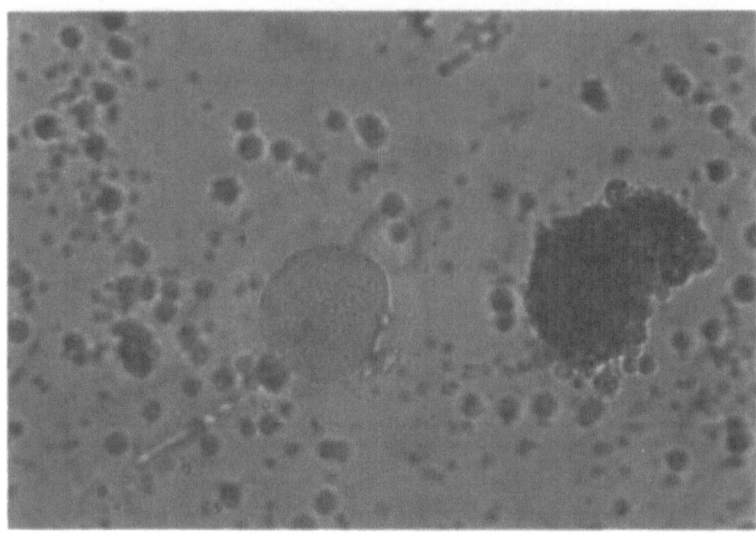

Fig. 3.   Human oocyte after cumulus dispersal with hyaluronidase.
           Note the presence of a polar body in the perivitelline
           space.

requires 24 hr. This time period is apparently extended in IVF-ET as
the normal time between hCG administration and surgical intervention
is 34-36 hr. Several preovulatory oocyte classification systems have
been suggested as summarized in Table I (Testart et al., 1983; Veeck,
1985). In the event that difficulties are encountered in evaluating
the relative state of maturation, hyaluronidase (0.1%) dispersal of
the cumulus can be used to improve visualization of the oocyte and
its perivitelline space (Figure 3). In this case, the presence of a
first polar body in the perivitelline space can be used as an
unequivocal measure of the maturation of the oocyte to metaphase II.

3.   PROCEDURES

     The details of the laboratory techniques for IVF-ET in the
University of Texas-Houston program have been published previously
(Sokoloski and Wolf, 1984). Here I have also incorporated variations
currently in use in the program at Oregon Health Sciences University.
The timing of laboratory events relative to the administration of hCG
is depicted in Table II.

     3.1. Materials

          a.   Ham's F-10 or HTF (human tubal fluid medium quality-
                control tested and available through Irvine
                Scientific, Irvine, CA) culture medium supplemented

Table II. Schedule for laboratory aspects of IVF-ET

Procedure:				
Patient receives hCG	Culture media preparation	Egg collection and fertilization	Evaluation of fertilization	Embryo transfer
p.m.		a.m. + p.m.	a.m.	
Monday	Tuesday	Wednesday	Thursday	Friday
Tuesday	Wednesday	Thursday	Friday	Saturday
Wednesday	Thursday	Friday	Saturday	Sunday
Thursday	Friday	Saturday	Sunday	Monday
Friday	Saturday	Sunday	Monday	Tuesday
Saturday	Sunday	Monday	Tuesday	Wednesday
Sunday	Monday	Tuesday	Wednesday	Thursday

       with heat inactivated maternal serum or equivalent, i.e., cord serum

b. Dulbecco's PBS (serum optional)

c. Tissue culture grade plastic petri dishes (Falcon 1007, 3001, Becton Dickinson, Oxnard, CA)

d. Dissecting microscope with heated stage warmer

e. Circulating temperature controlled water pump for plexiglass stage warmer (Aquamatic K Module, Model K-20, American Medical Systems, Cincinnati, OH)

f. Sterile Pasteur pipets/PiPump

g. 1 cc disposable tuberculin syringes with 25-gauge needles

h. Tissue culture grade organ culture dishes (Falcon, 3037, Becton Dickinson, Oxnard, CA)

i. Incubator, 2 or 3 gas control (Note: modified pediatric isolettes are becoming popular with the advent of outpatient egg pickup procedures and cryopreservation programs)

j. Disposable plastic 5/10 ml pipets

k. Marking pen, watchmakers forceps, burner

3.2. Laboratory Setup

a. On the day following hCG administration, prepare
    culture media. For details see Chapter 4, this
    volume, or Sokoloski and Wolf (1984). Label all media
    by patient name with color coded tape.

b. Preequilibrate media overnight--in loosely capped
    flasks and tubes--in 5% CO_2 in air incubated at 37°C.
    Remember to check and record incubator temperature and
    gas composition at least daily.

c. Approximately 1 hr before surgery, prepare individual
    organ culture dishes. Add 0.9 ml of insemination
    medium to center well and $\geq$ 2 ml of medium to moat.
    Label each dish with patient's name. During surgery,
    add information on egg origin as it becomes available.

d. Prepare work area: turn on slide warmer, circulating
    water for warming microscope stage, etc. Check inven-
    tories of disposable ware.

3.3. Egg Recovery

a. Upon receipt of a follicular fluid sample from the
    operating room, label the tube with the patient's
    name, source of the aspirate and nature of sample if
    not already done so by operating room personnel
    (aspirate, flushing, irrigant). Measure volume by
    comparison with a calibrated empty tube. Record the
    surgeon's estimate of follicular size.

b. Pour the aspirate into a sterile dish (Falcon 1007 or
    3001, depending on the volume) on the stage of the
    dissecting microscope and look for the cumulus mass
    containing the egg. Often the mass can be seen with
    the unaided eye, usually as a clear to white fluffy
    aggregate, like a small piece of clear to white
    gelatin. Under the dissection microscope, this mass
    is clearly distinct from the darker, more dense granu-
    losa sheets or plaques (Figure 2). When the straw-
    colored follicular fluid is contaminated with blood,
    pick out for further examination all suspicious
    looking cellular particulates and transfer to a dish
    containing medium. Confirm the presence of an egg.
    If, after looking through both the follicular aspirate
    and needle flush, no egg is found, inform the surgeon
    so that the follicle can be irrigated. Check the
    irrigated fluid above and repeat if necessary. If an
    oocyte is recovered from a bloody irrigant, be sure to

wash in medium before transferring to the organ
culture dish. Adherent blood clots or granulosa cell
plaques can be removed from the cumulus with 26-gauge
needles attached to tuberculin syringes.

c.   Eggs recovered from follicular aspirates either with
or without dilution in Dulbecco's PBS are rinsed in
culture medium containing 7.5% maternal serum, transferred to individual organ culture dishes and placed
in the incubator at 37°C in a humidified atmosphere of
5% CO_2, 5% O_2, 90% N_2 or 5% CO_2 in air.

d.   Semen is collected by masturbation and a routine semen
analysis is performed following at least 30 min of
liquefaction at 37°C (for details, see Chapter 6). We
have encouraged husbands to collect semen prior to
surgery in an effort to reduce anxiety levels.
However, the most common protocol utilized by IVF
programs involves semen collection after egg pickup
with a brief sperm preincubation prior to sperm-egg
mixing.

e.   Immediately prior to sperm addition (see Table I for
appropriate length of in vitro maturation), an
analysis is conducted on the enriched population of
spermatozoa and the volume required to deliver 5-
10 x 10^5 motile sperm per organ culture dish is
calculated, usually 3-15 µl. With oligospermic males,
efforts are made to reconstitute sperm pellets in
small volumes, thereby increasing the final sperm
concentration. For oligospermic males, inseminations
are conducted at 50 x 10^5 motile sperm per ml (Wolf et
al., 1984).

f.   Following approximately 14 hr of sperm-egg interaction, eggs are removed from the incubator for evaluation. At this time, the presence or absence of motile
sperm should be recorded. The cumulus should be
completely dispersed and appear as a nearly confluent,
circular layer of cells surrounding the egg. The
tightly adherent corona cells can be dispersed so that
the presence of pronuclei within the vitellus can be
confirmed. Alternatively, eggs can be transferred to
petri dishes for direct observation with an inverted
phase contrast microscope where pronuclei can often be
visualized.

g.   A series of hand pulled micropipets are prepared and
sized by breaking back with watchmakers' forceps to
give internal diameters at the tip on the order of

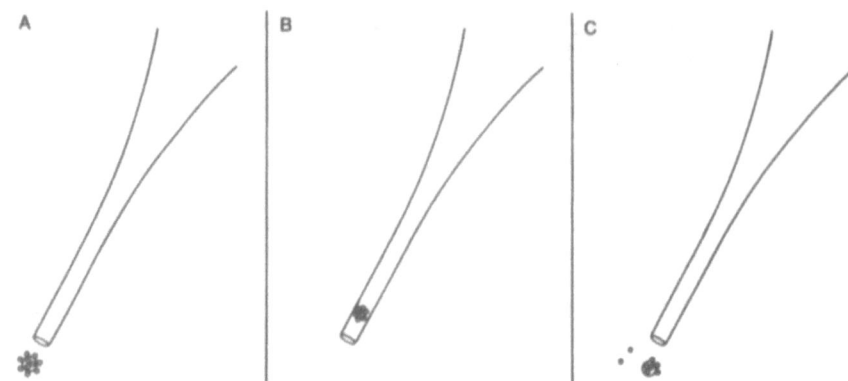

Fig. 4.   Mechanical removal of corona cells from human oocytes.
           A) Appearance of appropriately sized micropipet prior to
           oocyte aspiration.   B) Aspiration of oocyte–corona complex
           into the micropipet.   C) Expulsion of oocyte with dislodge-
           ment of corona cells.   The shearing force of corona cells
           against the walls of the micropipet is usually sufficient
           to dislodge them.

           100–200 μ.   It is the large portion, not the small tip
           of the pulled pipet, that is retained for this use.

    h.   Oocytes are dislodged from the bottom of organ culture
           dishes by a gentle stream of culture medium delivered
           from an unmodified Pasteur pipet. Corona cells are
           dispersed by gently pipetting the egg in and out of a
           pulled micropipet (Figure 4). Usually, only a portion
           of the corona cell complement must be dislodged.
           Occasionally a rubbery cumulus–corona complex
           surrounds the egg, precluding the use of a micropipet.
           Any time micropipetting would appear to jeopardize the
           egg, needles should be used. Often stretching a
           rubbery mass with two 25–gauge needles attached to 1-
           cc tuberculin syringes will result in corona removal
           or "popping" of the egg from the mass.

    i.   After confirming the presence of 2 pronuclei, eggs are
           transferred to organ culture dishes containing growth
           medium for continued culture.

        In the advent of failed or abnormal fertilization, several of
the more obvious explanations along with appropriate responses are
itemized in Table III. For a photographic record of human eggs and
embryos, both normal and abnormal, see Chapter 10.

Table III. Troubleshooting human oocyte fertilization

Outcome		Probable reasons	Response
Failed fertilization	– no motile sperm; no cumulus dispersal	bad medium – male factor	check medium pH, osmolarity; sperm survival or fertility potential by penetration assay
	– no PN	bad sperm or eggs	reinseminate with husband or perhaps donor sperm
	– motile sperm; cumulus dispersal	post- or immature eggs – male factor	reinseminate; screen couple for chlamydia
	– agglutinated sperm; cumulus dispersal	antisperm antibody	collect sperm in culture medium to minimize antibody contact
Polyspermic fertilization	– 3 or more PN	high inseminating motile sperm concentration – immature or aged oocytes	culture but do not transfer; possible freeze for future transfer if normal appearing blastocysts result
Fragmentation	– 0 or 1 PN + premature and often irregular cleavage	activation – spontaneous in aging oocytes	culture but do not transfer
Cracked zonae		excessive stress during aspiration; postmature oocytes	check negative pressure used during aspiration – should be approx. 100 mm Hg; reevaluate follicular recruitment and timing of hCG administration

4.   REFERENCES

Quigley, M. M., Wolf, D. P., Maklad, N. F., Dandekar, P. V., and
    Sokoloski, J. E., 1982, Follicular size and number in human in
    vitro fertilization, Fertil. Steril. 38:678–681.
Sokoloski, J. E. and Wolf, D. P., 1984, Laboratory details in an in
    vitro fertilization and embryo transfer program, in: Human In
    Vitro Fertilization and Embryo Transfer (D. P. Wolf and M. M.
    Quigley, eds.), Plenum Press, New York, pp. 275–296.
Testart, J., Frydman, R., De Mouzon, J., Lassalle, B., and Belaisch,
    J. C., 1983, A study of factors affecting the success of human
    fertilization in vitro. I. Influence of ovarian stimulation
    upon the number and condition of oocytes collected, Biol.
    Reprod. 28:415–424.
Veeck, L. L., 1985, Extracorporeal maturation: Norfolk, 1984, Ann.
    N.Y. Acad. Sci. 442:357–367.

Veeck, L. L., Wortham, J. W. E., Jr., Witmyer, J., Sandow, B. A.,
    Acosta, A. A., Garcia, J. E., Jones, G. S., and Jones, H. W.,
    Jr., 1983, Maturation and fertilization of morphologically
    immature human oocytes in a program of in vitro fertilization,
    Fertil. Steril. 39:594-602.
Wolf, D. P., Byrd, W., Dandekar, P., and Quigley, M. M., 1984, Sperm
    concentration and the fertilization of human eggs in vitro,
    Biol. Reprod. 31:837-848.

9

ANALYSIS OF EMBRYONIC DEVELOPMENT

Don P. Wolf

1.   INTRODUCTION

A comprehensive review of mammalian embryo culture in vitro is presented in Chapter 18. Additionally, a discussion of embryo quality as it relates to embryo freeze-thaw survival can be found in Chapter 16. This chapter, along with Chapter 8, is designed to supplement Chapter 10.

While initial attempts at culturing human embryos were carried out in media used for the development of mouse embryos (Edwards et al., 1970), both complex media like Ham's F-10 or Menezo's B$_2$ and simple media such as Earle's, Whittingham's T-6 or HTF (Quinn et al., 1985) supplemented with human or fetal cord serum are commonly employed. Reports by Menezo et al. (1984) and Feichtinger et al. (1986) indicate that serum is not absolutely essential as both human IVF and embryo culture can be conducted in medium containing only 1% human serum albumin. Indeed, successful fertilization, embryo development and pregnancy has been reported using a simple, chemically defined medium with no added protein (Caro and Trounson, 1986). However, a second study contrasting the presence and absence of patient serum in Ham's F-10 medium concluded that embryo cleavage was

superior in serum containing medium (Kruger et al., 1987). Despite
the availability of adequate media and culture conditions, further
research and development is desirable to optimize conditions that
support embryonic growth in vitro (see Chapter 18). This point is
exemplified in man by the observation of Fehilly et al. (1985) that
only a small percentage of in vitro fertilized oocytes develop to
expanded blastocysts when cultured under controlled environmental
conditions in Earle's medium supplemented with sodium pyruvate and
inactivated human serum.

The ultimate measurement of embryonic viability is the establishment of a pregnancy after transfer. Short of this indicator, the
rate of cleavage and the morphological quality of the embryo is used.
Biochemical markers of embryo viability would be helpful and, in this
regard, an embryo-derived platelet activating factor capable of
inducing transient maternal thrombocytopenia has been proposed as
both an in vitro and in vivo marker of embryo viability (O'Neill et
al., 1985a,b; Gidley-Baird et al., 1986; Roberts et al., 1987).

Since many IVF programs transfer embryos at the 2- to 4-cell
stage of development, cleavage rate alone is not highly discrimina-
tory in quality evaluation. The culturing of embryos for additional
time periods to improve quality assessment does not appear warranted
at the present time unless for cryopreservation purposes, since
pregnancy rates for postinsemination day 2 versus day 3 transfers are
comparable. Moreover, quality evaluation is further complicated by
the fact that variation in development rates can be seen within
embryos from a single patient cultured under identical conditions,
e.g., at 40 hr postinsemination embryos can range from 2- to 6-cell
stage (see Figures 26 and 27 in Chapter 10). Despite such intra-
patient variation, retarded cleavage rates probably reflect reduced
embryo viability (Trounson, 1982; Trounson et al., 1982; Mohr et al.,
1983; Fishel et al., 1985; Cummins et al., 1986). Conversely, the
suspicion has been expressed that, within limits, faster cleavage
rates are associated with higher clinical pregnancy rates.
Certainly, fragmenting (extremely rapid rate) or nondividing embryos
can be excluded on the basis of developmental rates. Growth rate
graphs have been determined for the human embryo and used in embryo
rating with limited success (Cummins et al., 1986). Increased
accuracy from growth rate evaluation can, of course, be obtained by
increasing the frequency of embryo observation, an undertaking which
is not without risk.

The morphological quality of human embryos, which are destined
to be transferred to the ovum donor, must be evaluated, of course, by
nondestructive techniques. Thus, approaches which rely upon transmission electron microscopic observations are superfluous to the
routine morphological evaluation of embryos produced in an IVF-ET
program. Extensive published records of normal cleavage stage human
embryos are available (Edwards, 1977, 1980; Trounson et al., 1982;

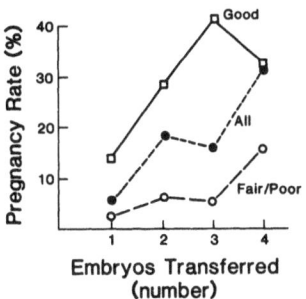

Fig. 1.   Relationship between embryo quality and pregnancy rate in
           an IVF-ET program.   (Redrawn from duPlessis et al., 1985.)

Veeck, 1986; see Chapter 10). Approaches to embryo assessment, which
are nondestructive but have not been exploited in the human, involve
fluorescein diacetate (Mohr et al., 1983; Hoppe and Bavister, 1984)
and the Hoechst DNA-binding dye, 33258 (Ebert et al., 1985). Fluorescein diacetate can be used to establish cellular integrity in
individual blastomeres and would most appropriately be applied to
abnormal embryos while the Hoechst dye has been used to facilitate
nuclei counting. While the relationship between embryo quality and
the establishment of pregnancy has not been rigorously documented, it
is assumed that higher pregnancy rates are associated with higher
quality embryos, a tenet established in bovine embryo transfer
studies. Preliminary reports with human embryos support this notion
(Figure 1; Cummins et al., 1986). Some groups (Speirs et al., 1983)
transfer embryos only if they appear healthy on the basis of dividing
blastomeres of equal size which occupy most of the space within the
zona. Moreover, cell division must be progressive and within established time limits. On the other hand, the totipotency of 2- to 8-
cell blastomeres would dictate the transfer of any embryo with at
least one healthy blastomere, the latter determined by at least the
confirmation of a nucleus (Zeilmaker and Alberda, 1982). Mohr et al.
(1983) reported that the transfer of "atypical" embryos containing a
few anucleate fragments and otherwise normal blastomeres or blastomeres which appeared multinucleated (at low magnification) was compatible with subsequent normal development and implantation. Another
argument in favor of transferring "atypical" embryos is the fact that
dramatic and rapid morphological changes occasionally occur during
normal development as evidenced by the embryo that scores poorly
early in the day of transfer and looks nearly perfect at transfer.

     A uniform grading system for human embryos is highly desirable.
We are currently grading embryos on a scale of 1-5 as suggested by
Mohr et al. (1983) (see also Cummins et al., 1986), taking into
consideration cell size and shape, the presence of fragments or other
debris in the perivitelline space, the gross quality of the ooplasm

Table I. Grading human embryos[a]

Grade	Characteristics
1	Very uneven blastomeres, i.e., one or more fragmenting blastomeres; lots of perivitelline space fragments; granular or otherwise unhealthy ooplasm; in a word: "bad."
2	Significantly distorted blastomeres or major unevenness in size; lots of perivitelline space fragments; otherwise healthy appearing.
3	Slightly distorted blastomeres or some unevenness in size; slight granularity; "healthy" ooplasm; modest perivitelline space fragments.
4	Even blastomeres; slight granularity and "healthy" ooplasm; a few perivitelline space fragments.
5	Even blastomeres; very clean; no perivitelline space fragments: "healthy" ooplasm.

[a]All should be $\geq$ 2 cell within appropriate time windows (normally 38–40 hr postsperm addition).

and the rate of cleavage (Table I; Figure 2; Chapter 10). In considering the rate of cleavage, the scores in Table I are adjusted up or down by 1/2 a point depending on whether or not the embryo is at 4-cell at 42 hr postinsemination. For more precise numbers, Cummins et al. (1986) report "ideal" growth rates of 24.7 hr for the pronuclear embryo, 33.6 hr, 45.5 hr and 56.4 hr for the 2-, 4- and 8-cell stage, respectively.

2.   DETAILS OF PROCEDURE

    2.1. Materials

        2.1.1.   Dissecting microscope
        2.1.2.   Inverted phase contrast microscope
        2.1.3.   Video or photographic equipment
        2.1.4.   Sterile Pasteur pipets/PiPump
        2.1.5.   Tissue culture grade organ culture dishes
        2.1.6.   Tissue culture grade plastic petri dishes
        2.1.7.   Growth medium

2.2. Methods

a. Remove organ culture dishes from the incubator indi-
    vidually and grade; evaluate the timing and quality of
    development; photograph or videotape as desired.
    Transfer the embryo to a petri dish for optimal photo-
    graphic records using differential interference
    microscopy.

b. Record the time of observation and the appearance of
    each embryo. Return each embryo to the incubator
    immediately if embryo transfer will follow within 4
    hr. Transfer to fresh growth medium in new organ
    culture dish if prolonged incubation will occur.

c. Make arrangements for embryo transfer (see Chapter
    11).

Table II. Troubleshooting early embryonic development. Observations
at 40-hr postinsemination.

Outcome	Probable reason	Response
2-6 cell	Successful fertili-zation and early cleavage	Prepare for ET
16-cell or greater; unequal blastomeres	Activation without sperm penetration	Discard; re-evaluate follicular recruit-ment schedule if frequent and persis-tent
2 PN - failed cleavage; cleavage: arrest	Heat or temperature shock; abnormal oogenesis; unknown causes	Optional for ET
3 or more PN - with cleavage	Polyspermic penetra-tion	Culture in vitro and cryopreserve for subsequent ET if normal blastocyst results

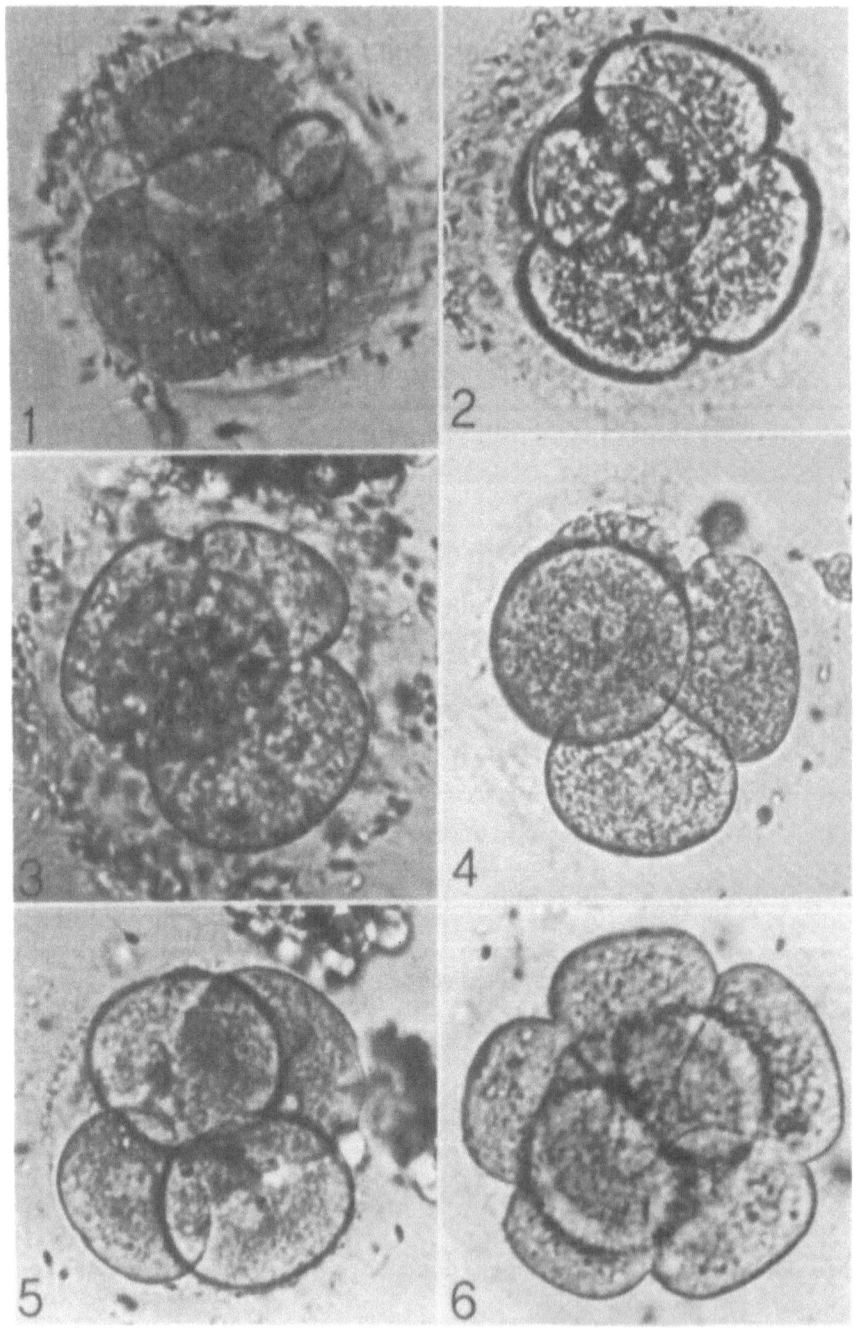

Some of the more common problems associated with early embryonic development in vitro are itemized in Table II along with suggestions for appropriate action.

3.   REFERENCES

Caro, C. M. and Trounson, A., 1986, Successful fertilization, embryo development, and pregnancy in human in vitro fertilization (IVF) using a chemically defined culture medium containing no protein, J. In Vitro Fert. Embryo Transfer 3:215-217.

Cummins, J. M., Breen, T. M., Harrison, K. L., Shaw, J. M., Wilson, L. M., and Hennessey, J. F., 1986, A formula for scoring human embryo growth rates in in vitro fertilization: Its value in predicting pregnancy and in comparison with visual estimates of embryo quality, J. In Vitro Fert. Embryo Transfer 3:284-295.

duPlessis, Y., Bowne, H., Abramczak, J., and Lopata, A., 1985, Morphological assessment of human embryos in relation to the success of in vitro fertilization, Abstract presented at the Fourth World Conference on In Vitro Fertilization, Melbourne, Australia.

Fig. 2.   Photographs of fresh unmounted human embryos. (1-5) Four-cell human embryos at various grades of development; (1) Poor. Note the misshapen blastomeres and fragments; (2) Fair to poor. Note single large fragment; (3) Fair. Note slightly uneven blastomere size and some small fragments; (4) Fair to good; (5) Good. Note the large spherical blastomeres of even density and size; (6) A good eight-cell embryo. (Reproduced with permission from Mohr et al., 1983.)

Ebert, K. M., Hammer, R. E., and Papaioannou, V. E., 1985, A simple method for counting nuclei in the preimplantation mouse embryo, Experientia 41:1207-1209.

Edwards, R. G., 1977, Early human development: from the oocyte to implantation, in: Scientific Foundations of Obstetrics and Gynaecology (E. E. Philipp, J. Barnes, and M. Newton, eds.), Heinemann Medical Books, Chicago, pp. 175-252.

Edwards, R. G., 1980, Conception in the Human Female, Academic Press, New York.

Edwards, R. G., Steptoe, P. C., and Purdy, J. M., 1970, Fertilization and cleavage in vitro of preovulator human oocytes, Nature (London) 227:1307-1309.

Fehilly, C. B., Cohen, J., Simons, R. F., Fishel, S. B., and Edwards, R. G., 1985, Cryopreservation of cleaving embryos and expanded blastocysts in the human: a comparative study, Fertil. Steril. 44:638-644.

Feichtinger, W., Kemeter, P., and Menezo, Y., 1986, The use of synthetic culture medium and patient serum for human in vitro fertilization and embryo replacement, J. In Vitro Fert. Embryo Transfer 3:87-92

Fishel, S. B., Cohen, J., Fehilly, C., Purdy, J. M., Walters, D. E., and Edwards, R. G., 1985, Factors influencing human embryonic development in vitro, Ann. N.Y. Acad. Sci. 442:342-356.

Gidley-Baird, A. A., O'Neill, C., Sinosich, M. J., Porter, R. N., Pike, I. L., and Saunders, D. M., 1986, Failure of implantation in human in vitro fertilization and embryo transfer patients: the effects of altered progesterone/estrogen ratios in humans and mice, Fertil. Steril. 45:69-74.

Hoppe, R. W. and Bavister, B. D., 1984, Evaluation of the fluorescein diacetate (FDA) vital dye viability test with hamster and bovine embryos, Anim. Reprod. Sci. 6:323-335.

Kruger, T. F., Stander, F. S. H., Smith, K., Van der Merwe, J. P., and Lombard. C. J., 1987, The effect of serum supplementation on the cleavage of human embryos, J. In Vitro Fert. Embryo Transfer 4:10-12.

Menezo, Y., Testart, J., and Perrone, D., 1984, Serum is not necessary in human in vitro fertilization, early embryo culture, and transfer, Fertil. Steril. 42:750-755.

Mohr, L. R., Trounson, A. O., Leeton, J. F., and Wood, C., 1983, Evaluation of normal and abnormal human embryo development during procedures in vitro, in: Fertilization of the Human Egg In Vitro (H. M. Beier and H. R. Lindner, eds.), Springer-Verlag, Berlin, pp. 211-221.

O'Neill, C., Gidley-Baird, A. A., Pike, I. L., Porter, R. N., Sinosich, M. J., and Saunders, D. M., 1985a, Maternal blood platelet physiology and luteal-phase endocrinology as a means of monitoring pre- and postimplantation embryo viability following in vitro fertilization, J. In Vitro Fert. Embryo Transfer 2:87-93.

O'Neill, C., Pike, I. L., Porter, R. N., Gidley-Baird, A. A.,
    Sinosich, M. J., and Saunders, D. M., 1985b, Maternal recogni-
    tion of pregnancy prior to implantation. Methods for monitoring
    embryonic viability in vitro and in vivo, Ann. N.Y. Acad. Sci.
    442:429-439.
Quinn, P., Kerin, J. F., and Warnes, G. M., 1985, Improved pregnancy
    rate in human in vitro fertilization with the use of a medium
    based on the composition of human tubal fluid, Fertil. Steril.
    44:493-498.
Roberts, T. K., Adamson, L. M., Smart, Y. C., Stanger, J. D., and
    Murdoch, R. N., 1987, An evaluation of peripheral blood platelet
    enumeration as a monitor of fertilization and early pregnancy,
    Fertil. Steril. 47:848-854.
Speirs, A. L., Lopata, A., Gronow, M. J., Kellow, G. N., and
    Johnston, W. I. H., 1983, Analysis of the benefits and risks of
    multiple embryo transfer, Fertil. Steril. 39:468-471.
Trounson, A. O., 1982, Factors influencing the success of fertiliza-
    tion and embryonic growth in vitro, in: Human Conception In
    Vitro (R. G. Edwards and J. M. Purdy, eds.), Academic Press,
    New York, pp. 201-205.
Trounson, A. O., Leeton, J. F., and Wood, C., 1982, In vitro fertili-
    zation and embryo transfer in the human, in: Follicular Matura-
    tion and Ovulation (R. Rolland, E. V. van Hall, S. G. Hillier,
    K. P. McNatty, and J. Schoemaker, eds.), Excerpta Medica,
    Amsterdam, pp. 313-322.
Veeck, L., 1986, Atlas of the Human Oocyte and Early Conceptus,
    Williams and Wilkins, Baltimore.
Zeilmaker, G. H. and Alberda, A. Th., 1982, Some observations on
    human oocyte maturation and fertilization and a consideration of
    the risks involved in employing the technique of in vitro
    fertilization, in: Follicular Maturation and Ovulation (R.
    Rolland, E. V. van Hall, S. G. Hillier, K. P. McNatty, and J.
    Schoemaker, eds.), Excerpta Medica, Amsterdam.

10

MORPHOLOGY OF HUMAN EGGS AND EMBRYOS

Don P. Wolf, Marybeth Gerrity, and Gregory S. Kopf

1.   INTRODUCTION

Photographic or video micrographic records of human eggs and
embryos are useful for patient record keeping, quality control,
research purposes and teaching. The morphology of human oocytes,
eggs and embryos has been described extensively over the past several
years and the readers' attention is drawn to the publication by
Lucinda Veeck (Atlas of the Human Oocyte and Early Conceptus, Lucinda
L. Veeck, Williams and Wilkins, Baltimore, 1986). Photo or video
micrography must be a simple and rapid process within the context of
an IVF-ET program since exposure of eggs or embryos to changes in

temperature, light and pH may be damaging. Ideally, photo or video
recordings should be made with a microscope in an enclosed environmental chamber. Since many programs use an inverted microscope for
the routine examination of eggs and embryos, a standard through-the-
lens system made available by the manufacturer is usually the most
practical and cost-effective approach. For convenience a system
should be chosen that is parfocal with the microscope and permits
through-the-lens viewing and recording. Systems are available with
fully automated light metering capabilities. Records can also be
made with a dissecting microscope (see Figures 4 and 11). Advantages
include the fact that the eggs/embryos do not have to be transferred
back and forth between microscopes. However, the limited resolving
capability of these microscopes greatly detracts from their usefulness, especially true with lower quality microscopes.

     Film choice will vary. For color slides, useful in teaching,
Ectachrome 200 daylight film or equivalent is satisfactory. Be
aware, however, that color slides do not always reproduce well into
black and white prints. For the latter, Kodak Technical Pan 2415 or
Tri-X Pan is recommended. This is an inexpensive approach for daily
record keeping. Negatives, prints or a proof sheet may be kept on
file. Black and white prints are also useful if the ultimate goal
includes publication.

     An alternative approach, albeit more expensive, to a 35 mm
format is available in Polaroid® films, color as well as black and
white. A major advantage in this case is rapid development. A high
contrast black/white film (ASA 400) is used for documentation at the
Hospital of the University of Pennsylvania. These films are inferior
in quality to conventional 35 mm films with regard to both color and
clarity (due to the high speed, it tends to be grainy). Under normal
conditions, the slides prepared from this film have a finite life.
However, they may be made archival by adding some relatively simple
wash steps at the end of the development series.

     It must be emphasized that because of the properties of the
plastic and the design of the organ culture dishes (that is, a dish
that slopes toward the center), it is essential to transfer the
material to be photographed from organ culture dishes to petri dishes
for optimal results. Photography of eggs/embryos in organ culture
dishes may result in distortion and light scattering, thus resulting
in a poorer quality image.

     Some IVF programs document their laboratory procedures using
video tape. The major advantage to this type of documentation is
that monitoring can be done on a real time basis. Disadvantages are
that the videotapes break easily, are expensive and the clarity of
the picture may be compromised unless high quality equipment is used.
Record keeping using tapes (e.g., storage, retrieval of specific
segments) is considerably more difficult than with photographs.

The following atlas is comprised of micrographs from three different IVF-ET programs utilizing several of the photographic combinations described above. The optical magnifications vary from 50 to 500X.

2.   NOMENCLATURE

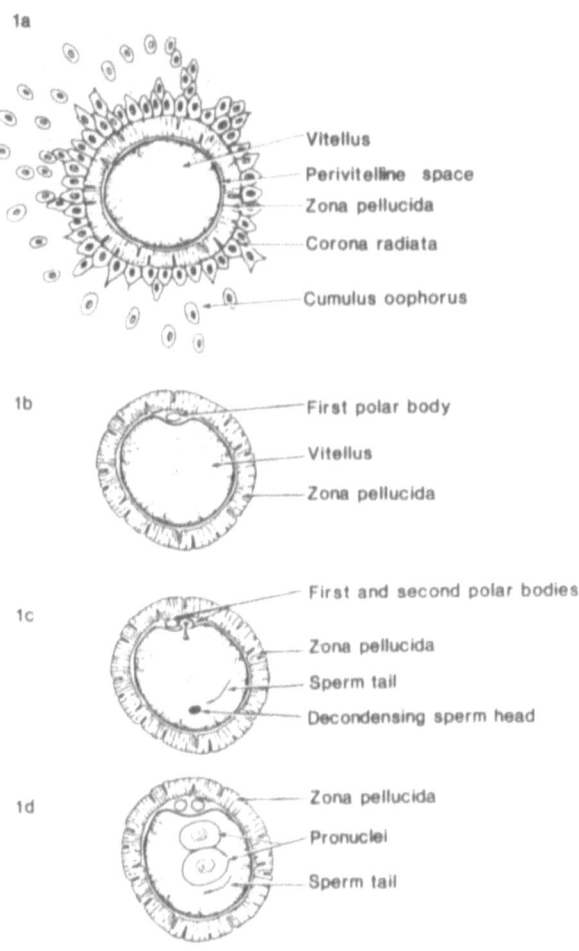

Fig. 1.   Schematic representation of a cumulus-enclosed human egg
           before (1a) and after (1b) cumulus and corona dispersal,
           sperm-egg fusion (1c), and pronuclear formation (1d). The
           relevant structures have been identified to facilitate the
           characterization of eggs and embryos depicted in this
           chapter.

3. NORMAL CASES

3.1. Immature Oocytes

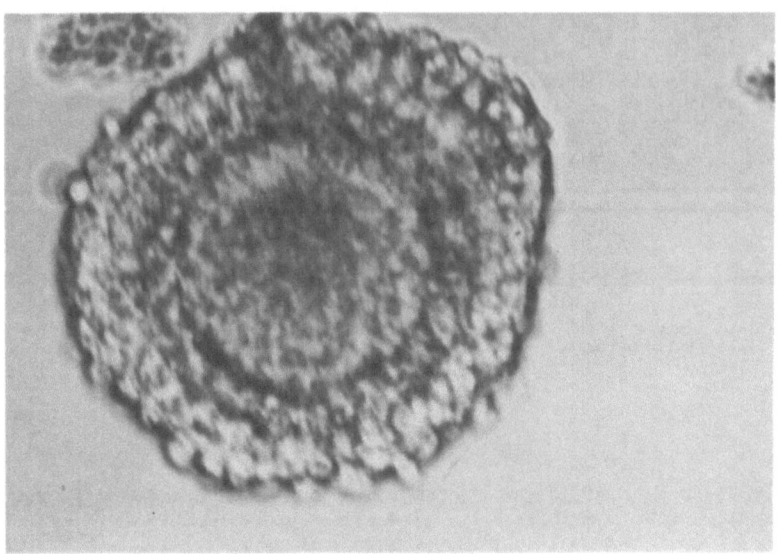

Fig. 2.   Immature oocyte; Prophase I.

Description: The vitellus is surrounded by several tightly packed
layers of granulosa cells. This oocyte probably contains an intact
germinal vesicle, although it cannot be seen in this micrograph.
Immature oocytes like this are not associated with a cumulus oophorus
and are usually recovered from the accidental or inadvertent aspiration of a small follicle (< 8 mm diameter). Incubation for prolonged
time periods (> 24 hr) under IVF conditions may result in oocyte
maturation. However, the prognosis for fertilization, cleavage and
implantation after ET is poor. One of 5 oocytes obtained from a
patient hyperstimulated with hMG (2 ampules/day, days 3–9; hCG, day
10). This oocyte never matured sufficiently to even attempt fertilization.

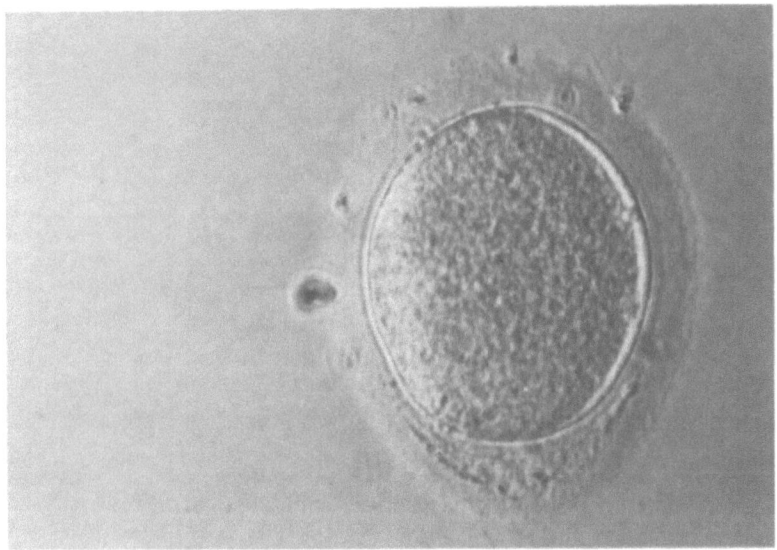

Fig. 3.   Immature oocyte; between Prophase I and Metaphase II.

Description: This oocyte without any associated cumulus or corona
cells was recovered from a preovulatory follicle. Since it did not
contain a visible germinal vesicle or a polar body, it was presumed
to have resumed meiosis with nuclear progression between Prophase I
and Metaphase II. Note also the "tight" zona with minimum perivitelline space. This oocyte extruded a first polar body after 24 hr in
culture and fertilized (polyspermic); it was not transferred.

3.2. Mature Preovulatory Oocytes

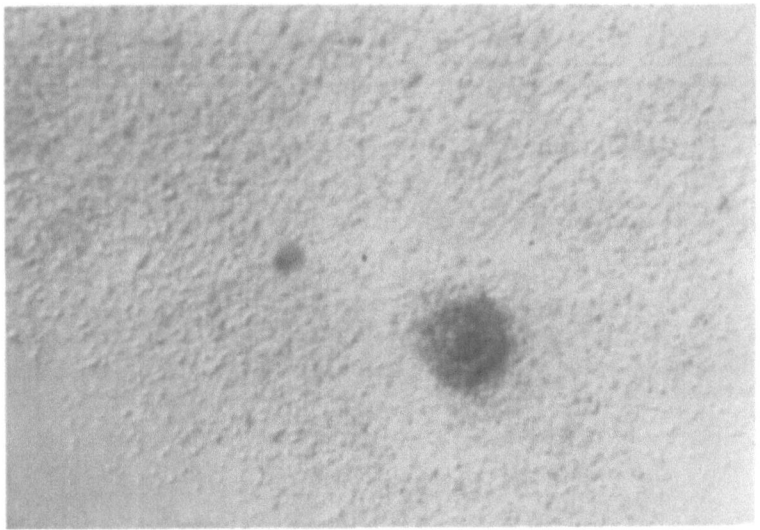

Fig. 4.   Mature preovulatory oocyte; Metaphase II.

<u>Description</u>: The cumulus oophorus associated with this oocyte is
large and expanded but the corona radiata is still slightly dense.
The first polar body is very difficult to see in this photomicrograph
taken with a dissecting microscope. Insemination of eggs at this
stage of maturity should be within 4–6 hr of culture. This oocyte
was recovered from a patient stimulated with a combination of clomiphene citrate and hMG.

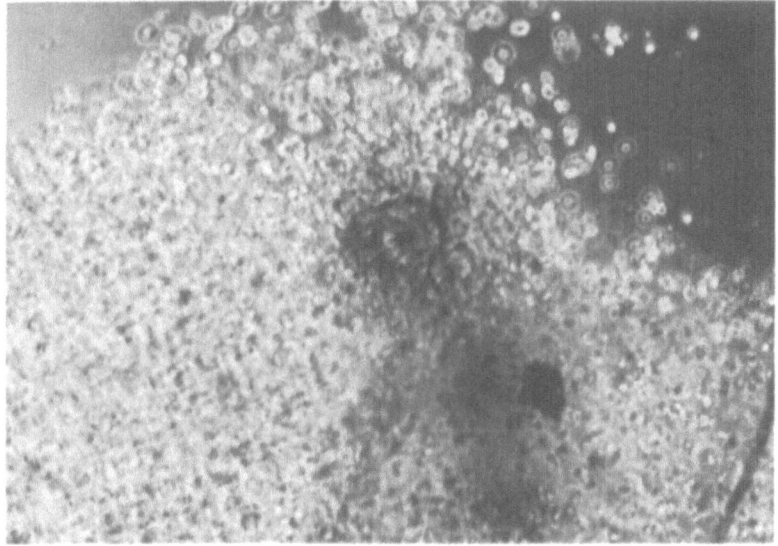

Fig. 5.   Mature preovulatory oocyte; Metaphase II.

Description: The cumulus oophorus is expanded.   The corona radiata
is still somewhat tightly compacted around the oocyte and is uneven.
Recovered from a patient stimulated with clomiphene citrate only.
Additional incubation of this oocyte type is required to allow for
greater corona expansion.

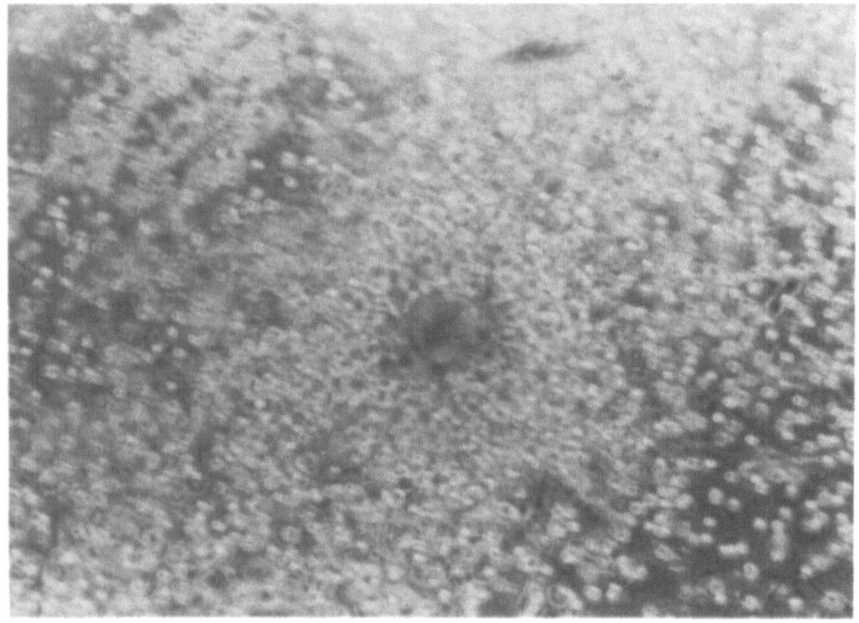

Fig. 6.   Mature preovulatory oocyte; Metaphase II.

Description: The cumulus oophorus is expanded. The corona radiata
displays a radiating appearance and is slightly more expanded than
that seen in Figure 5.

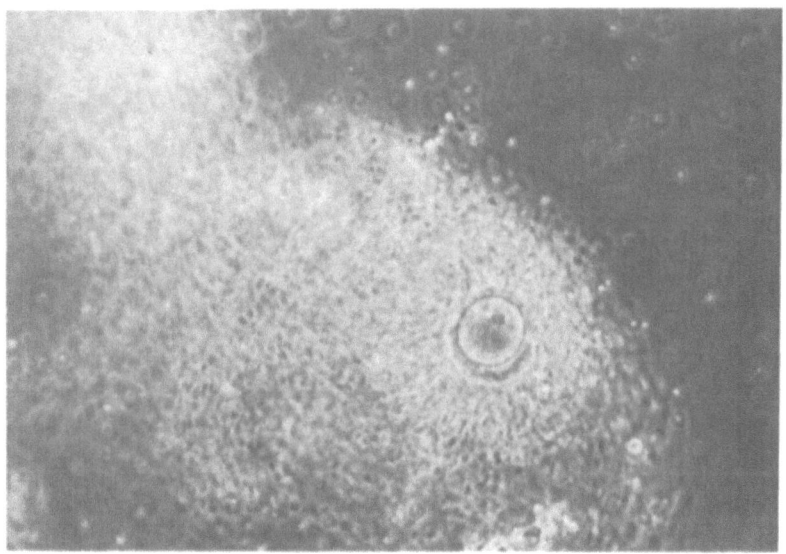

Fig. 7.   Mature preovulatory oocyte; Metaphase II.

Description: This oocyte is enclosed in an expanded cumulus oophorus
and a radiating corona. Recovered from a hMG-stimulated cycle in
which a spontaneous LH surge was detected. Following fertilization,
this was 1 of 4 embryos (see Figure 22) transferred resulting in a
singleton pregnancy.

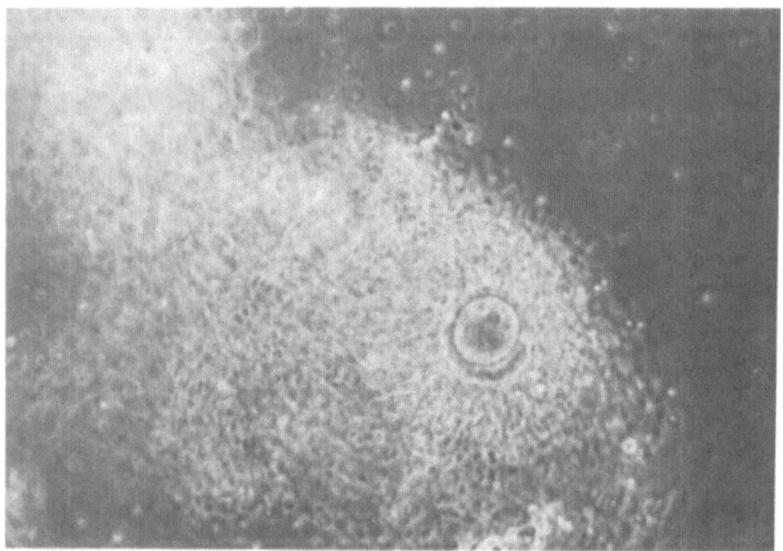

Fig. 8.   Mature preovulatory oocyte; Metaphase II.

Description: This oocyte is characterized by an extremely expanded
cumulus oophorus. The corona radiata has assumed a clumpy, unevenly
distributed appearance suggesting postmaturity.

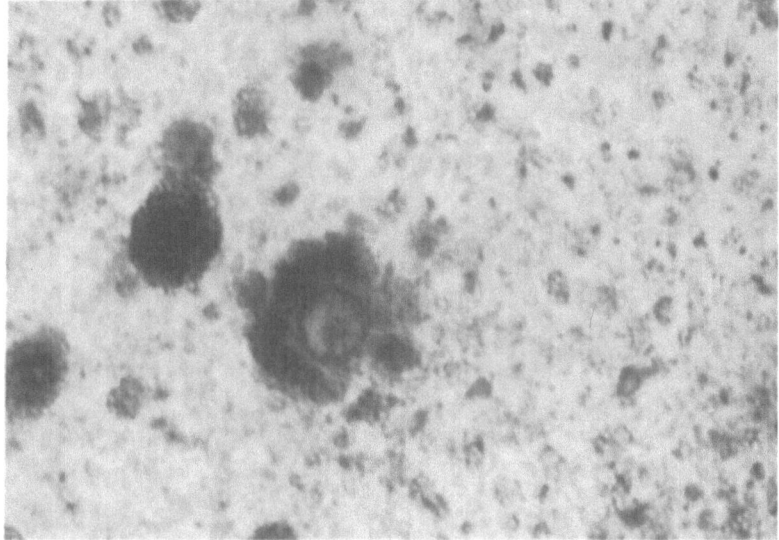

Fig. 9.   Postmature preovulatory oocyte; Metaphase II.

Description: This oocyte is surrounded by a clumpy dark cumulus and
corona. Note also the uneven distribution of corona around the
oocyte. Recovered from a patient stimulated with clomiphene citrate
and hMG. hCG was given on a falling E_2 curve.

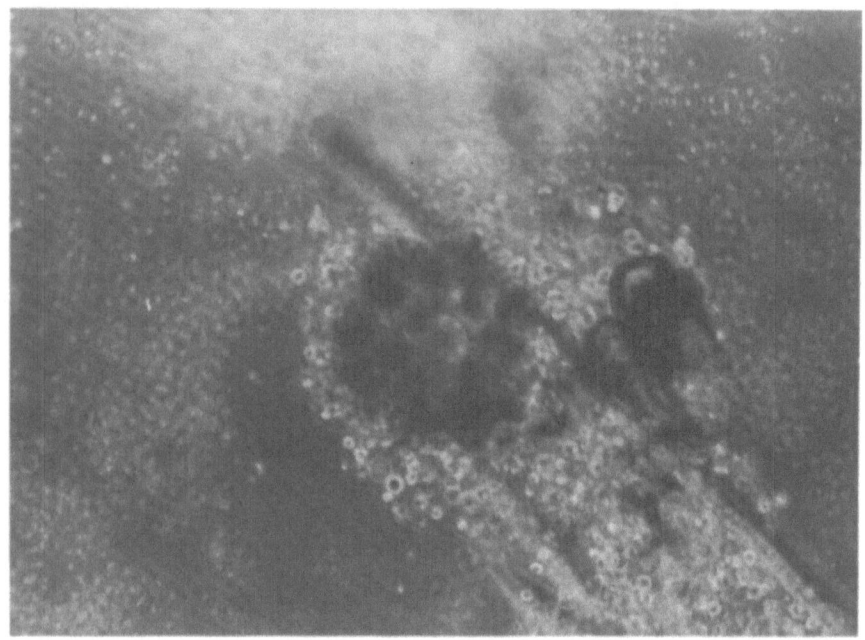

Fig. 10.   Postmature preovulatory oocyte; Metaphase II.

Description: Note the sparse but very clumpy cumulus mass and dense, beaded corona radiata.   This oocyte was inseminated within 3 hr of recovery but did not fertilize.   Five eggs were retrieved in this hMG-stimulated cycle and all 5 were bordering on the postmature stage (hCG was administered on a falling E_2 curve).   Only 1 of the 5 fertilized normally and developed to a 2-cell.   No pregnancy occurred.

3.3. Pronuclear Eggs/Zygotes

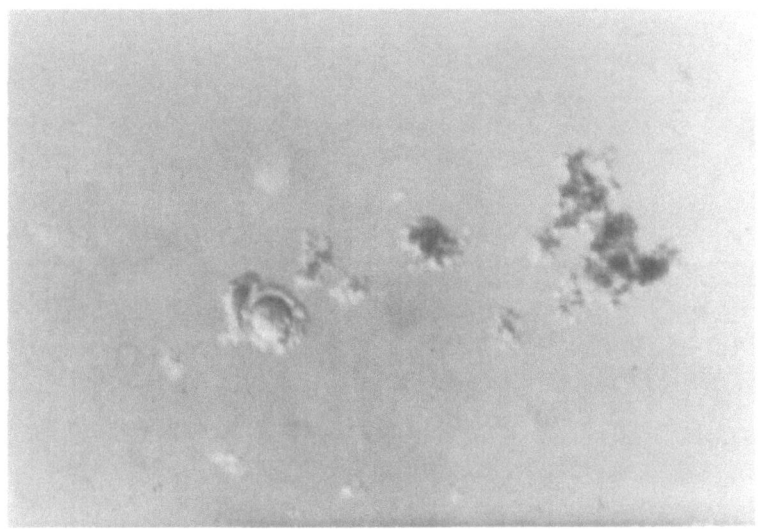

Fig. 11. Pronuclear egg:zygote; 18 hr postinsemination.

Description: Two pronuclei can be seen with a dissecting microscope
after the corona radiata is dispersed by micropipetting or by teasing
with needles. In this micrograph, the pronuclei are indicated by the
arrows. The dispersed corona radiata is also visible. Corona cells
remaining adherent to the zona will often propagate over 24 hr of
culture and form a monolayer of cells enclosing the embryo and
"trapping" it against the plastic.

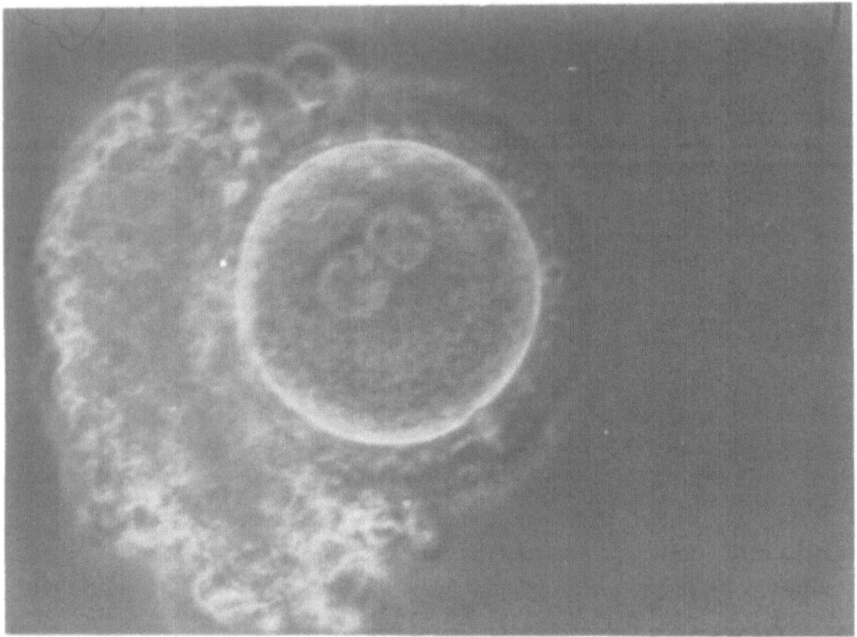

Fig. 12.   Pronuclear egg:zygote; 20 hr postinsemination.

Description: Two centrally located pronuclei are very obvious in
this fertilized egg recovered from a clomiphene citrate-stimulated
cycle. The corona has been dislodged only partially and will remain
intact through embryo transfer. Uniform slight granularity of
ooplasm characterizes a healthy cell which cleaved to a high-quality,
2-cell embryo by the time of transfer.

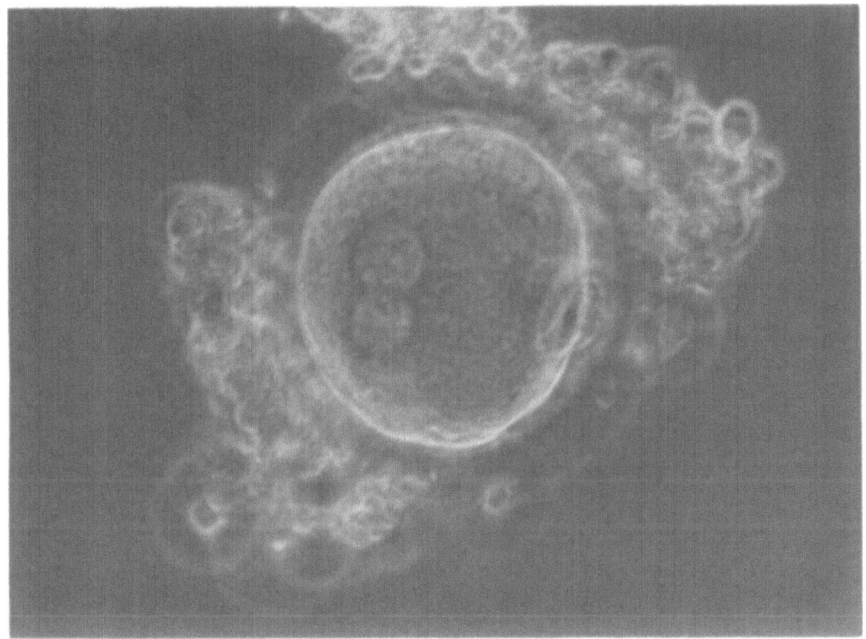

Fig. 13.   Pronuclear egg:zygote; 22 hr postinsemination.

Description: This was a mature preovulatory oocyte at retrieval from
a hMG-stimulated cycle which had a visible first polar body and was
inseminated 4 hr after retrieval.   Note the presence of a polar body
at 3 o'clock in the photograph.   This zygote developed to a 6-cell
embryo and was transferred with 1 other embryo.   No pregnancy
occurred.

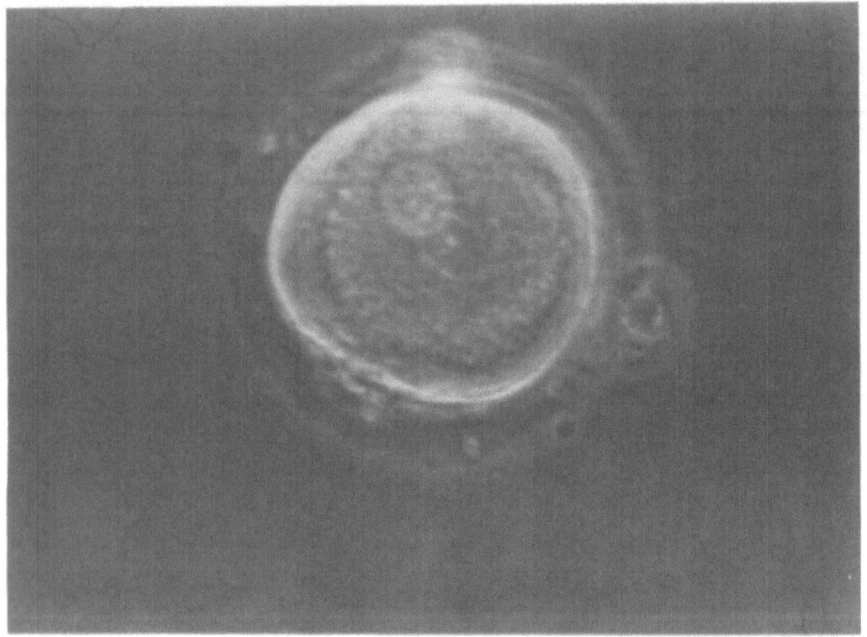

Fig. 14. Pronuclear egg:zygote; 21 hr postinsemination.

Description: This pronuclear egg is of relatively low quality as
evidenced by the dense, granular central cytoplasm and slightly
misshapen vitellus. Recovered from a hMG-stimulated cycle.

3.4. Cleaving Embryos

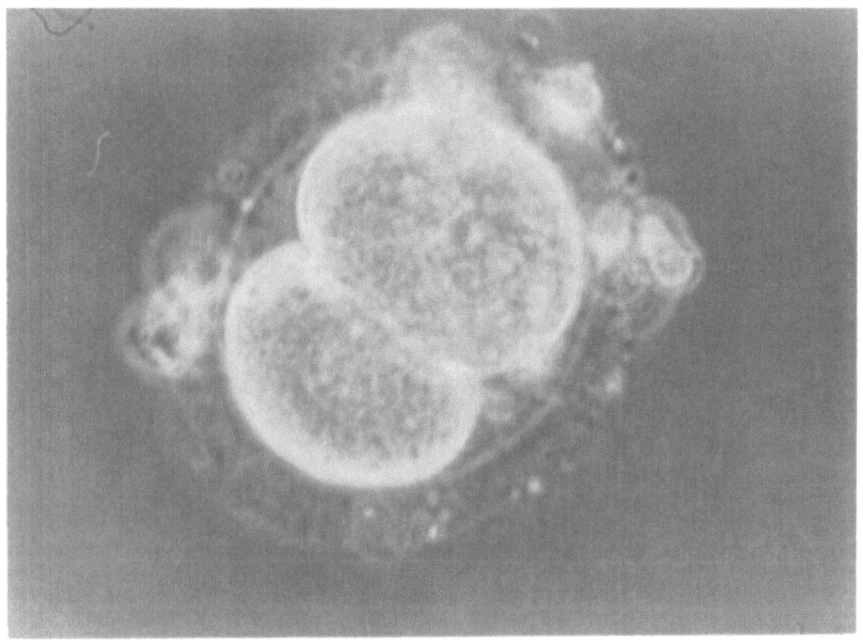

Fig. 15. Two-cell embryo; 41 hr postinsemination.

Description: This embryo deserves a high grade because of the
uniform size and shape of its blastomeres, the relatively debris-free
perivitelline space and the healthy appearance of the ooplasm. At
retrieval from a clomiphene citrate-stimulated cycle, this was a
mature, preovulatory oocyte which was inseminated within 3 hr of
recovery. This was 1 of 4 embryos transferred without the establishment of a pregnancy.

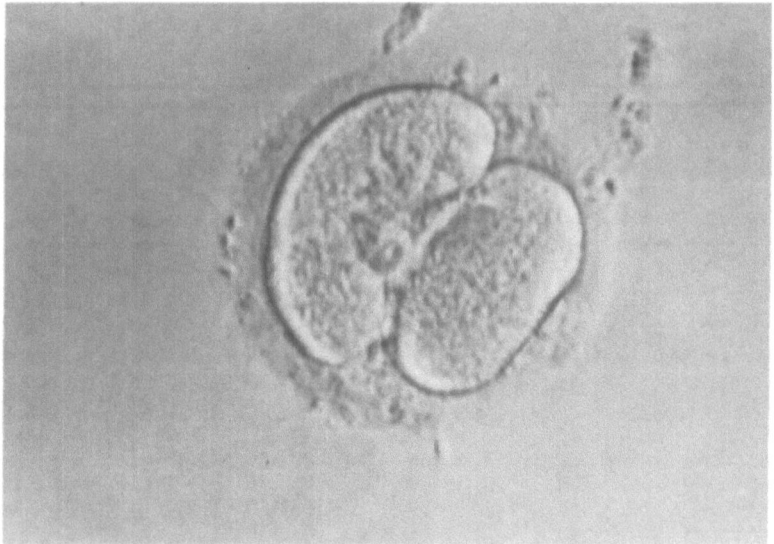

Fig. 16. Two-cell embryo.

Description: Typical low quality embryo (Grade 2) characterized by
uneven size and shape of blastomeres, granularity of ooplasm and
debris-laden perivitelline space. This embryo was transferred alone
and resulted in a term pregnancy.

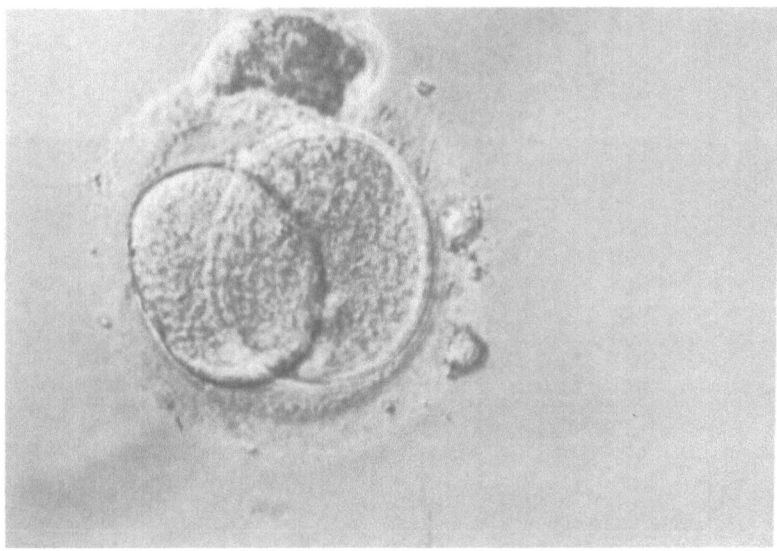

Fig. 17.   Two-cell embryo.

Description: This is another low quality embryo (Grade 2-3) showing
extreme granularity. Sperm can be seen in/on the zona pellucida and
a cluster of corona cells remains at the top of this micrograph.

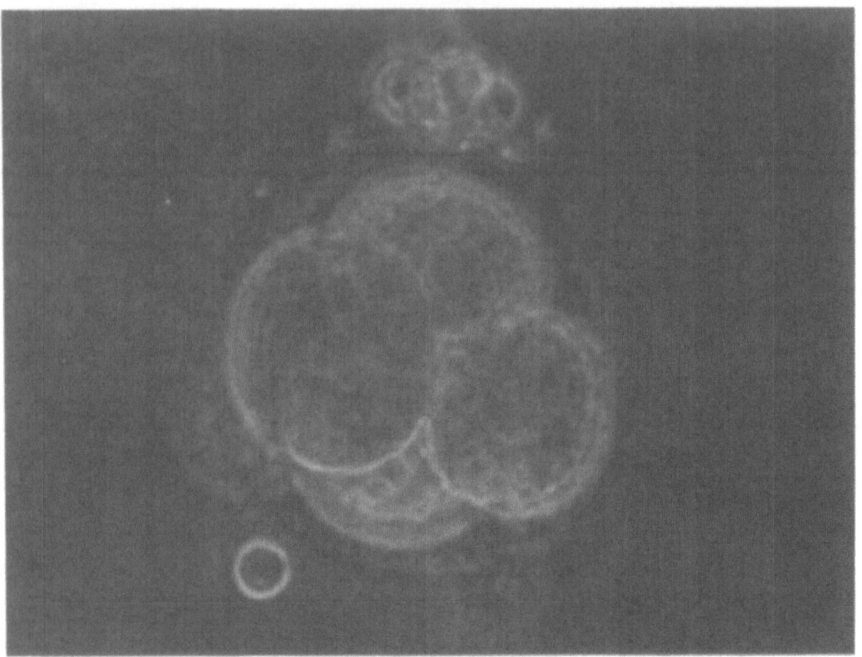

Fig. 18. Four-cell embryo; 46 hr postinsemination.

Description: High-quality, cleaving embryo (Grade 4-5) characterized
by even size and shape of blastomeres. Individual nuclei should be
visible under differential interference optics. At retrieval from a
hMG-stimulated cycle, this was a mature, preovulatory oocyte which
was inseminated within 3 hr of recovery. One of 5 embryos transferred which resulted in a twin term pregnancy.

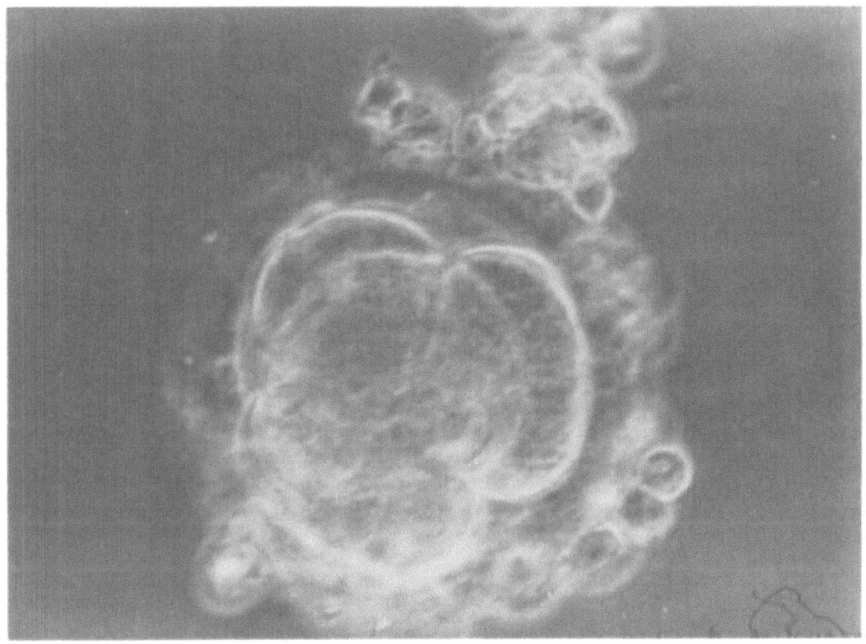

Fig. 19.   Four-cell embryo; 41 hr postinsemination.

Description: This embryo is of intermediate quality (Grade 3-4).
Note that the 4 blastomeres are uneven. A sperm bound to or in the
zona pellucida can be seen at 11 o'clock.   Cells of the corona
radiata are still present. At retrieval from a clomiphene citrate-
stimulated cycle, this was a mature, preovulatory oocyte and was
inseminated within 3 hr. This was 1 of 4 embryos transferred. No
pregnancy was established.

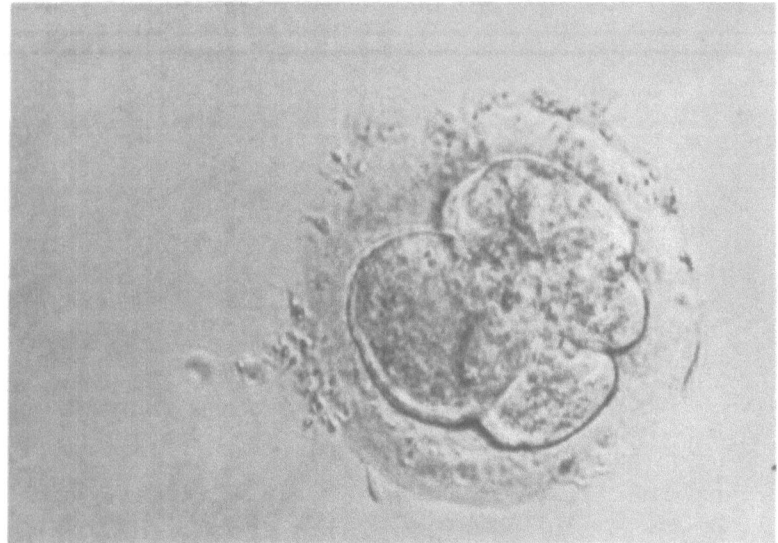

Fig. 20.   Four-cell embryo.

<u>Description:</u>   This low to intermediate quality (Grade 1-2) embryo is 1 of 11 recovered from a hMG-stimulated cycle.

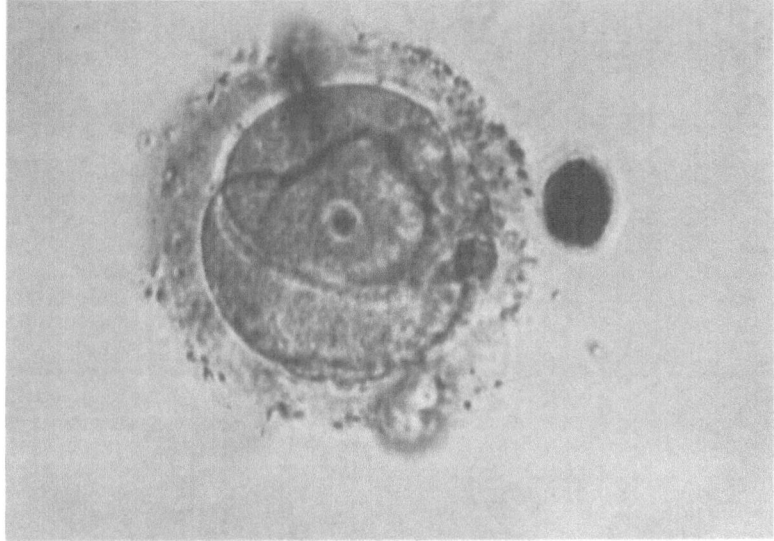

Fig. 21. Four-cell embryo.

Description: This embryo is very low in quality (Grade 1) based upon
the uneven size and shape of its blastomeres. Recovered from the
cycle described in Figure 20. Note the small, dark structure
(refractile body?) in this embryo.

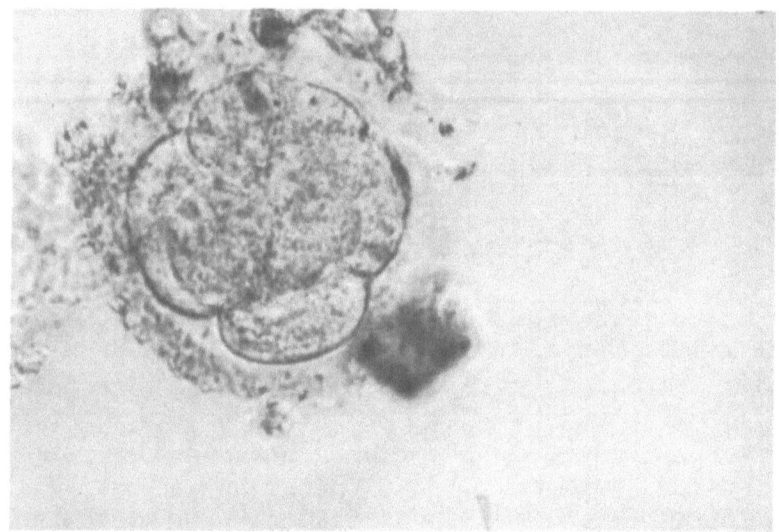

Fig. 22. Five-cell embryo; 43 hr postinsemination.

Description: The corona radiated remnants on the zona and the
unevenness of the blastomeres detract from the appearance of this
high-quality embryo (Grade 4). This embryo resulted from insemination of the egg depicted in Figure 7.

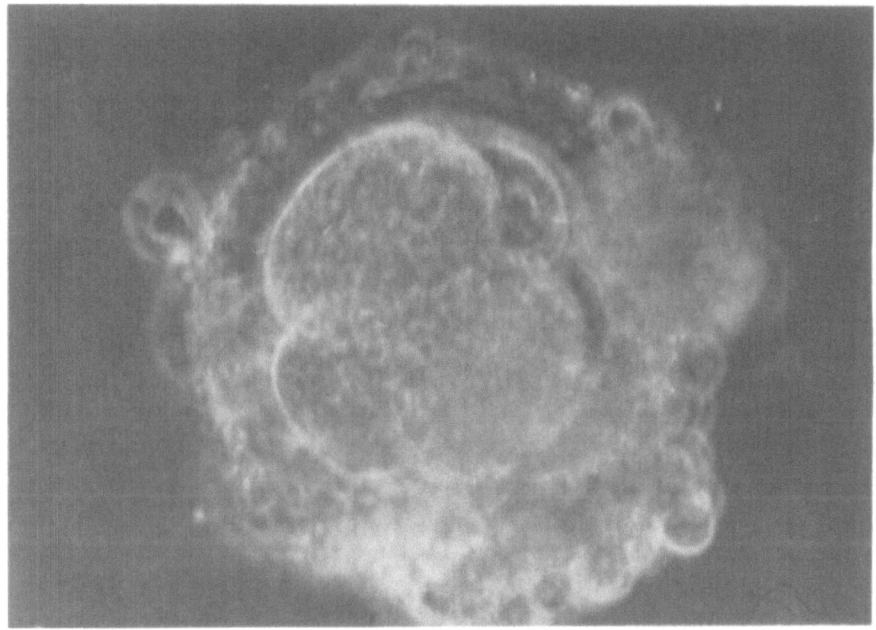

Fig. 23. Five- or six-cell embryo; 41 hr postinsemination.

Description: A high quality embryo (Grade 4) with slightly uneven
blastomere size. Sperm bound/penetrating the zona are seen between
10 and 1 o'clock. At retrieval from a hMG-stimulated cycle, this was
graded as a mature/postmature preovulatory oocyte. An additional 4-
cell embryo was also transferred with this embryo. No pregnancy was
established.

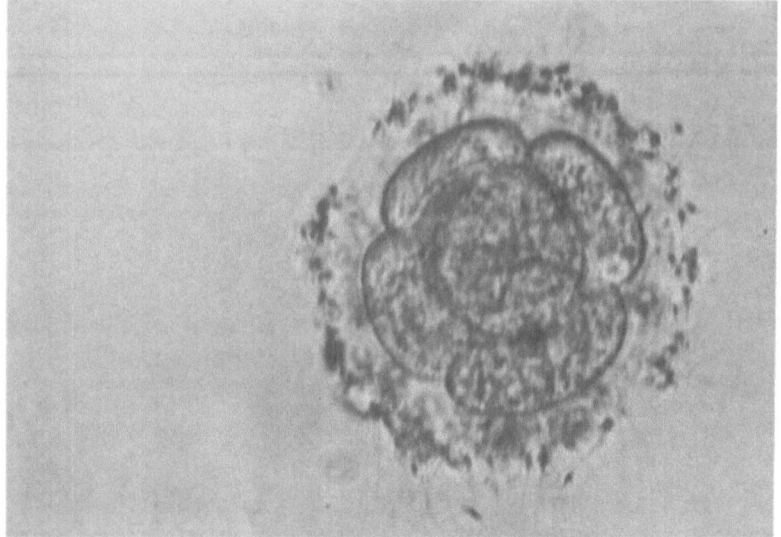

Fig. 24. Six-cell embryo; 41 hr postinsemination.

Description: Many sperm heads are visible at the outer margin of the zona pellucida in this near perfect (Grade 5) embryo. This embryo was transferred along with 3 others to produce a transient β-hCG rise but not a clinical pregnancy.

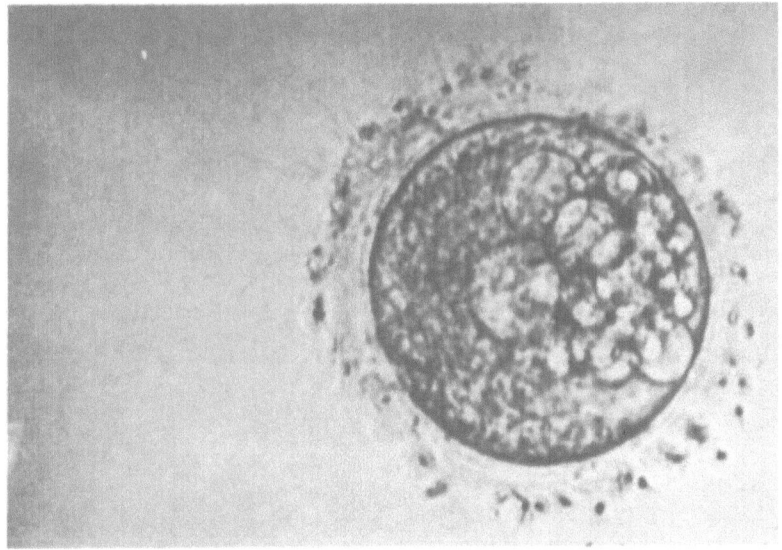

Fig. 25. Early blastocyst; 92 hr postinsemination.

Description: This blastocyst was derived from 1 of 13 oocytes
recovered from and inseminated in a hMG-stimulated cycle. Note the
unfused blastocoelic cavities (see also Figure 27).

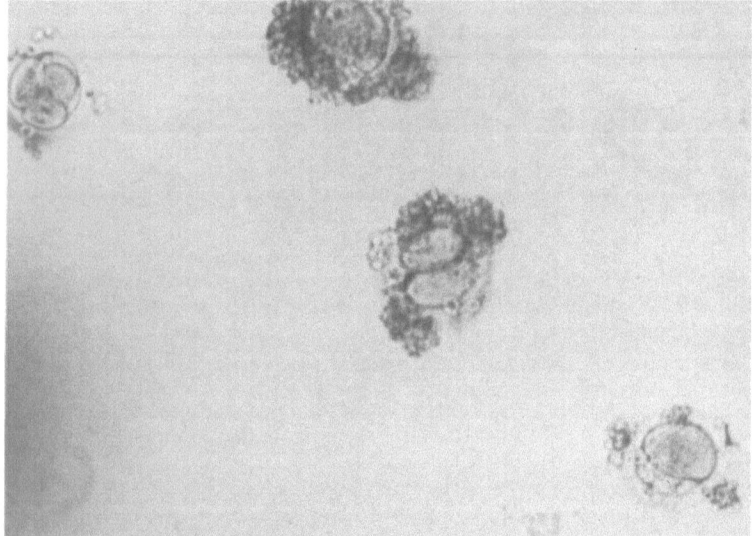

Fig. 26.   Human embryos for transfer; 42 hr postinsemination.

Description: An example of the variation in stage and quality of
embryos cultured under similar conditions from the same patient.
Transfer of these 4 embryos resulted in a twin term pregnancy.

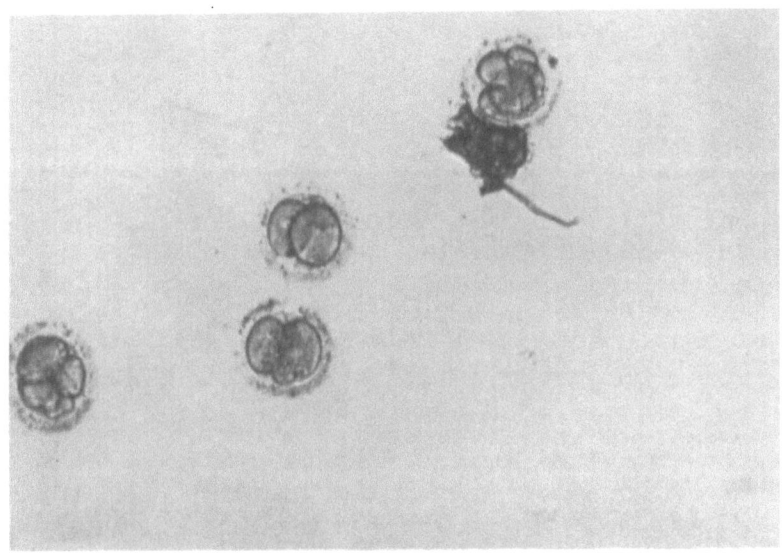

Fig. 27. Human embryos for transfer.

Description: Another example of the variations in embryos from a
single patient cultured under similar conditions in vitro. The stage
of embryonic development varies from 2- to 6-cell. Thirteen oocytes
were recovered from this patient (see also Figure 25).

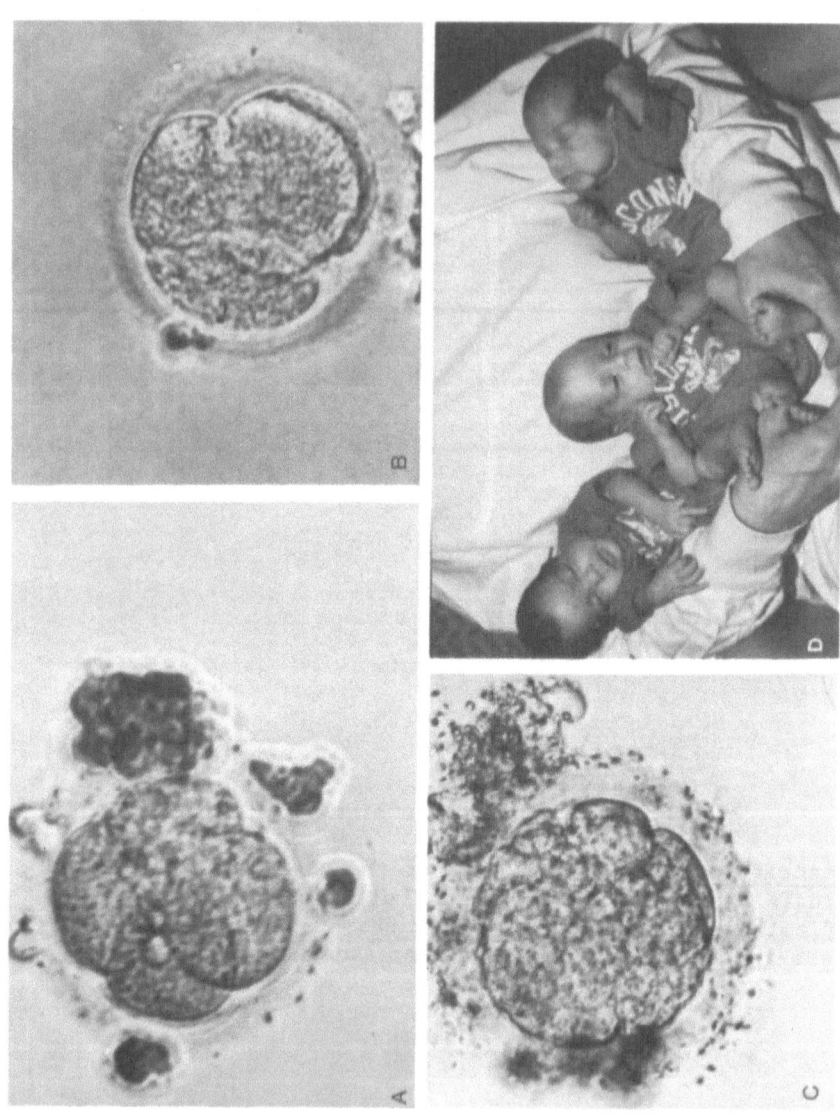

Fig. 28.   Three of 5 embryos (A,B,C) transferred to a 37-year-old patient with infertility of unknown etiology (9 years). Note the blebby appearance of the embryos, indistinct blastomeres and granular appearance of the cytoplasm. The outcome of this transfer is depicted in (D). Five embryos continued development posttransfer with an empty sac visualized by ultrasound at 7 weeks and an ectopic implantation that did not require surgical intervention found at the time of cesarean delivery.

4.   SPECIAL CASES

4.1. Spontaneously Activated Oocyte

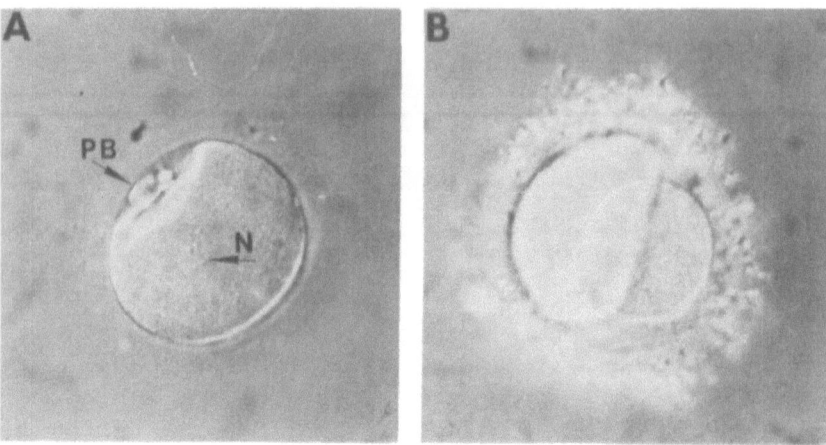

Fig. 29.   Spontaneously-activated oocyte; parthenogenote.

Description: Panel A depicts a spontaneously-activated oocyte as it
appeared approximately 4 hr after placing in culture. Note the
presence of a polar body(s) (PB) and a single membrane enclosed
nucleus (N) with individual nucleoli.   Panel B depicts the same
oocyte undergoing premature cleavage within 16 hr of recovery (from
Byrd, W. and Wolf, D. P., 1984, in: Human In Vitro Fertilization and
Embryo Transfer [D. P. Wolf and M. M. Quigley, eds.], Plenum Press,
New York).

4.2. Shrunken Oocyte

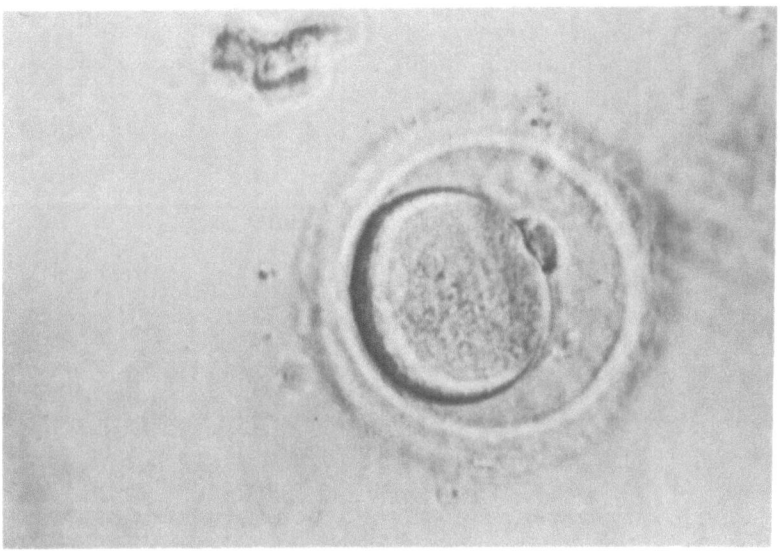

Fig. 30.   Shrunken oocyte; Metaphase II.

Description: At retrieval, this oocyte was shrunken but contained a polar body. It did not fertilize and must be considered both post-mature and degenerative.

4.3. Fractured Eggs/Zonae

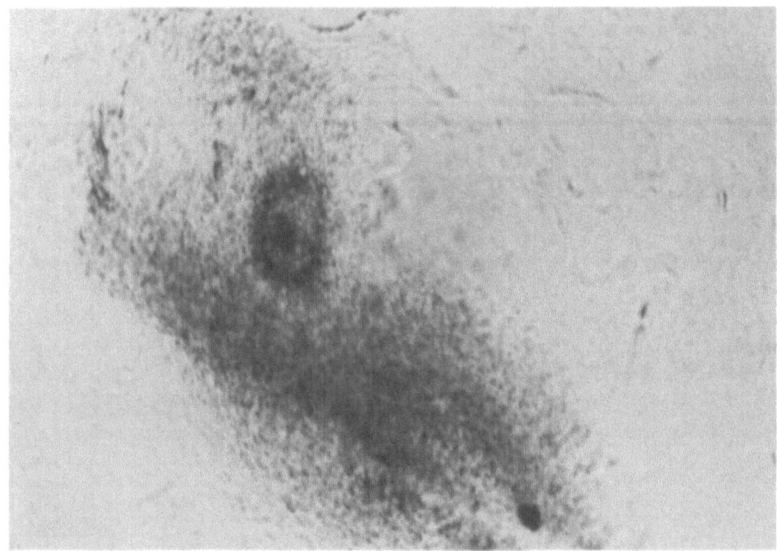

Fig. 31.   Fractured egg/zona.

Description: This oocyte was recovered from a clomiphene citrate
hMG—stimulated cycle where a 2—day "coast" occurred between the last
hMG and hCG.   Note the dark cumulus and corona.

4.4. Atretic/Degenerated Oocyte

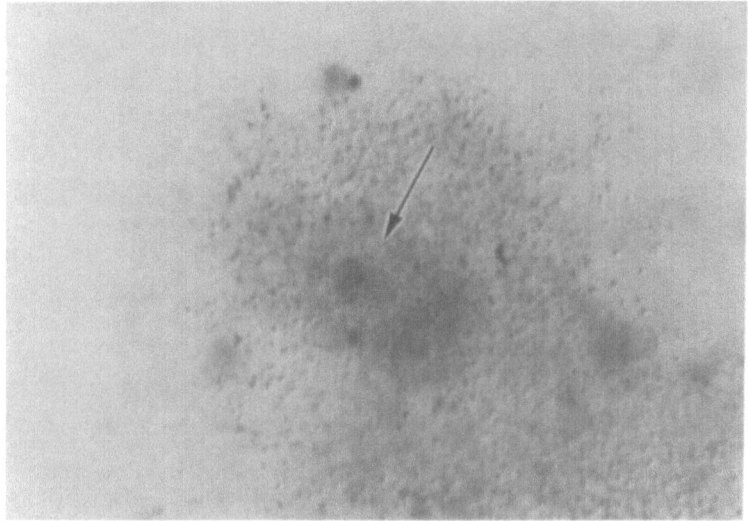

Fig. 32. Atretic/degenerated oocyte.

Description: An apparently normal cumulus mass recovered from a
preovulatory follicle. An expanded corona surrounds an empty zona
pellucida (arrow). Although considered unlikely, oocyte loss could
have been induced during aspiration.

4.5. In Vitro Aged Oocyte

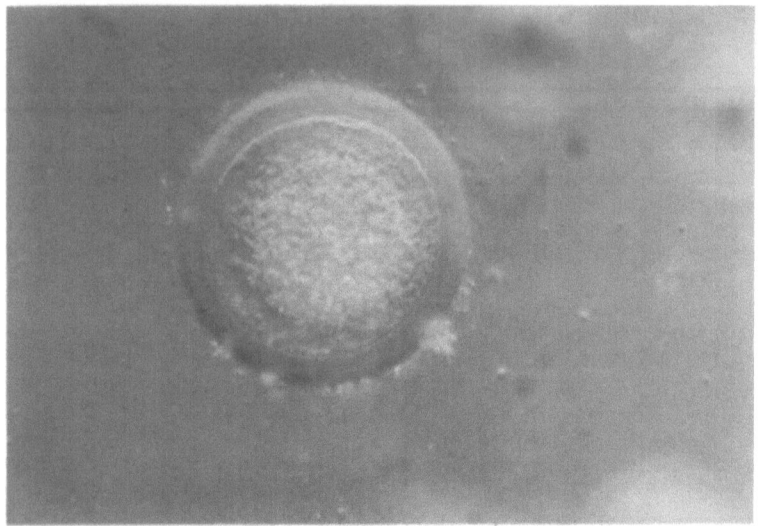

Fig. 33.   In vitro aged, preovulatory oocyte.

Description: This oocyte was inseminated with very low quality sperm
(< 10% motile) and did not fertilize.   After 4 days in culture,
obvious degeneration has occurred including increased granularity of
ooplasm and lysis completely filling the perivitelline space.

4.6. Fragmenting Oocyte

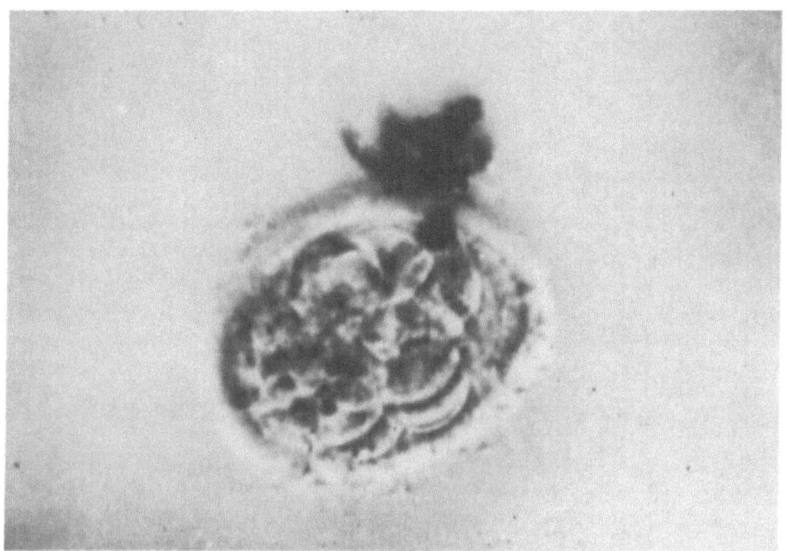

Fig. 34.   Fragmenting oocyte.

<u>Description</u>: This structure contained 12–16 fragments at 24 hr postinsemination. Recovered from a clomiphene citrate-stimulated cycle.

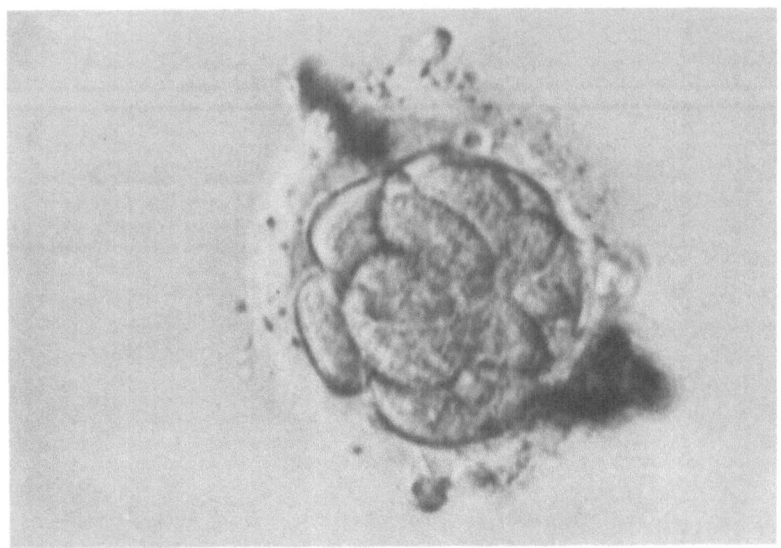

Fig. 35. Fragmenting oocyte.

Description: At 48 hr postinsemination, this structure could be
mistaken for a cleaving embryo. Examination at 16 hr postinsemination was negative for polar bodies and pronuclei. Thereafter,
division was irregular and out of phase temporally. Recovered from a
clomiphene citrate hMG-stimulated cycle.

4.7. Polypronuclear Egg

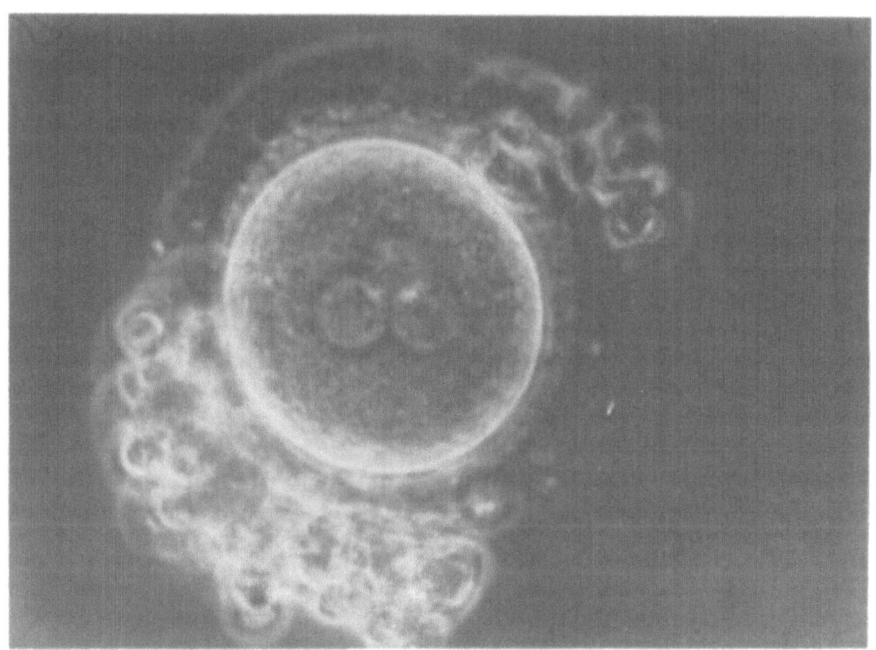

Fig. 36. Polypronuclear egg; polyspermic.

Description: Three pronuclei are obvious in this micrograph taken 19
hr postinsemination. This cell cleaved to an apparent high-quality,
4-cell embryo (Figure 39). At retrieval from a hMG-stimulated cycle,
it was an apparently mature, preovulatory oocyte.

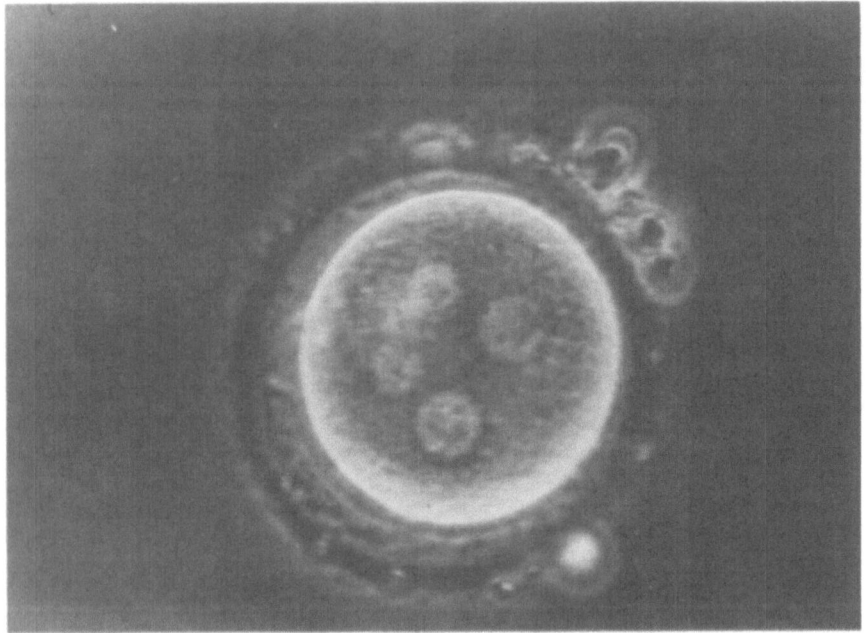

Fig. 37. Polypronuclear egg.

Description: Five pronuclei are visible in this oocyte 20 hr after insemination; cleavage to a 5-cell "embryo" occurred. Derived from a mature, preovulatory oocyte in a hMG-stimulated cycle.

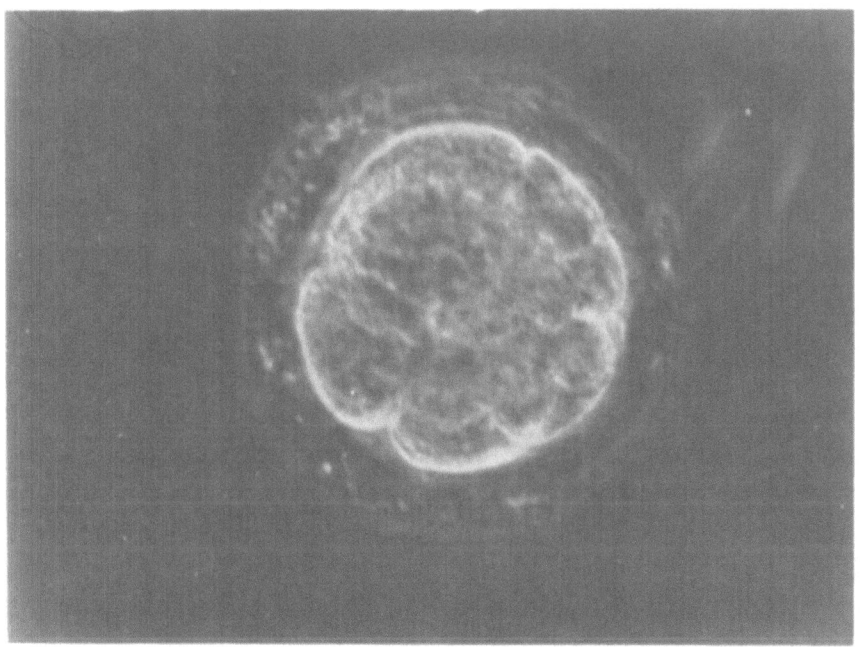

Fig. 38. Polypronuclear egg undergoing division.

Description: Three pronuclei were visible in this egg at 20 hr
postinsemination. By 44 hr postinsemination, 6 individual cells plus
a number of blebs or anucleated fragments were apparent. Mature,
preovulatory oocyte at retrieval from a hMG-stimulated cycle. Three
other normally developing embryos were transferred and a pregnancy
was established giving rise to twin males.

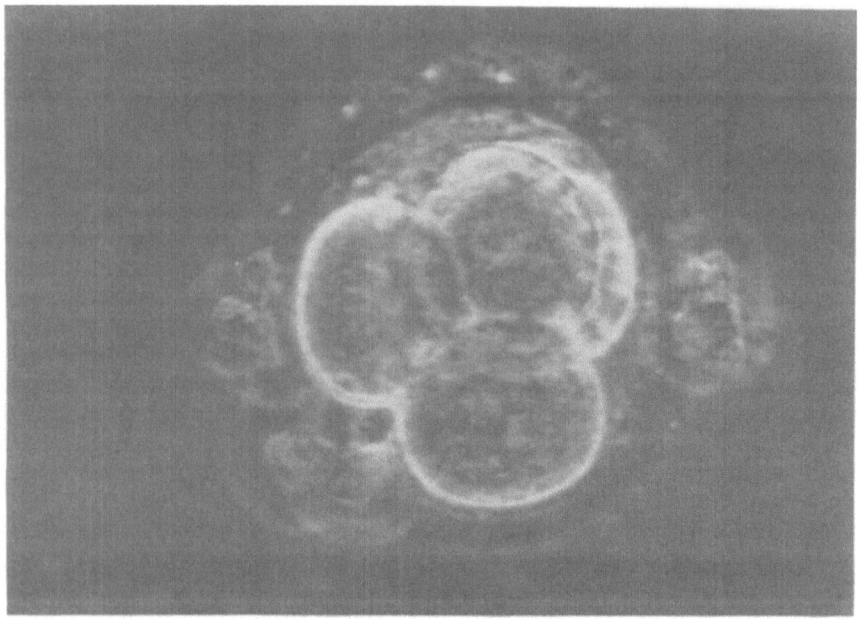

Fig. 39. Dividing polypronuclear egg.

Description: An apparent high quality 4-cell embryo at 42 hr post-insemination resulting from the polypronuclear egg depicted in Figure 36.

11

EMBRYO TRANSFER

Marybeth Gerrity and John S. Rinehart

1.   INTRODUCTION

The goal of embryo transfer is the placement of embryos in the uterus with minimal trauma to both embryos and recipient. Embryo transfer in the human is accomplished in as small a volume of medium as is possible. The purpose of this chapter is to describe both the equipment that is required for embryo transfer as well as the procedure itself.

2.   CATHETER TYPES

In 1982, Leeton et al. described the technique for the transfer
of human embryos. They compared the success rate of embryo transfer
using both end opening and side opening embryo transfer catheters.
Their results indicated that there was no significant difference in
the pregnancy rates obtained with the two types of catheters.
However, if biochemical pregnancies were excluded, end opening
catheters offered a slight advantage. Since that time, however, the
choice of an embryo transfer catheter in each IVF program has been
based largely on the personal preferences of the surgeon and embryologist as well as on the gold standard of the specific program that
they choose to emulate.

End opening catheters, the first to be put into widespread use,
as the name implies, have a straight open end. They are available in
several forms. Cook Company (Australia) manufactures an end opening
catheter (METS-1) designed by the IVF program at Monash University.
This particular catheter is available with a flexible introducer
(Figure 1). Another popular variation on the end opening catheter is
the Tom Cat catheter (Monoject, St. Louis, MO). This catheter has
received widespread use in recent years because it is easily available and is disposable (Kerin et al., 1981). It is simple to use,
atraumatic and does not require an introducer.

Side opening catheters have been widely endorsed because, it is
felt, they are easier to use in a nulligravid woman (Leeton et al.,
1982). These catheters have a machined end which is plugged with
teflon and a side hole for expelling the embryos at a right angle to
the catheter length (Figure 1). These catheters are often favored
because of the reduced risk of mucus plugging during cervical
passage. They are somewhat more difficult to load and practice with
mouse embryos is recommended. There are several types of side
opening catheters available. The Norfolk type has been widely used
and was described by Jones et al. (1983b) (Horace Barnaby, Baltimore,
MD). This particular catheter can be used with an inflexible metal
guide for introduction through the cervical os. The catheter is not
supplied with a syringe connection and a 27-gauge needle must be
inserted to provide a connection for a syringe. The rigid introducer
is a decided disadvantage in performing atraumatic transfers. Cook
Company also markets a side opening catheter (METS-2) designed by
Monash University's IVF program. As with the Monash end opening
catheter, this catheter is available with a flexible guide which can
be molded to suit the anatomy of the embryo recipient.

In selecting a suitable catheter, several considerations are
paramount: the type of connection between catheter and syringe,
availability, and the desirability of having a catheter that is
disposable. See Figure 1 for examples of several.

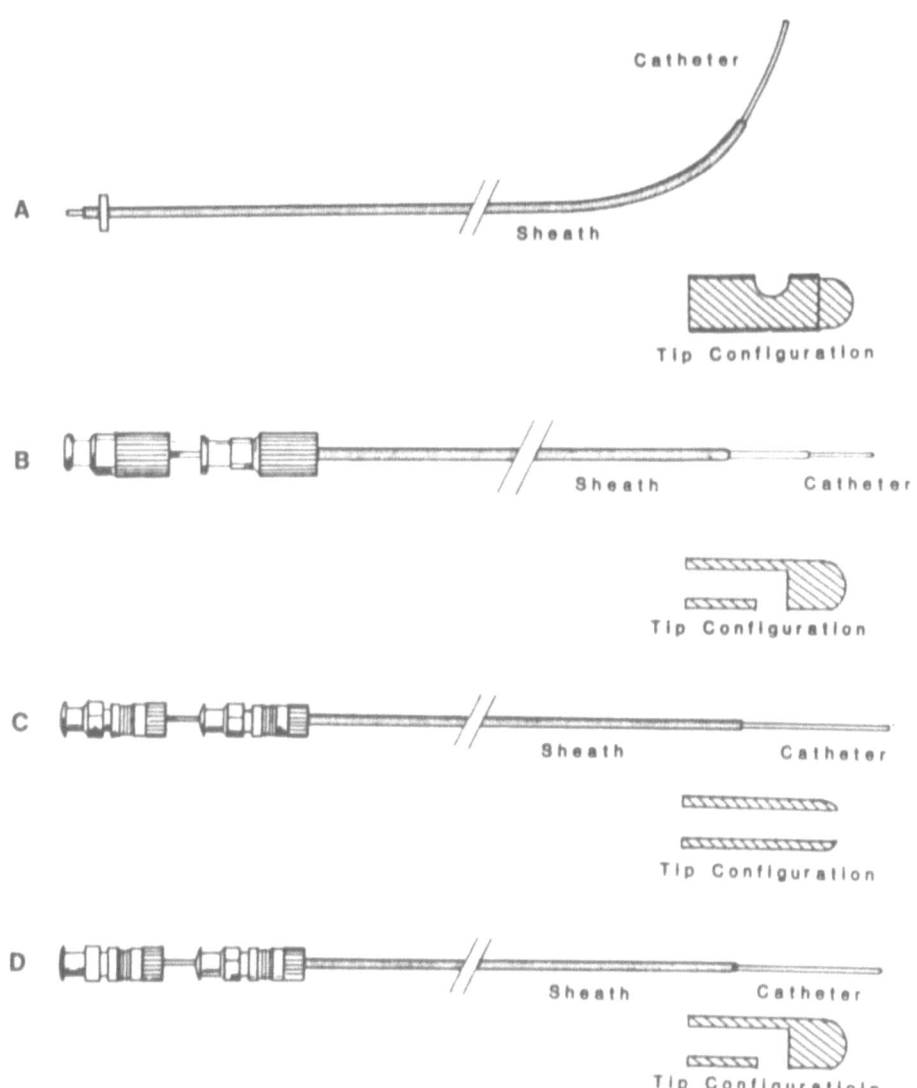

Fig. 1.   Embryo transfer catheters utilized for human IVF-ET.
           A) Side-opening teflon catheter with stainless introducer.
           B) Royal Women's Hospital ET Set; side opening (JLLS-4).
           C) Monash ET Set; end opening (METS-1).   D) Monash ET Set;
           side opening (METS-2).   B.C.D., William A. Cook, Australia
           PTY. LTD.

3.   SYRINGES

Regardless of what choice is made for an embryo transfer catheter, a syringe must be attached to draw up and expel the embryos. There are three main types which are used in in vitro fertilization programs to attach to the embryo transfer catheter.

a.   The first is the plastic disposable tuberculin syringe. This syringe is widely available, cheap and disposable. Accurate measurement of volumes with a plastic syringe is difficult. It is essential to rinse this type of syringe thoroughly before allowing media that contains embryos to come into contact with it. Plastic disposable syringes often contain a lubricant on the plunger which helps the plunger move smoothly. This same lubricant, however, is toxic to mouse embryos, so care must be taken to prevent medium from coming into contact with the end of the plunger itself.

b.   The second type of syringe which is available is a glass disposable syringe, which is manufactured by several companies (e.g., Becton-Dickinson, Glaspak). As with the plastic disposable syringe, these syringes are relatively inexpensive and disposable. However, the plunger slides so easily against the glass surface that it is also possible to expel embryos inadvertently. If this type of syringe is used, extreme care must be exercised to prevent premature plunger movement.

c.   The third type is the Hamilton syringe. These syringes are available in a gas-tight version which minimizes gas exchange in the transfer catheter. Accurate measurement of volumes with Hamilton syringes is easy. The gliding action of the plunger in the syringe is unsurpassed. These syringes are made of glass and are not disposable. They are also relatively expensive (approximately $15 ea.). Therefore they must be cleaned and sterilized for use at each procedure. There are two types of plungers available with this syringe: one with a micrometer end and one with a straight plunger. Manual dexterity and individual preference will dictate which will be selected.

4.   MEDIUM

There have been no well-controlled studies which compare successful embryo transfer as a function of the type of embryo transfer medium used. In many cases the choice of medium will depend upon the configuration of the physical facilities available to the IVF program.

There are two general types of transfer media to choose from: air-buffered medium and bicarbonate-buffered medium. Air-buffered medium such as Dulbecco's phosphate buffered saline (PBS) or Hepes buffered medium are not widely used for embryo transfers. The advantage of air-buffered media is that the physiologic pH of the medium is maintained in room air; therefore, the length of time the embryos are exposed to air will not be as critical. Centers where handling and transport of embryos is a prolonged process or takes place over long distance may opt for air-buffered media. Bicarbonate-buffered medium is far more widely used in embryo transfer. The medium used is generally the same as that used in the IVF process (e.g., Ham's F-10).

The concentration of protein used in the transfer medium varies considerably between IVF programs. Medium containing 15-20% serum (growth medium) has been used successfully, as has high serum (75-90%) medium. Users of high serum transfer medium cite as advantages: ease of expelling embryos from the catheter, decreased tendency for the embryos to adhere to the end of the transfer catheter, and an increased pregnancy rate (Feichtinger et al., 1983). However, Meldrum's group has suggested that the nonphysiological acid pH of high serum transfer medium may actually be detrimental to the embryos (Chetkowski et al., 1985).

5.   PREPARATION AND CLEANING OF INSTRUMENTS USED IN EMBRYO TRANSFER

The responsibility for cleaning and sterilizing embryo transfer catheters should lie with the IVF embryo lab, and not in the operating room or central services of the hospital. Some manufacturers of embryo transfer catheters provide detailed descriptions of how these catheters should be cleaned and sterilized (e.g., Cook, Australia). After each use, the catheter should be thoroughly cleaned with 7 X tissue culture detergent (Linbro, Flow Labs, Arlington, VA) followed by ultrapure water. Excess water can be cleared with filter-sterilized air and catheters packaged in individual peel packs for sterilization. The catheter can be sterilized in two ways: heat or gas sterilization. If the catheter used is suitable for heat sterilization, it should be placed in a drying oven at 275°C for 2 hr. If gas sterilization is used, the catheter should be sterilized using ethylene oxide according to a standard sterilization protocol. The catheters must then be allowed to degas for a minimum of 48 hr before use. In addition, the catheters must be washed vigorously with tissue culture medium before use to remove any ethylene oxide residues which are toxic to mouse embryos (Schiewe et al., 1984). Quality control for this procedure should include use of sterile indicator markers. If Hamilton syringes are used for embryo transfer and they come into contact with medium, they should be washed in the same way described, packed individually and sterilized according to manufacturer's directions.

Steam sterilization of embryo transfer catheters should be avoided
since steam can leave a mineral deposit on all surfaces. There have
been anecdotal reports of formation of microbubbles with some types
of cleansing protocols. When one is filling the embryo transfer
catheter, it is therefore important to examine the catheter walls for
uniform filling. Drying of these microbubbles on the walls may
prevent uniform filling. Removal of microbubbles and other residues
can be accomplished using sonication.

    If the chosen catheter is supplied by the manufacturer with an
introducer (e.g., Monash catheters), then these should be treated as
a matched pair. Etching the hub of both the transfer and external
catheter (introducer) with a number or letter will prevent separation
of the pair. Catheters should be inspected after each use for signs
of damage, crimping, etc. Replacing catheters after a certain number
of uses is also a good policy.

6.   EQUIPMENT AND SUPPLIES

    In addition to embryo transfer catheters and syringes, a
quantity of embryo transfer medium should be available for rinsing
the catheters and syringes following the transfer. Additional
syringes, petri dishes and a dissecting microscope are also required.

7.   TIMING OF EMBRYO TRANSFER

    Embryo transfer is generally accomplished 48–54 hr after egg
retrieval. This timing, of course, assumes that mature preovulatory
oocytes have been obtained. In cases where only immature oocytes
have been aspirated, embryo transfer may be delayed. In situations
where there is a mix of mature and immature oocytes, embryo transfer
usually takes place as dictated by the fastest developing embryos.
Therefore, it might be possible to transfer a 4-cell embryo along
with a pronuclear embryo. Since it is known that embryo development
slows in culture and that embryo transfer media do not reproducibly
support embryo development to the preimplantation stage in the human
(Fishel et al., 1985; Cummins et al., 1986), there is no advantage to
holding embryos for extended periods of time in the laboratory; in
fact, this practice may be detrimental to continued development. On
the other hand, patients with high serum estradiols may benefit from
a delayed embryo transfer to allow the serum E_2 to decrease causing a
more favorable P/E_2 ratio (Gidley-Baird et al., 1986).

8.   EMBRYOS

    The decision concerning which embryos should be transferred will
depend largely on ethical and legal restrictions in the state in

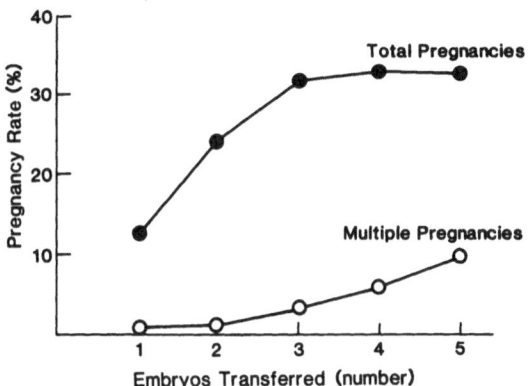

Fig. 2.   The relationship between pregnancy rate and the number of
           embryos replaced in human IVF-ET. Data from Byrd and Wolf
           (1984).

which the transfer is taking place as well as the availability of
alternate means of embryo preservation. While it has been shown that
the success of embryo transfer (measured in percent pregnancies)
increases as a function of the numbers of embryos transferred (Wood
et al., 1981), there is no further increase in pregnancies with
transfers of four or more embryos (see Figure 2). Furthermore,
transfer of multiple embryos increases the risk of multiple pregnancy
with its accompanying medical risks to both the mother and child
(Kerin et al., 1983). Therefore, most centers set a limit as to the
number of embryos that will be transferred to the uterus during the
treatment cycle. There are several options for use of any extra
embryos: they can be frozen for transfer in a future cycle, they can
be donated to another recipient or they can be continued in culture
until growth stops. Destruction or experimentation on these embryos
is restricted in many states. Some centers have eliminated the
problem of the extra embryo by limiting the number of oocytes which
are retrieved or fertilized at the time of the IVF procedure. There
are distinct disadvantages to both of these approaches. Limiting the
number of eggs retrieved may leave the best eggs behind and could
result in hyperstimulation syndrome. Limiting the number of eggs
fertilized forces a very subjective decision on the embryologist
concerning which eggs are best. As will be discussed below, it is
difficult to predict which egg or embryo will result in a pregnancy.
Fertilizing the eggs permits some selection by the sperm. It is
essential that a policy be developed at the institution to deal with
the issue of extra eggs or embryos before the situation arises. In
an era where many programs have a waiting list of couples needing
donor eggs, wasting these materials is unwarranted.

Examination of embryos before transfer often sheds little light on which embryo will most likely result in a clinical pregnancy. The embryologist should identify the approximate cleavage stage of the embryo. Examination of the embryos can be facilitated by transferring them from the organ culture dish (Falcon 3095) to a regular flat bottom petri dish (Falcon 3001). The organ culture dishes often distort the embryo optically and they become difficult to examine in detail. It is not uncommon to see blebbing or indistinct blastomeres in human embryos (see Chapters 9 and 10 for additional details). It is also not unusual to examine an embryo and conclude that it is fragmenting or has ceased development only to observe several hours later that extensive reorganization has occurred resulting in a regular-appearing embryo. Human embryos which are entering cleavage divisions often look their worst. A number of investigators have demonstrated that successful embryo transfer is highly correlated with the rate of development; therefore, the fastest developing embryos are best selected for embryo transfer (e.g., Cummins et al., 1986). The judgement of the embryologist will also come into play in this decision. If the IVF center has a cutoff of four embryos for transfer and there are five embryos but one appears to have halted development, it is often best to replace all five of the embryos. In all cases, the patient and her husband should be fully apprised of the fate of all embryos and given the opportunity to participate in the decision-making process.

There are still very few objective clues as to which embryos will result in pregnancies and which will not. Recent reports of an embryo-derived, platelet-activating factor that can be detected in the growth medium surrounding a viable embryo (using a bioassay in splenectomized mice; O'Neill et al., 1985) may signal a new phase in which objective measurements of embryo development may, at last, be available. Nevertheless, there are many examples of embryos which appeared to be less than optimal that have apparently resulted in pregnancies. While it is probably no longer acceptable to transfer every embryo that fertilizes and cleaves, when in doubt it is best to return the embryo to the recipient.

9.   EMBRYO TRANSFER PROCEDURE

Two different procedures for embryo transfer are described in this section. They illustrate the division of labor as well as the sequential steps required.

Method 1. The side opening catheter with introducer (Figure 1).

Materials

       a.   Embryo transfer catheter and introducer (METS-2, Cook Australia)

      b.   Ham's F-10 with 75-90% Fetal Cord Serum (FCS).
      c.   35 mm (Falcon) petri dish
      d.   Dissecting microscope
      e.   Sterile Pasteur pipets/Pi-pump (blue)
      f.   Syringes (1 cc)
      g.   Needles (18 gauge or any available size)
      h.   Speculum
      i.   Sterile 4 x 4 gauge and ring forceps
      j.   Single tooth tenaculum
      k.   Sterile tape

Preparation of the Patient and Catheter for the Transfer

a.   Uterine depth should be measured using a sound during the ovulation induction phase of the cycle (i.e., not at egg retrieval or embryo transfer).

b.   The embryo catheter will be inserted to a depth of 1 cm below the top of the fundus. The catheter should be measured so that the tip of the catheter reaches approximately 1 cm beyond the end of the embryo transfer catheter. For example, if the uterus is sounded at 8 cm, then the introducer should be marked at 6 cm and the transfer catheter marked at 7 cm using sterile tape.

c.   At the time of embryo transfer, the patient should be positioned with her uterus in a dependent position (knee-chest for an anteverted uterus and lithotomy for a retroverted uterus).

d.   A speculum is placed in the vagina, the cervix clearly visualized and cleaned with a sterile 4 x 4 forceps with saline or culture medium.

e.   The external catheter (introducer) is inserted to the internal os as measured.

Transfer Procedure

a.   The embryos are prepared for the transfer procedure in the embryo lab. As already described, visualization and photography will be easier if the embryos are transferred from the organ culture dish to a flat bottom petri dish. After photography and/or assessment, all of the embryos are combined in a single petri dish containing the transfer medium.

b.   In order to insure that as small a volume as possible is used for transfer, it is easiest to draw up all of

the embryos in a small volume of medium using a
sterile pipet.

c.   Place all of the embryos in a single drop of medium in
      an empty 35 mm petri dish. Place this dish on the
      stage of the dissecting microscope and focus.

d.   Put on sterile gloves and take the premarked transfer
      catheter with one ml syringe attached from the assis-
      tant. Usually the assistant will be the physician who
      is performing the embryo transfer. Care must be taken
      from this point on to use sterile procedures.

e.   Rinse the catheter with large volumes of transfer
      media and expel all of this media from the catheter.

f.   Take up a small volume (approximately 20 µl of medium)
      followed by 30 µl of air, followed by the embryos in
      as small a volume of medium as possible, followed by
      30 µl of air, followed by 20 µl of medium. Care must
      be taken to be sure that a column of medium is
      included between the embryos and the syringe in order
      to assure that the entire catheter will be cleared.
      To predetermine these volumes, it is simplest to use a
      Hamilton syringe while drawing up the amounts
      described above. A cruder but effective alternative
      is to rehearse the transfer using an extra transfer
      catheter. Place 20 µl of medium in an empty petri
      dish and draw it up into the transfer catheter.
      Measure the height of the fluid column. After some
      practice, visual inspection will permit use of proper
      volumes.

g.   When the tip of the catheter is removed from the petri
      dish, check to insure that the air bubbles do not
      move. This would indicate a loose connection or air
      leak and embryos may be lost.

h.   At this point, the physician should have placed the
      external catheter (introducer) into the cervix to the
      point of the internal os as measured and as already
      described. The embryologist should now approach the
      patient and run the catheter into the external
      catheter to the marked depth.

i.   Push the plunger on the syringe emptying the contents
      of the catheter into the uterus. While keeping the
      transfer catheter in place, remove the syringe, fill

with a small volume of air and inject an additional 50
μl of air.

j.   Wait 30 seconds and withdraw the catheter with a
     slight twisting motion.

k.   Flush both the internal and external catheters with
     medium and verify that there are no retained embryos
     using the methods described below (flushing from top
     to bottom). Use needles to tease apart any mucus
     masses. Allow the flushings to settle and re-examine.

l.   Once it has been verified that there are no retained
     embryos, the physician can remove the speculum.

m.   In some cases, it may be difficult to pass the intro-
     ducer to the point of the internal os. Traction on
     the lip of the cervix with a single tooth tenaculum
     may aid this process.

n.   The patient remains in the prone position for approxi-
     mately 30 min after the embryo transfer procedure and
     is then transferred to a cart which can be placed in
     Trendelenburg. In cases where the transfer has been
     performed in the knee-chest position, the patient will
     be transferred to the cart on her stomach. In cases
     where the transfer was performed in dorsal lithotomy
     position, the patient will be transferred to the cart
     on her back. The period of recovery after embryo
     transfer varies according to the center but is usually
     about 3 hr. After that period of time, the patient is
     transferred to her car in a reclining position and
     told to limit her activities by remaining in bed for
     the next 24 hr.

Method 2.   The Tom Cat Catheter

Materials

a.   Tom Cat catheter (3 1/2 Fr, 5 1/2" open; Monoject,
     St. Louis, MO)
b.   Speculum
c.   Tissue culture medium (Ham's F-10 and 15% FCS)
d.   4-0 silk with cutting needle/needle holder
e.   Sterile 4 x 4 gauze and ring forceps
f.   35 mm (Falcon 3001) petri dish
g.   Dissecting microscopes
h.   Sterile Pasteur pipets/Pi-pump (blue)
i.   Syringe

Preparation of the Patient and the Catheter

a.   All transfers using Tom Cat catheters are done in the
     dorsal lithotomy position.

b.   The depth and position of the uterus is measured
     during ovulation induction using a Tom Cat catheter.
     Since these catheters are disposable, "rehearsed"
     transfers are feasible for each woman.

c.   The speculum is placed and the cervix is washed off
     with culture medium using sterile 4 x 4 sponges.

d.   The Tom Cat catheter is attached to a disposable
     syringe. The physician molds the catheter to the
     shape of the uterus which was previously determined
     and hands it off to the embryologist for loading.

Transfer Procedure

a.   The embryos are placed in a single drop of growth
     medium (Ham's F-10 and 15% FCS) in the same way as
     described for the side opening catheter.

b.   10 µl of air, then 10 µl of fluid containing the
     embryos, then 10 µl of air are loaded in the catheter
     which is handed off to the physician.

c.   The physician gently inserts the catheter into the
     cervix/uterus to the measured depth and injects the
     embryos.

d.   The physician withdraws the catheter and hands it off
     to the embryologist for flushing (refer to procedure
     described below).

e.   If placement of the catheter is difficult, gentle
     traction on the lip of the cervix may facilitate the
     process. To do this atraumatically, the physician
     makes a single pass through the lip of the cervix
     using a 4-0 silk suture with a cutting needle. The
     two "tails" of the suture are used to apply traction.
     After the transfer is complete, the suture is removed.

f.   The patient remains on her back with feet elevated for
     30 min. At the end of 30 min, she is transferred to a
     cart placed in Trendelenburg for 3 hr. She is placed
     so that her uterus will be in a dependent position--on
     her stomach for anteverted, on her back for a retro-
     verted uterus.

10. FLUSHING

After embryo transfer, it is essential to flush the transfer
catheter to rule out the possibility of a retained embryo. Flushing
of both transfer catheter and introducer should be accomplished from
the top down. This is done by filling a syringe with medium and
attaching it to the catheter, flushing down. If you flush from the
bottom up, it is possible to force the retained embryo further up the
catheter, missing it. It is also useful to have several needles
attached to syringes available for teasing apart mucus or any other
materials that may be flushed from the catheter. In this way, a
retained embryo will not be missed in the mucus mass.

11. RETAINED EMBRYO

After flushing the embryo transfer catheter and introducer or
external catheter, a retained embryo may be found. This embryo
should be retransferred to the recipient. We have had personal
experience with cases of pregnancies resulting after more than one
embryo transfer, that is, retransfer of a retained embryo. Anecdotal
information indicates that other groups have had similar experiences.
Therefore, it does not appear that this process precludes pregnancy.
The embryo should be rinsed in culture medium and a new embryo
transfer catheter should be used.

12. MEDICATIONS USED

Based on factors that affect ET success in cattle, doxycycline
is administered to most patients receiving embryo transfer (Trounson
et al., 1978), beginning on the day of laparoscopy and continuing for
five days at a dose of 100 mg twice a day. In cattle, doxycycline
use is based on observations that surgical transfers are more
successful than nonsurgical ones; bacterial contamination of the
catheter in the vagina was thought to be the cause. Many candidates
for IVF have a history of tubal or pelvic inflammatory disease. If a
specific pathogen or other infectious disease of the reproductive
tract can affect the outcome of IVF, antibiotic prophylaxis may be
merited (Rowland et al., 1985). On the other hand, specific treatment of an identified pathogen when possible may be preferable to a
"shotgun" type of prophylaxis (Barriere et al., 1985). A recent
review by Axelrod and Talbot (1986) suggests that a battery of infectious disease screens and cultures including Chlamydia, mycoplasma,
gonorrhea, syphillis, hepatitis B surface antigen and HTLV-III should
be included as part of the pre-IVF workup. These authors contend
that these pretreatment infectious disease screens will decrease the
probability of contamination of laboratory personnel and equipment as
well as reducing the risk of ascending infection of the reproductive
tract at the time of embryo transfer. Since some of these diseases

have disastrous implications for the newborn, the risk of congenital
infection can also be reduced in this way.

Inhibition of prostaglandin synthesis has been used in the
period prior to embryo transfer without influence on outcome in the
case of ibuprofen (Wood et al., 1981). However, any manipulation
that decreases uterine motility may enhance success of embryo
transfer. For this reason, the use of a tenaculum is avoided and/or
drugs such as Motrin, progesterone, valium and even the β-adrenergic
receptor agonist, ritordrine are employed (Katayama et al., 1985) to
reduce uterine smooth muscle activity.

Exogenous administration of progesterone and/or hCG to rescue
the corpus luteum is also a source of controversy. The evidence is
unclear as to whether or not ovulation induction causes a luteal
phase defect. Garcia et al. (1984) have shown that the endometrium
in hMG-treated IVF cycles is "more advanced" in some patients than in
others. The authors felt that an advanced endometrium enhanced the
probability of implantation. Since this advanced endometrium is
associated with increased serum progesterone levels, the authors
advocated the use of progesterone supplementation for IVF cycles.
Leeton et al. (1985) recently performed a randomized prospective
study to evaluate the effect of progesterone supplementation on
pregnancy rate in Clomid/hMG-induced cycles. They found no significant difference in pregnancy rate between progesterone-treated and
nontreated cycles.

Some groups have also advocated hCG to support the corpus
luteum. Mahadevan et al. (1985) reported on a prospective study of
the use of low dose hCG during the luteal phase in IVF patients
stimulated with Clomid/hMG. Three treatment groups were included:
hCG-treated, untreated (both assigned randomly) and hCG-treated
patients with a suspected luteal phase defect. Their results
indicated that hCG-treated patients experienced longer luteal phases
than did untreated patients (18 vs. 12.5 days); therefore, hCG
prolonged the life of the corpus luteum. However, none of the
patients treated with hCG (n = 10) became pregnant while four of the
patients in the untreated group (n = 10) established pregnancies.
While the sample size in this study was small, a tentative conclusion
is that hCG may affect embryo viability or implantation. At the very
least, the administration of hCG during the luteal phase precludes
early serum pregnancy testing.

The role of the progesterone (P)/estradiol (E_2) ratio may
actually be the key to these apparent contradictions. Gidley-Baird
et al. (1986) reported that failure of implantation in IVF patients
with viable embryos (as assessed by platelet response) occurs in
patients where E_2 is high and P is low and that the P/E_2 ratio is
more critical than the absolute value of either hormone. They have
shown that high E_2 levels block implantation in the mouse and that

this effect can be reversed by progesterone administration. P/E_2 ratios may provide objective criteria for progesterone supplementation in each patient as well as an index for adjusting dose and treatment route (suppository vs. injection).

13. FACTORS INFLUENCING SUCCESS

A recent study by Englert et al. (1986) compared a number of factors in embryo transfer that were thought to contribute to pregnancy rate. They found that the duration of the procedure and the person performing it did not affect outcome. They confirmed earlier work by Leeton et al. (1982) which indicated that the ease or difficulty of the transfer was very predictive of outcome (difficult transfers resulted in fewer pregnancies) and that blood on the cervix, catheter or in the uterus greatly decreased the pregnancy rate. Most importantly, they found no effect of position used for transfer on pregnancy rate. Since most gynecologists are more proficient with procedures performed in the dorsal lithotomy position and there is less need for manipulation and use of instruments, which may cause bleeding, one could argue that dorsal lithotomy is a preferred position. In this position, rigid introducers offer no advantage (Englert et al., 1986). Finally, Bellinge et al. (1986) have shown that insemination of IVF patients on the day of laparoscopy significantly increased the implantation rate. This effect could not be attributed to IVF independent pregnancies since these patients all had documented tubal factors.

Ultrasound visualization of the process of embryo transfer was first described by Strickler et al. (1985). The authors reported that this method made catheter and embryo placement easier and more reproducible. Ultrasound also provided a means of monitoring fluid (if not embryo) ejection from the transfer catheter. The series reported with this method was quite small and wider testing is needed. Ultrasound measurement of endometrial thickness (Fleischer et al., 1986) indicates that current methods do not differentiate between the endometrium of patients during conception versus nonconception cycles.

14. PREGNANCY TEST

Generally, the pregnancy test is performed 10-12 days following embryo transfer. In IVF patients, this test takes the form of both a qualitative pregnancy test and a quantitative β-hCG determination. If the test is positive, β-hCG titers are measured at regular intervals in order to evaluate doubling time and assess the normalcy of the pregnancy. The subject of what constitutes a pregnancy in IVF has been described by Jones et al. (1983a). Pregnancy should be determined by rising β-hCG levels as well as ultrasound confirmation.

The risk of ectopic pregnancy in IVF patients is reported to be approximately 10%; therefore, patients should be closely followed until intrauterine pregnancy is established.

15. REFERENCES

Axelrod, P. and Talbot, G. H., 1986, Infection control considerations for in vitro fertilization and embryo transfer programs, Infect. Control 7:373-378.

Barriere, P., Lopes, P., Boiffard, J. P., L'Hermite, A., and Lerat, M. F., 1985, An unusual cause of failure of in vitro fertilization: report of a case, J. In Vitro Fert. Embryo Transfer 2:170-171.

Bellinge, B. S., Copeland, C. M., Thomas, T. D., Mazzucchelli, R. E., O'Neill, G., and Cohen, M. J., 1986, The influence of patient insemination on the implantation rate in an in vitro fertilization and embryo transfer program, Fertil. Steril. 46:252-256.

Byrd, W. and Wolf, D. P., 1984, Oogenesis, fertilization and early development, in: Human In Vitro Fertilization and Embryo Transfer (D. P. Wolf and M. M. Quigley, eds.), Plenum Press, New York, pp. 213-273.

Chetkowski, R. J., Nass, T. E., Matt, D. W., Hamilton, F., Steingold, K. A., Randle, D., and Meldrum, D. R., 1985, Optimization of hydrogen-ion concentration during aspiration of oocytes and culture and transfer of embryos, J. In Vitro Fert. Embryo Transfer 2:207-212.

Cummins, J. M., Breen, T. M., Harrison, K. L., Shaw, J. M., Wilson, L. M., and Hennessey, J. F., 1986, A formula for scoring human embryo growth rates in in vitro fertilization: its value in predicting pregnancy and in comparison with visual estimates of embryo quality, J. In Vitro Fert. Embryo Transfer 3:284-295.

Englert, Y., Puissant, F., Camus, M., Van Hoeck, J., and Leroy, F., 1986, Clinical study on embryo transfer after human in vitro fertilization, J. In Vitro Fert. Embryo Transfer 3:243-246.

Feichtinger, W., Kemeter, P., and Szalay, S., 1983, The Vienna program of in vitro fertilization and embryo-transfer--a successful clinical treatment, Eur. J. Obstet. Gynecol. Reprod. Biol. 15:63-70.

Fishel, S. B., Cohen, J., Fehilly, C., Purdy, J. M., Walters, D. E., and Edwards, R. G., 1985, Factors influencing human embryonic development in vitro, Ann. N.Y. Acad. Sci. 442:342-356.

Fleischer, A. C., Herbert, C. M., Sacks, G. A., Wentz, A. C., Entman, S. S., and James, A. E., Jr., 1986, Sonography of the endometrium during conception and nonconception cycles of in vitro fertilization and embryo transfer, Fertil. Steril. 46:442-447.

Garcia, J. E., Acosta, A. A., Hsiu, J. G., and Jones, H. W., Jr., 1984, Advanced endometrial maturation after ovulation induction with human menopausal gonadotropin/human chorionic gonadotropin for in vitro fertilization, Fertil. Steril. 41:31-35.

Gidley-Baird, A. A., O'Neill, C., Sinosich, M. J., Porter, R. N.,
    Pike, I. L., and Saunders, D. M., 1986, Failure of implantation
    in human in vitro fertilization and embryo transfer patients:
    the effects of altered progesterone/estrogen ratios in humans
    and mice, Fertil. Steril. 45:69-74.
Jones, H. W., Jr., Acosta, A. A., Andrews, M. C., Garcia, J. E.,
    Jones, G. S., Mantzavinos, T., McDowell, J., Sandow, B. A.,
    Veeck, L., Whibley, T. W., Wilkes, C. A., and Wright, G. L.,
    Jr., 1983a, What is a pregnancy? A question for programs of in
    vitro fertilization, Fertil. Steril. 40:728-733.
Jones, H. W., Jr., Acosta, A. A., Garcia, J. E., Sandow, B. A., and
    Veeck, L., 1983b, On the transfer of conceptuses from oocytes
    fertilized in vitro, Fertil. Steril. 39:241-243.
Katayama, P., Roesler, M., Gunnarson, C., Halverson, G., and Meyer,
    M., 1985, In vitro fertilization and embryo transfer, Wis. Med.
    J. 84:9-11.
Kerin, J. F., Quinn, P. J., Kirby, C., Seamark, R. F., Warnes, G. M.,
    Jeffrey, R., Matthews, C. D., and Cox, L. W., 1983, Incidence of
    multiple pregnancy after in-vitro fertilization and embryo
    transfer, Lancet 2:537-540.
Kerin, J. F. P., Jeffrey, R., Warnes, G. M., Cox, L. W., and Broom,
    T. J., 1981, A simple technique for human embryo transfer into
    the uterus, Lancet 2:726-727.
Leeton, J., Trounson, A., Jessup, D., and Wood, C., 1982, The tech-
    nique for human embryo transfer, Fertil. Steril. 38:156-161.
Leeton, J., Trounson, A., and Jessup, D., 1985, Support of the luteal
    phase in in vitro fertilization programs: results of a con-
    trolled trial with intramuscular Proluton, J. In Vitro Fert.
    Embryo Transfer 2:166-169.
Mahadevan, M. M., Leader, A., and Taylor, P. J., 1985, Effects of
    low-dose human chorionic gonadotropin on corpus luteum function
    after embryo transfer, J. In Vitro Fert. Embryo Transfer 2:190-
    194.
O'Neill, C., Pike, I. L., Porter, R. N., Gidley-Baird, A. A.,
    Sinosich, M. J., and Saunders, D. M., 1985, Maternal recognition
    of pregnancy prior to implantation: Methods for monitoring
    embryonic viability in vitro and in vivo, Ann. N.Y. Acad. Sci.
    442:429-439.
Rowland, G. F., Forsey, T., Moss, T. R., Steptoe, P. C., Hewitt, J.,
    and Darougar, S., 1985, Failure of in vitro fertilization and
    embryo replacement following infection with Chlamydia
    trachomatis, J. In Vitro Fert. Embryo Transfer 2:151-155.
Schiewe, M. C., Schmidt, P. M., Bush, M., and Wildt, D. E., 1984,
    Effect of absorbed/retained ethylene oxide in plastic culture
    dishes on embryo development in vitro, Theriogenology 21:260
    (abstract).
Strickler, R. C., Christianson, C., Crane, J. P., Curato, A., Knight,
    A. B., and Yang, V., 1985, Ultrasound guidance for human embryo
    transfer, Fertil. Steril. 43:54-61.

Trounson, A. O., Rowson, L. E., and Willadsen, S. M., 1978, Non-
    surgical transfer of bovine embryos, Vet. Rec. 102:74-75.
Wood, C., Trounson, A., Leeton, J., Talbot, J. M., Buttery, B., Webb,
    J., Wood, J., and Jessup, D., 1981, A clinical assessment of
    nine pregnancies obtained by in vitro fertilization and embryo
    transfer, Fertil. Steril. 35:502-508.

12

ANIMAL MODELS FOR THE STUDY OF FERTILIZATION AND EARLY DEVELOPMENT

IN VITRO

Barry D. Bavister

1.   INTRODUCTION

The development and successful application of human IVF-ET has grown out of pioneering work using animal models, particularly the mouse, rabbit and hamster (Yanagimachi and Chang, 1964; Gwatkin, 1977; Hartmann, 1983; Biggers, 1987). The successful adaptation of these procedures to human gametes and embryos, from the first unequivocal demonstration of IVF (Bavister et al., 1969; Edwards et al., 1969) to the growth of IVF blastocysts in vitro, took only a few years (Edwards, 1980) and the first human IVF birth occurred in 1978 (Steptoe and Edwards, 1978). Since that time, there have been striking improvements in the clinical procedures for human IVF, but the same cannot be said for the laboratory aspects of the technique. Due to a variety of constraints placed upon the use of human embryos, relatively little research can be done in this species. One would expect that relevant new knowledge might come from animal IVF studies. However, although this is true in a limited sense, direct extrapolation from animal studies to the practice of human IVF is not possible. There are two main reasons for this. One is the obvious fact that there are substantial between-species differences in the details of fertilization and early embryonic development. For example, it is well known that mouse embryos do not require exogenous amino acids for growth to the blastocyst stage in vitro, but there is an absolute amino acid requirement for development of rabbit blasto- cysts (Kane and Foote, 1970). Indirect evidence indicates that primate embryos also need amino acids for development beyond the cleavage stage in vitro (Boatman, 1987). Secondly, few animal models are available that can provide information which is directly useful to human IVF: only in the mouse, rabbit, and a few nonhuman primates has IVF with subsequent development of viable embryos in vitro been routinely accomplished. As a result, the practice of human IVF is presently taking place without a sufficient supporting infrastructure of knowledge derived from animal models. It is up to animal scientists to bridge this gap as quickly as possible and to provide new data on basic mechanisms involved in fertilization and early development, as well as on the design of new culture media for supporting IVF and normal embryo growth that can be applied to the human clinical situation.

It might be argued that, with the incidence of human IVF in the 80-90% range and high frequencies of cleavage to the 4- or 8-cell stage, there is little need for improvement in procedures for IVF and embryo culture (EC). However, this conclusion may be untenable. Abnormalities of oocyte maturation, fertilization or early cleavage may not be manifested until later stages, following embryo transfer when it is no longer possible to visually monitor the progress of peri-implantation development. Using IVF monkey embryos, substantial differences in developmental capability can be exhibited after the 4- to 8-cell stage has been reached (Bavister et al., 1983). The same may well be true for human embryos, which has implications for the practice of IVF. It is well known that primate blastocysts secrete substantial amounts of chorionic gonadotropin. If the majority of human IVF embryos succeeded in developing to the blastocyst stage following embryo transfer, then most pregnancy failures should be associated with positive β-hCG results (so-called "biochemical pregnancies"). Since this is usually not the case, it seems likely that defective embryonic development prior to implantation is one of the prime causes of failure in IVF-ET. This contention is consistent with the demonstrated increase in the (singleton) pregnancy success rate when two or more embryos are transferred (Edwards and Steptoe, 1973; Fishel et al., 1985).

What are the possible causes of losses of IVF embryos? These fall into two main categories: problems with the maternal milieu and intrinsic embryonic defects. In the second category, defective oocyte maturation (leading to abnormal fertilization and/or cleavage), anomalies of fertilization and inadequate embryo culture conditions could all contribute to embryonic death. These complex problems need to be approached separately. From the research view-point, addressing the problems exclusively with human gametes is difficult, to say the least. Although the total number of human oocytes that are recovered annually and fertilized in vitro is very large, most of the resulting embryos are committed to be transferred back to the donors. Moreover, most practitioners of human IVF-ET are unwilling or unable, for a variety of reasons, to "tinker" with an established regimen for the sake of research. For example, prolong-ing the duration of embryo culture (usually ~ 2 days) to 3 or 4 days prior to embryo transfer could provide new information on the inherent viability of embryos and on the suitability of different culture media. This approach has been useful with nonhuman primate embryos (see Section 4). However, such a practice (at present) carries a high risk of embryonic loss due to unsatisfactory culture conditions and has been followed by few human IVF laboratories.

In summary, the practice of human IVF-ET seems to have reached a point at which the established laboratory protocols are hardly being improved, because of insufficient data from either human or animal studies to formulate new protocols that will substantially increase the incidence of successful term pregnancies. Because of the obvious constraints (practical, ethical and legal) on human IVF research, the major burden would seem to fall on increased animal IVF research to point the way towards improvements in the application of IVF-EC in humans. If such efforts are to be meaningful, appropriate animal models need to be selected and carefully designed experiments conducted. Detailed studies of gametes and embryos in species related to humans (e.g., rhesus monkeys, baboons) can be rewarding in terms of defining, for example, parameters for successful oocyte maturation and embryo development. Aside from this, the performance of IVF-EC in animals can provide invaluable training experiences for the laboratory personnel. Finally, several animal models have been devised that indirectly aid the performance of human IVF, e.g., mouse embryo culture for testing culture media and the hamster zona-free test for evaluating human sperm fertilizing ability; these tests are described in Chapters 5 and 7, respectively.

The following sections discuss the use of different animal
models for studying IVF-EC, and their appropriateness for extrapolation to the human situation (see also Chapter 18).

2.   OOCYTE MATURATION

Normal embryonic development can only take place after
fertilization of a normal (mature) oocyte. In primates, there is
insufficient understanding of the regulation of oocyte maturation
within the preovulatory follicle. This hinders our ability to mature
oocytes under controlled in vitro conditions. There are practical
consequences to this situation. The practice of IVF, whether for
research or for treatment of infertility, is compromised by lack of
knowledge about the control of oocyte maturation. Incomplete or
otherwise abnormal oocyte maturation may contribute to embryonic
losses during normal pregnancy and failure of pregnancy following
transfer of IVF embryos. In primates and in domestic species, at
least a few immature oocytes are likely to be recovered during
follicular aspiration following ovarian stimulation with exogenous
gonadotropins. Unless these oocytes can be brought to full maturity
in vitro, they represent a loss of efficiency in the treatment cycle.

An enormous literature exists on the regulation of oocyte
maturation. The great majority of studies have used only superficial
morphological criteria of oocyte maturity, nearly always involving
the breakdown of the so-called "germinal vesicle" and extrusion of
the first polar body (PB1), which are primarily signs of nuclear
maturation. However, it is crucial to examine the question of
cytoplasmic maturation, which is frequently abnormal in oocytes
matured in vitro. Cytoplasmic maturation defects associated with
developmental anomalies have been noted in a number of studies using
eggs from humans and animals. These defects range from (1) inability
to decondense the nuclei of penetrating sperm, resulting in failure
to complete fertilization (Thibault and Gerard, 1970; Leibfried and
Bavister, 1983); (2) errors in cortical granule distribution and/or
response, resulting in either polyspermy (leading to embryonic death)
or premature zona block (leading to failure of sperm penetration);
and (3) decreased viability and/or retarded development of embryos
subsequent to apparently normal fertilization, resulting in
embryonic/maternal asynchrony, thus compromising pregnancy. The last
of these defects might be particularly difficult to diagnose in
circumstances where IVF embryos are transferred soon after fertilization.

In order to adequately evaluate the normality of oocytes matured
in vitro, the development of embryos subsequent to fertilization must
be examined. Several laboratories have begun to examine the control
of oocyte maturation in relation to the functional capability of

oocytes after fertilization. Most of this work has been done using
the mouse, sheep, bovine and nonhuman primate.

2.1. Mouse

The study of oocyte maturation control in the mouse has
advanced to examination of the chemical interactions between cumulus
cells and oocyte metabolism. This topic is discussed by Schultz in
Chapter 17. In a recent study, germinal vesicle stage oocytes from
inbred mice were matured and fertilized in vitro, then grown to the
morula stage before transfer to recipients; about 30% of these
embryos produced offspring (Schroeder and Eppig, 1984). The inclusion of serum in the culture medium was found to be necessary to
prevent "zona hardening" and to permit fertilization of in vitro
matured oocytes (Eppig and Schroeder, 1986). Of particular interest
to the human IVF situation is that human cord serum could not support
fertilization in these experiments, whereas a variety of sera from
other species could do so.

2.2. Sheep

Moor and his colleagues have successfully matured sheep
oocytes in vitro, as demonstrated by embryo development to the
blastocyst stage following fertilization in vivo, and by the birth of
young after transfer of some of these blastocysts (Staigmiller and
Moor, 1984). Some of the requirements for oocyte maturation in vitro
have been worked out by Moor and his colleagues, including the need
for an intact corona cell layer and coculture with supplementary
granulosa cells. Serum, gonadotropins and estradiol were also
routinely incorporated in the culture milieu for sheep oocyte maturation, and an agitating (nonstatic) system was also found to be
important (Moor and Trounson, 1977; Staigmiller and Moor, 1984). An
important finding is that sheep oocytes are irreversibly damaged by
exposure to temperatures of < 32°C (Moor and Crosby, 1985). In view
of this, in experiments with rhesus monkeys, oocytes are maintained
at or above 32°C during collection and manipulation (Boatman, 1987).

2.3. Cattle

Bovine oocytes can undergo maturation in vitro, using
culture conditions similar to those described by Moor and his
colleagues for sheep oocytes (Critser et al., 1986a,b). Bovine
oocytes for in vitro maturation studies and IVF are usually obtained
by aspirating 1 to 5 mm follicles of ovaries from slaughtered
animals. Such oocytes have tightly condensed cumulus oophorus and
usually display a prominent nucleus or so-called "germinal vesicle"
(Leibfried and First, 1979). These oocytes provide good research
opportunities for the study of regulation of oocyte maturation.
Human oocytes are sometimes aspirated in an immature condition and it
is desirable to complete oocyte maturation in vitro. Knowledge of

procedures for achieving this in primates is presently lacking.
Bovine oocytes may be more useful models for study than rodent
oocytes; the timing of oocyte maturation in vitro for bovine oocytes
(24 to 36 hr) is more comparable to the situation in primates than,
for example, the mouse which needs only 16 to 18 hr (Eppig and
Schroeder, 1986).

Following IVF of in vitro matured oocytes, embryos have been
transferred to sheep oviducts, where they developed into blastocysts
(Critser et al., 1986a,b). In vitro matured bovine oocytes after IVF
can also develop in culture as far as the 8-cell stage (F. Barnes,
personal communication). In those locations where bovine oocytes are
readily obtainable, they can provide a useful model for examining the
regulation of maturation, and the information so gained might be
applicable to primates.

2.4. Hamster

Since IVF procedures are well worked out in this species
(Bavister, submitted), studies on in vitro maturation of hamster
oocytes could become another useful comparative approach. In one IVF
study, although the majority of germinal vesicle stage hamster
oocytes were able to undergo nuclear maturation and to extrude PB1 in
vitro, only 2% of the penetrated oocytes were able to decondense the
sperm nucleus, compared to 98% of control oocytes (Leibfried and
Bavister, 1983). This model has not been developed further.

2.5. Nonhuman Primate

Ultimately, the regulation of oocyte maturation needs to be
examined in a primate model. Studies using rhesus monkey oocytes
have been initiated using two main approaches. In one, the
"indirect" or in vivo approach, follicular fluid (FF) parameters are
retrospectively correlated with the developmental capacity of oocytes
following IVF (Bavister, in press). In the second approach, complete
maturation of oocytes from excised ovaries is examined in vitro.

Pregnant mare serum gonadotropin (PMSG) or human menopausal
gonadotropin (hMG, Pergonal) was used to stimulate follicular growth
in rhesus monkeys (Bavister et al., 1983; Boatman et al., 1986). The
response to stimulation was monitored in some animals by ultrasonography (Morgan et al., 1987a). The developmental competence of rhesus
monkey oocytes was examined following aspiration from follicles of
gonadotropin-stimulated animals (Boatman et al., 1986; Boatman,
1987). In one series of experiments, rhesus oocytes were classified
as (1) mature (M) if they extruded PB1 in culture within 8 hr
following aspiration, or (2) as completing maturation in vitro (MIV),
if more than 8 hr were required for PB1 formation (Figure 1). When
the M and MIV ova were subjected to IVF, no significant difference
was found between these two groups of eggs in terms of their ability

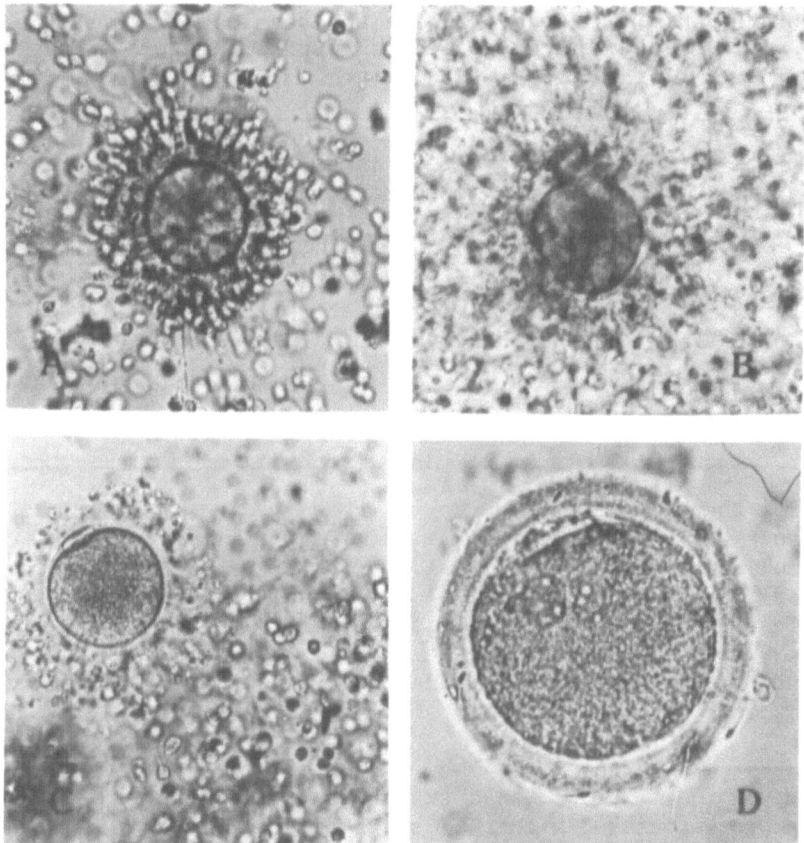

Fig. 1.   Completion of oocyte maturation in vitro. (A) Immature egg
           with unexpanded corona radiata. (B) Mature egg with
           expanded cumulus oophorus and corona radiata. (C) Mature
           egg with first polar body. (D) Fertilized egg (13 hr
           postinsemination [PI]) with two pronuclei near syngamy and
           two polar bodies (slightly out of focal plane). Optical
           magnification: A to C, 50X; D, 100X. (From Boatman,
           1987.)

to undergo fertilization (Table I). However, there was a significant
difference ($p \leq 0.001$) in the ability of these two classes of ova to
reach the 6- to 8-cell stage in vitro following IVF. These data
illustrate that oocyte defects may not show up until later stages of
embryo development.

In order to obtain information on oocyte maturation in the
rhesus monkey, follicular fluid (FF) steroid concentrations were
examined in association with different classes of oocytes: "mature"
and "maturation completed in vitro," as defined above, and

Table I. Comparison of fertilization and development of rhesus
          monkey oocytes classified as initially mature versus
          matured in vitro[a]

Class	No. ova fertilized/ no. inseminated (%)	No. $\geq$ 6- to 8-cells in vitro (% of ova fertilized)
Mature (M)[b]	40/61 (65.6)	35/37[c,d] (94.6)
Maturation completed in vitro (MIV)[e]	35/49 (71.4)	21/35[d] (60.0)

[a]Data from Boatman (1987).
[b]Incubated < 8 hr prior to insemination, first polar body (PB1)
 confirmed after cumulus dispersal.
[c]Excludes 2 embryos transferred (1 live birth [Boatman et al., 1986]
 and 1 lost prior to 6- to 8-cell stage).
[d]x^2, p $\leq$ 0.001.
[e]Incubated > 8 hr prior to insemination, PB1 confirmed after cumulus
 dispersal.

nonmature/atretic. There was a significant elevation of FF
progesterone associated with the mature oocyte category compared with
the others, but there were no significant differences in estradiol,
testosterone, or dehydrotestosterone (Morgan et al., 1986; Bavister,
in press). In addition, there were no significant differences for
any of the steroids when the data were re-analyzed to compare in
vitro fertilized (n = 35) versus unfertilized (n = 12) oocytes.
Thus, although oocytes which are (nearly) mature at recovery have
substantially more developmental capacity as a group than oocytes
which complete a considerable part of their maturation in vitro, no
significant differences in FF steroids between these two oocyte
categories were detected, other than an increase in P for the mature
category. In contrast to these preliminary data, Carson et al.
(1982) reported that the FF concentration of E_2 and the ratio of E_2
to P were positively correlated with the fertilizability and developmental capacity of human oocytes. The difference between these two
studies may be due in part to the difference in the gonadotropin
preparations used for follicular stimulation. Since any gonadotropin
stimulation is likely to disturb the normal chemical balance of the
FF milieu, comparative data are needed from unstimulated cycles.

     A small number of oocytes have been recovered by aspirating
follicles in unstimulated rhesus monkeys. Fully mature oocytes are

difficult to obtain from unstimulated monkeys because of the difficulty of precisely timing the completion of oocyte maturation (there
is no rapid RIA for LH in monkeys). Moreover, only one oocyte is
obtained per cycle. In preliminary experiments on spontaneous cycles
(Morgan et al., 1987b), significant differences (p < 0.05) were found
in both FF E_2 and P associated with oocytes that were mature (M) at
the time of collection (E_2, 153.2 ± 31 ng/ml; P, 3.2 ± 0.2 µg/ml) and
immature oocytes (IM) that required up to 24 hr in culture to extrude
PB1 (E_2, 772.6 ± 291 ng/ml; P, 4.7 ± 0.6 µg/ml). However, the IM
oocytes were collected earlier in the cycle, when E_2 levels are
higher. Both M and IM oocytes were able to undergo IVF. There was
no significant difference in FF E_2 values associated with M oocytes
in spontaneous cycles versus M oocytes in stimulated cycles (153.2 ±
31 ng/ml and 137 ± 14 ng/ml, respectively). These data suggest that,
in the rhesus monkey, FF E_2 by itself is not an indicator of the
capacity of oocytes to complete maturation in vivo or in vitro.

An alternative approach to the study of oocyte maturation
control in primates is to subject normal-looking, immature oocytes to
different treatments in vitro to obtain information about regulatory
mechanisms, then to test their capacity to undergo fertilization and
embryo growth in vitro and/or in vivo, as in the nonprimate studies
cited earlier. Rhesus oocytes having > 2 cell layers of condensed
cumulus oophorus were collected from excised ovaries and cultured in
simple medium (TALP) or complex medium (CMRL); calf serum, FSH-P
(Burns-Biotec, Omaha, NE) and hCG were also tested. When judged to
be mature, oocytes were inseminated in vitro (Figure 2). Significantly more (p < 0.025) oocytes cultured in the simple medium
containing serum and FSH underwent nuclear breakdown (83%) compared
to other treatments (mean of 53 ± 3%). Overall, the simple medium
was better than the complex medium for maturation of oocytes in vitro
(Morgan et al., in press). The reasons for this difference are not
yet known. Caution should be used when selecting a culture medium
for in vitro maturation of primate oocytes. For example, maturation
of squirrel monkey oocytes in Ham's F-10 medium was found to be very
limited compared to medium 199 (Kuehl and Dukelow, 1979). A possible
explanation for this difference is that hypoxanthine, which inhibits
maturation of mouse oocytes in vitro (Downs et al., 1985), is present
in Ham's F-10 at almost 7 times its concentration in medium 199.
However, medium CMRL-1066, which was less suitable for rhesus monkey
oocyte maturation (Morgan et al., in press), does not contain
hypoxanthine.

Some of the immature rhesus oocytes that were obtained from
excised ovaries and completely matured in vitro underwent IVF,
ranging from 2% of oocytes cultured with serum only to 22% of oocytes
cultured with serum, FSH-P and hCG. Fifty-seven percent of the
fertilized oocytes cleaved to the 6- to 8-cell stage in vitro (Morgan
et al., in press). These data show that immature ("germinal
vesicle") primate oocytes can complete both nuclear and cytoplasmic

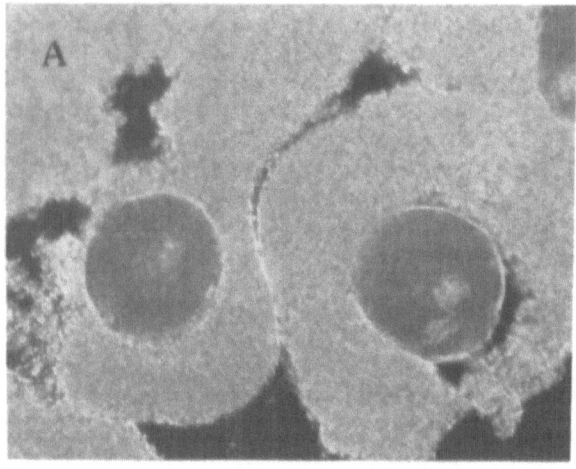

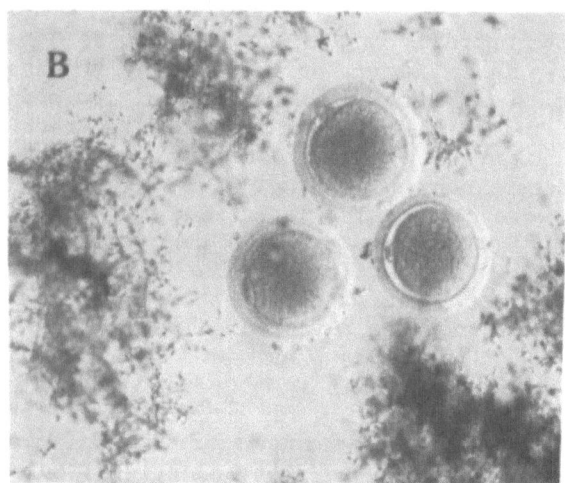

Fig. 2.   (A) Immature oocytes immediately after aspiration from
            follicles of excised rhesus monkey ovaries. Germinal
            vesicles are clearly seen in center of each oocyte. 52X.
            (B) Oocytes following in vitro maturation, several hours
            after insemination; each oocyte has a well-defined polar
            body. Large masses are clumped sperm from in vitro
            fertilization system. 36X. (From Boatman and Bavister,
            1984, with permission.)

maturation in vitro, although it remains to be shown that the embryos
are viable. Now the factors controlling maturation in vitro need to
be examined in more detail.

     The combination of these two approaches (analysis of the FF
environment associated with normal oocyte maturation, and studies on

conditions needed for complete oocyte maturation in vitro) seems the
most productive way to obtain new information on the control of
oocyte maturation. Information obtained from the study of oocyte
maturation in nonhuman primates should be helpful both for understanding the control of oocyte maturation and for making the best use
of available oocytes for research and/or for infertility treatment.
The maturation of oocytes in vitro followed by IVF is one of the most
promising approaches for increasing the supply of early embryos for
nonhuman primate research. However, new approaches for stimulating
multiple follicular development need to be assessed because of the
production by nonhuman primates of antibodies against foreign gonadotropins, rendering animals refractory to further treatment cycles
(Bavister et al., 1986).

3.   IN VITRO FERTILIZATION

The basic events of fertilization are similar in animals and in
humans. Experimental studies using suitable animal species not only
can provide basic data on fertilization events, but can also be a
useful training experience. Rodent species (mouse, rat and golden
hamster) are the easiest to use, but the details of fertilization in
nonhuman primates more closely resemble those in humans. In several
respects, bovine IVF may provide the most practical analogies with
IVF in primates.

3.1. Rodent

There are several advantages of studying fertilization in
rodents. (1) IVF is easy to accomplish. Procedures for IVF that
yield consistently good IVF results have been described by a number
of authors (see reviews by Gwatkin, 1977; Rogers, 1978; Hartmann,
1983). A detailed procedure that yields consistently high IVF
results using hamster gametes has recently been described (Bavister,
submitted). (2) Eggs are readily available in substantial numbers
(30 to 60 per donor) from superovulated animals. Animals are usually
superovulated by injection of pregnant mare serum gonadotropin (PMSG)
followed by human chorionic gonadotropin (hCG) (e.g., Leibfried and
Bavister, 1982). (3) The egg cytoplasm is relatively translucent, so
that details of fertilization can easily be visualized in living eggs
by phase contrast or by interference contrast microscopy. The two
pronuclei and both polar bodies can easily be seen. The tail of the
fertilizing sperm is usually hard to see in mouse eggs, especially
several hours after sperm penetration, but in the golden hamster, the
sperm tail is very large and persistent so that it can be seen in the
egg cytoplasm as late as the 2-cell stage (Marston et al., 1964; see
also Bavister, 1980). This is useful for verifying that IVF has
taken place.

There are some disadvantages of using rodent species. These include: (1) differences in morphological details of fertilization in the rat and hamster compared with those of human eggs. For example, the fertilizing sperm tail is not visible by light microscopy in human eggs because the sperm tail rapidly disintegrates following penetration into the egg cytoplasm (Bavister et al., 1969; Soupart and Strong, 1974). It may be difficult to detect both polar bodies in fertilized primate eggs, since PB1 sometimes disintegrates. By contrast, at least in hamster eggs, both polar bodies are routinely found (Bavister, submitted). Another feature of rodent eggs is that the cumulus-corona layer readily disperses during IVF, whereas in primates this layer usually persists through fertilization. Thus, the only easily and consistently visible morphological feature common to both rodent and primate eggs is the presence of the pronuclei. (2) Culture conditions used for rodent species are nearly always simple balanced salt solutions, such as modified Krebs-Ringer bicarbonate for mouse gametes (Biggers, 1987), or modified Tyrode's solution for the hamster (Bavister and Yanagimachi, 1977). For human IVF, complex media are often used so that it becomes difficult to extrapolate conditions devised for animal IVF to those best suited for human gametes. However, some human IVF clinics routinely use simple media for IVF, and it appears that results are indistinguishable from, if not better than, those obtained with complex media (Purdy, 1982; Trounson, 1983; Quinn et al., 1985; Caro and Trounson, 1986). A modified balanced salt solution developed for hamster IVF was used for the earliest successful human IVF experiments (Bavister et al., 1969; Edwards et al., 1969). A similar culture medium supports IVF and cleavage in the rhesus monkey (Bavister et al., 1983; Boatman et al., 1986; Boatman, 1987). Differences in culture media formulations for IVF in different species may thus be more imagined than real. (3) IVF with rodent eggs nearly always involves using sperm obtained from the excised vas deferens or cauda epididymidis. It is difficult, although possible, to collect freshly ejaculated sperm from these animals. This limitation has some drawbacks for research.

In summary, IVF with rodent gametes provides a useful training exercise for those with little or no previous experience in IVF. The advantages associated with using rodents for IVF make these animals preferred species for research on sperm capacitation and fertilization.

3.2. Rabbit

IVF was first demonstrated using rabbit gametes (Thibault et al., 1954; Chang, 1959). Following the discovery of procedures for capacitating rabbit sperm in vitro (Brackett and Oliphant, 1975), improved methods for rabbit IVF have been developed (Brackett et al., 1982).

There are several advantages of using rabbit gametes for IVF.
(1) Rabbit eggs, recovered from follicles or from the surface of the
ovary, resemble primate eggs more closely than do rodent eggs.
Rabbit eggs are opaque due to the presence of cytoplasmic lipid and
cleaving rabbit eggs are similar to human eggs in their appearance
(Brackett and Oliphant, 1975).   (2) Semen can easily be collected
using an artificial vagina; the rabbit is one of the few species in
which both epididymal and ejaculated sperm can be conveniently
collected so that comparative data can be obtained on fertilizing
ability and other characteristics of these two categories of sperm.
(3) IVF rabbit eggs will develop in culture to the blastocyst stage
without showing any developmental blocks (Kane, 1987).   This property
is discussed in Section 4.

Disadvantages of using rabbits include the fact that rabbits are
expensive to purchase and to maintain.   The animals are very sensitive to stress and readily become pseudopregnant if mishandled.   In
addition, the block to polyspermy in rabbit eggs is entirely at the
level of the plasma membrane, unlike all other commonly used species.

On balance, the disadvantages of using rabbit IVF for training
purposes seem to outweigh the advantages.   However, the rabbit does
offer significant and, in some respects, unique advantages for
research and for studying embryonic development in vitro.

3.3. Cattle

Bovine gametes may seem an unlikely choice for IVF studies,
but there are some significant advantages associated with bovine IVF.
(1) Bovine oocytes for IVF are easily obtained, as described in
Section 2.   (2) Bovine semen (fresh or frozen) can easily be
obtained, providing some potentially useful opportunities for basic
research on capacitation.   Bovine sperm are temperature sensitive, so
that precautions against temperature shock need to be taken, which
also provides good practical experience.   The large numbers of
spermatozoa in a single bovine ejaculate provide plenty of cells for
comparing alternative methods of sperm preparation, e.g., sperm
washing or swim-up procedures.   (3) The fertilizing spermatozoon is
not visible by light microscopy in IVF bovine oocytes, a situation
comparable to that with human IVF.   The criteria for accomplishment
of fertilization in the bovine largely consist of the presence of two
pronuclei and/or cleavage.

There are some disadvantages associated with using bovine
gametes, assuming that these are readily available, which is not
always the case.   The presence of pronuclei in IVF bovine oocytes
must be demonstrated either by fixing and staining eggs, or by using
a fluorescent stain (e.g., Hoechst or DAPI stains).   Alternatively,
oocytes can be centrifuged to displace cytoplasmic lipid revealing
the presence of pronuclei (Minhas et al., 1984).

In summary, IVF with bovine oocytes represents a useful advanced training exercise and is also a valuable research tool. Data on the control of bovine oocyte maturation as well as on bovine IVF and embryo cleavage may be useful to the practice of human IVF.

3.4. Nonhuman Primate

A few laboratories are now routinely performing IVF studies using nonhuman primates. These animals are very useful both for research and for IVF training purposes. Repeatable procedures have been devised for accomplishing IVF in the rhesus monkey (Bavister et al., 1983; Boatman and Bavister, 1984; Boatman et al., 1986; Boatman, 1987). Among the advantages of using nonhuman primates are: (1) The eggs are very similar, if not indistinguishable, from human eggs (Boatman, 1987). (2) Eggs are expendable. They can be fixed and stained to examine details of fertilization and therefore provide an opportunity to learn about basic aspects of fertilization in primates. The ability to perform destructive tests on nonhuman primate embryos can provide unique feedback about practical aspects of IVF procedures, such as the suitability of different culture conditions. Information of this kind most likely can be directly extrapolated to human IVF-EC. (3) Nonviable oocytes with intact zonae pellucidae ("zona equivalent" eggs) have been used to develop procedures for sperm capacitation and to examine conditions required for sperm:zona interaction (Boatman and Bavister, 1984; Figure 3). This approach is extremely useful in a species in which relatively few viable oocytes can be obtained. Zona equivalent oocytes are usually collected from excised ovaries and stored frozen in DMSO. (4) Viable eggs are usually recovered by aspiration at laparoscopy from ovarian follicles in a similar manner to the recovery of human oocytes. Thus, the stage of maturity of these aspirated oocytes may be very close to their counterparts in the human.

Disadvantages of using nonhuman primates include the obvious fact that the animals are fairly scarce and very expensive to use. Rhesus monkey sperm are generally more delicate than human sperm, and conditions for sperm capacitation may not be the same (Bavister et al., 1983; Boatman and Bavister, 1984). In addition, IVF culture conditions as presently used may not be comparable to the human situation in which complex medium, usually Ham's F-10, is commonly used.

A problem already referred to is the production of an immune response by nonhuman primates treated with foreign (animal and human) gonadotropins. The inability to perform repeated follicular stimulation treatment cycles in the same monkey becomes a limiting factor in the progress of IVF research. Alternative stimulation methods need to be developed and tested, such as the use of releasing hormones or immunization against inhibin, as in the sheep (Bindon et al., 1986).

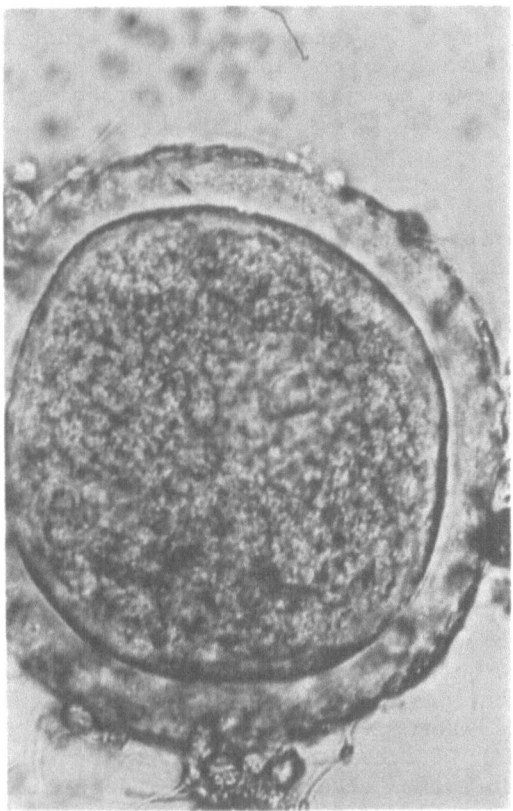

Fig. 3.   "Zona equivalent" egg, with poor quality cytoplasm and
           germinal vesicle. Zona pellucida of such eggs can serve as
           indicators of sperm penetrating ability; head of pene-
           trating sperm is deeply embedded in the zona pellucida at
           right. 206X (From Boatman and Bavister, 1984, with
           permission.)

    In summary, nonhuman primates, when available, can provide
invaluable training and research opportunities to complement human
IVF activities. The fertilized egg in rhesus monkeys is sufficiently
similar to human eggs that data obtained in this species should be
readily extrapolatable to the human.

4.   EMBRYO DEVELOPMENT IN CULTURE

    As mentioned earlier, animal species can within certain limits
provide useful models for the study of IVF-EC. As with IVF, the
study of embryo development in animal species has both advantages and

disadvantages.   It should be noted particularly that in vitro fertilized embryos often have lower developmental capacity than those
fertilized in vivo, even though no obvious morphological differences
are found (Brackett et al., 1982; Shalgi, 1984; Boatman, 1987).

4.1. Rodent

     The majority of present knowledge about development of
embryos in vitro, and about culture conditions for supporting embryogenesis, has come from experiments with the mouse (nearly always
inbred strains). However, there are some problems in extrapolating
this knowledge to primates: (1) As mentioned earlier, the culture
requirements for inbred mouse embryos are not the same as for embryos
of other species, including primates (Biggers, 1987; Boatman, 1987).
(2) A particular problem with most rodent embryos (rat, hamster,
outbred mouse) is that developmental blocks occur during cleavage in
vitro. Frequently, growth ceases entirely at the 2- or 4-cell stage
(Bavister, 1987; Biggers, 1987). Elimination of these in vitro
blocks to development is a desirable objective for improving the
usefulness of rodent species for research and for broad training in
embryo culture. (3) The regulation of embryonic development may
differ in rodents compared with other animals. Cleavage of rodent
embryos seems to be quite regular, with little or none of the cytoplasmic blebbing that can sometimes occur in primate embryos. In
spite of this, such abnormal-looking primate embryos can develop to
term (Boatman, 1987). Furthermore, asynchrony of early cleavage is
not commonly seen in normal development of rodent embryos, whereas
apparently asynchronous blastomere division has frequently been
observed in studies with monkey embryos (Bavister et al., unpublished
observations).

4.2. Rabbit

     In some important respects, rabbit embryos in culture
behave more like primate embryos than rodent embryos do. IVF rabbit
embryos can develop in culture to the blastocyst stage, i.e., there
is no block to development in vitro (Kane and Foote, 1970; Brackett
et al., 1982). Nutrient requirements for rabbit embryos are much
more demanding than for mouse embryos (Kane and Foote, 1970; Biggers,
1987; Kane, 1987). Thus, the rabbit embryo is probably the better
model for primate embryogenesis studies. Rabbit embryos may have
"growth factor" requirements (Kane, 1980, 1987). Elucidation of the
nature and mode of action of these factors could be very beneficial
for attempts to improve the viability of primate embryos in culture.
The rabbit is the only species other than the mouse in which systematic studies of culture requirements, including energy sources, have
been conducted. These data may prove to be valuable for understanding the needs of primate embryos.

4.3. Cattle

Recently, conditions for accomplishing cleavage of IVF bovine eggs have been established. IVF embryos have been transferred to the oviducts of sheep for cleavage and blastocyst development (Eyestone et al., 1985; Critser et al., 1986a,b). Approximately 50% of IVF oocytes will develop as far as the 8-cell stage in vitro, although these embryos are not viable as judged by embryo transfer (F. Barnes, personal communication).

An exciting development comparable to the discovery of growth factors for rabbit embryos has been the finding that trophoblast from developing blastocysts secretes growth factors that can sustain development of cleavage stage embryos (Camous et al., 1984; Heyman and Ménézo, 1987). It is not known if such factors are needed by primate embryos for optimal development and viability in vitro, but this is an exciting possibility. Analysis of the nature and mechanism of action of these factors in the bovine and rabbit could be helpful in attempts to formulate culture systems that are more appropriate for human embryos.

4.4. Nonhuman Primate

Embryos of nonhuman primates are very similar to those of man with respect to morphological appearance, timing of cleavage stages and culture requirements (Bavister et al., 1983; Boatman, 1987; see also Figure 4). Data obtained with nonhuman primate embryos are most likely to be helpful in improving culture conditions and procedures for human IVF-EC. An example of this is the testing of culture conditions for supporting growth of IVF embryos to the blastocyst stage, which was accomplished using rhesus monkeys (Boatman, 1987). As shown in Table II, although no significant differences were found between a simple balanced salt solution (TALP) and modified CMRL-1066 for cleavage of IVF embryos to the 6- to 8-cell stage, there was a striking difference in the ability of these media to support growth to or beyond the morula stage.

Culture conditions for growth of embryos to the blastocyst stage need improvement. Only 50% of the IVF rhesus embryos reached the early blastocyst stage, and less than half of these developed into expanded blastocysts (Boatman, 1987; see also Figure 5). Transfer of viable early blastocysts instead of early cleavage stages could substantially improve pregnancy success rates. If late-stage embryo transfer were done in the human, there could be several benefits. First, the patient would have had longer to recover from anesthesia, and secondly, less viable embryos would most likely have ceased development or slowed down, so that only the most vigorous embryos would be transferred.

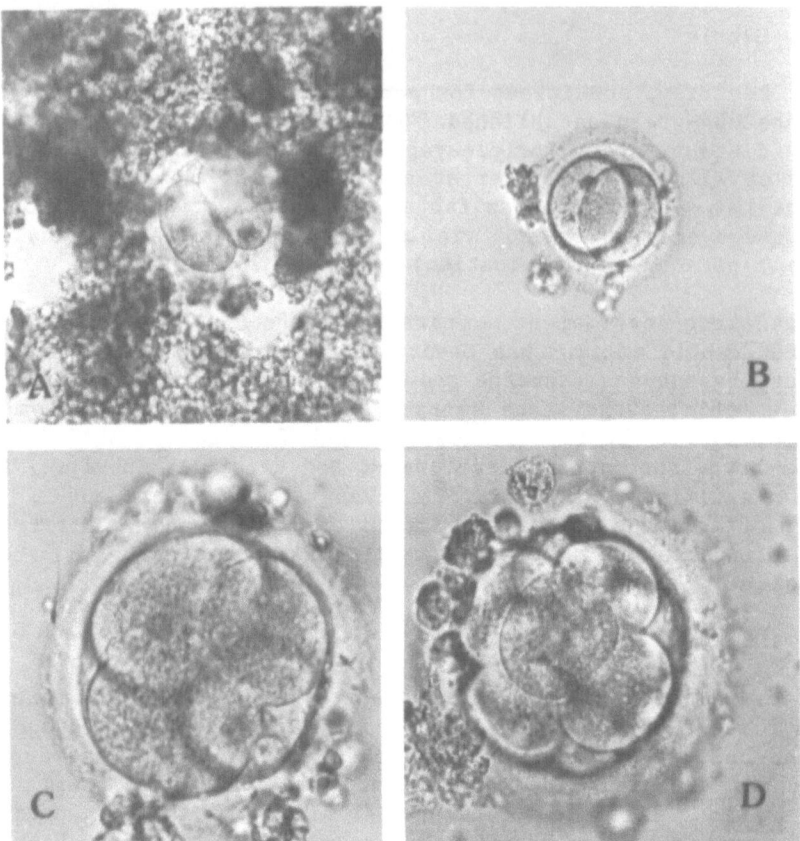

Fig. 4.   (A) Two-cell embryo (16 hr PI) with nucleated blastomeres
           and two polar bodies, in fertilization drop. (B) Same
           embryo as in A (19 hr PI) after removal of excess sperm and
           corona cells. Optical magnification A and B, 50X.
           (C) Four-cell embryo (same embryo as in A and B, 34 hr PI).
           (D) Eight-cell embryo (42 hr PI). Optical magnification C
           and D, 100X. Nucleated blastomeres can be seen in both
           embryos. (From Boatman, 1987.)

        Nonhuman primate embryos can also be of value when testing
potentially toxic experimental procedures, e.g., the use of fluorescent probes for evaluating embryo viability (Hoppe and Bavister,
1984). The application of such tests would be risky with human
embryos until adequate data on nonhuman primates are available. A
preliminary study with fluorescein diacetate was performed in the
rhesus monkey, but the results were inconclusive in terms of subsequent embryo development (Boatman et al., unpublished data). Another
use of nonhuman primate embryos would be to examine the feasibility

Table II. Comparison of development of IVF rhesus monkey embryos in two culture media[a]

	2-cells		≥ 6- to 8-cells		≥ morula	
Culture medium	No. (%)[b]	Time[c]	No. (%)[d]	Time[c]	No. (%)[d]	Time[c]
TALP[e]	23/29 (79)	26 ± 5	20/23 (87)	54 ± 14	3/23 (13)	88 ± 11
CMRL (modified)[g]	75/110 (68)	23 ± 5.5	56/72[f] (78)	59 ± 19	46/68[f] (68)	111 ± 16

[a]Data from Boatman (1987).
[b]Percent of inseminated oocytes.
[c]Time (hr) postinsemination (PI); mean ± S.D.
[d]Percent of fertilized ova.
[e]Modified Tyrode's solution with supplements including 3 mg/ml BSA (Bavister et al., 1983).
[f]Numbers of embryos reduced due to transfers to recipients.
[g]CMRL-1066 with supplements including 20% fetal (human, calf) or adult (rhesus, horse) serum (Boatman et al., 1986). Oocytes inseminated in TALP, then transferred to CMRL at 12 to 18 hr PI.

of embryo splitting and embryo sexing, and to develop optimal procedures for these techniques.

Prolonged culture of nonhuman primate embryos can be used to evaluate the functional significance of factors secreted by healthy developing embryos that may prove to be indicators of embryo viability. An example of this is the secretion of chorionic gonadotropin by developing baboon embryos (Pope et al., 1982). Chemical analysis of such factors and identification of the developmental stages at which they are secreted could lead to establishment of noninvasive tests of embryo quality. Useful basic information on the regulation of primate embryo development could be obtained by studies on the mechanism of action of these factors. It may become possible to enhance growth of primate embryos in vitro by providing such factors in the culture environment, similar to the studies done in the bovine (Heyman and Ménézo, 1987).

Another use for nonhuman primate embryos is to devise improved procedures for cryopreservation of embryos (Figure 6). Risks can be taken with nonhuman primate embryos that may well lead to their demise, which is not always advisable (or permissible) with human embryos. Development of reliable cryopreservation methods for nonhuman primate embryos should be a priority, partly because this technique would allow separation of the production of IVF embryos from the need to find synchronous surrogate recipients for embryo transfer (Boatman, 1987).

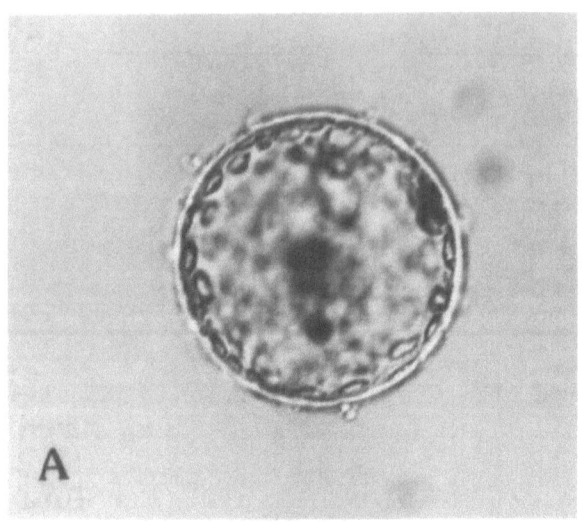

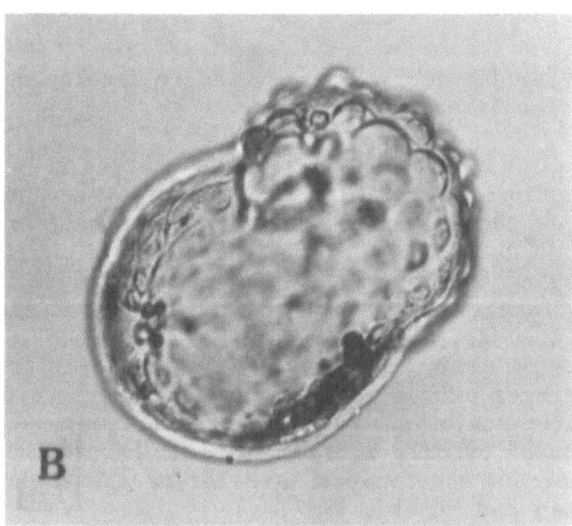

Fig. 5.   (A) Fully expanded zonal blastocyst (253 hr postinsemina-
           tion [PI] in vitro) showing numerous trophectodermal
           projections through the zona. (From Boatman, 1987.)
           (B) Early stage of hatching (same embryo as A, 265 hr PI).
           50X. (From Bavister, 1986.)

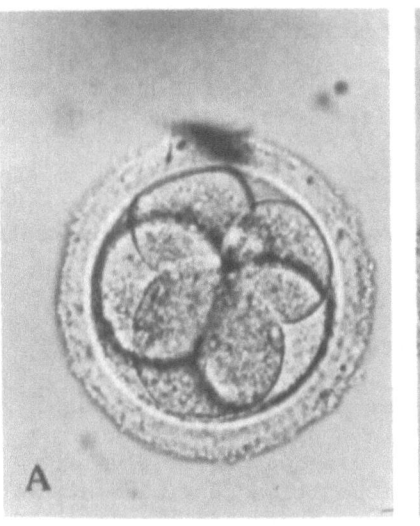
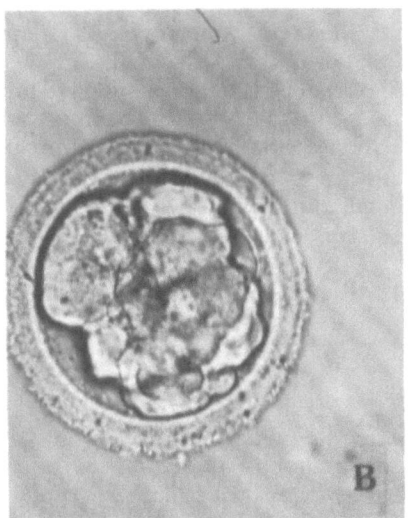
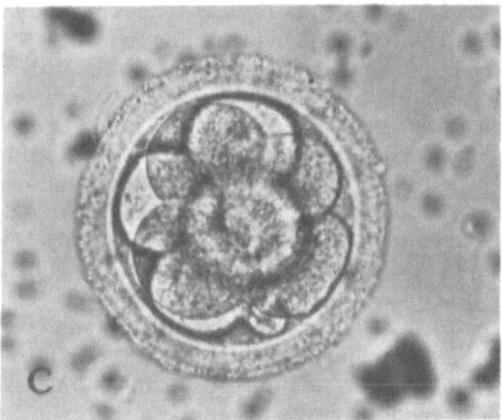

Fig. 6.   (A) Five-cell embryo (57 hr PI) prior to processing through
           steps of cryoprotectant. (B) Same embryo immediately after.
           thawing, in 1.5 M glycerol + PBS. (C) Same embryo (now 8-
           cell) postthaw following sucrose dilution steps and 25 hr
           in culture (37°C, 5% CO_2 in air). 100X. (From Boatman,
           1987.)

5.   TEST SYSTEMS USING ANIMAL GAMETES AND EMBRYOS

     In addition to the value of animal studies for basic research
into IVF-EC and for gaining expertise in gamete/embryo handling,
there are several tests involving animal gametes and embryos that may
be directly useful to the successful accomplishment of human IVF-EC.
These include tests of sperm fertilizing ability and quality-control
tests for culture media and apparatus.

5.1. Hamster Zona-Free Egg Penetration Test

The performance and value of this test for human sperm
fertilizing ability have been extensively discussed in Chapter 8.
There is no doubt that, under carefully controlled conditions, the
zona-free test can provide useful basic information on sperm capacitation and the acrosome reaction. Moreover, this test is unsurpassed
for allowing examination of the paternal chromosomal complement.

5.2. Quality-Control Tests for Culture Media

5.2.1.   Mouse embryo culture test

This test has been described in Chapter 5.
Consideration needs to be given to the sensitivity of the test. If a
strain of mouse is used whose embryos are capable of developing in
vitro even under poor culture conditions, then the value of the test
for human IVF-EC is questionable. Furthermore, if the stage of
embryo is selected for its high tolerance to the culture environment,
the value of test results is even more undermined. Using the
"standard" mouse 2-cell embryo culture test, one laboratory found no
difference in blastocyst development when tap-water or distilled
water was used to prepare the culture medium (Silverman et al.,
1987). It has been recommended that culture of 1-cell mouse embryos
should be used as the bioassay because of their substantially greater
sensitivity to culture conditions than 2-cell embryos (Quinn et al.,
1985; see also Spielmann et al., 1980). This would extend the
duration of the assay from 4 days to 5. Another recent study showed
that mouse IVF and culture of (in vivo fertilized) 1-cell, 2-cell and
even 8-cell embryos could all be used to detect differences in the
purity of water used in the preparation of culture media (Fukuda et
al., 1987). The most sensitive of these assays was embryo development to the hatched blastocyst stage. Using embryo culture from the
1-cell stage to expanded or hatched blastocyst, or from 2-cells to
hatched blastocyst, it was possible to discriminate between triple-
distilled water and reverse-osmosis/Milli-Q water used in the preparation of the culture medium (Fukuda et al., 1987).

5.2.2.   Sperm motility bioassay

Several laboratories engaged in human IVF-EC have
suggested using survival of human spermatozoa as an indicator of
culture medium quality (see Purdy, 1982). While this assay can be
completed in much less time than the mouse embryo assay, a drawback
of this sperm viability assay is that human sperm are remarkably
rugged; motility can persist in vitro for 48 hr or more, so the
sensitivity of this test is likely to be low.

A novel quality control test has been devised based on the
sensitivity of hamster spermatozoa to contaminants in culture media

(Bavister and Andrews, in press). The test is performed in protein-free medium so that the effect of contaminants is enhanced. The test can be completed in one day, and it is used routinely to check every batch of water for IVF-EC, as well as culture equipment. For example, using this assay, spermicidal contaminants leached from one brand of syringe-mounted filters were detected (Bavister and Andrews, in press). The test can also detect imminent failure of the cartridges in a Milli-Q water purification system before any change is detected in the resistance of the product water.

Whichever test is employed, there is no doubt of the advisability of checking the quality of culture media and apparatus used in IVF-EC with a reliable animal bioassay before subjecting valuable oocytes or embryos to in vitro conditions. Hopefully, development of more sensitive tests of this kind will help to improve not only the chemical purity of culture media but also their biological suitability for IVF and embryo development.

6.   SUMMARY

IVF-EC using appropriate animal species can be useful for several reasons: (A) Developing expertise in gamete handling and IVF-EC procedures, and in gaining knowledge of the morphology and physiology of embryogenesis. (B) Providing basic information that will lead to improvement of culture media for sustaining normal embryonic development. This is critically important if prolonged culture of embryos, i.e., to the early blastocyst stage, is desired prior to embryo transfer. (C) As tests of various kinds that are useful in the performance of different aspects of human IVF-EC. (D) Elucidating the nature of embryonic growth factors and their role in regulation of embryogenesis. This basic information may lead to the discovery of noninvasive tests for embryo viability, and of the means to sustain in vitro growth of less hardy embryos.

7.   REFERENCES

Bavister, B. D., 1980, Recent progress in the study of early events
    in mammalian fertilization, Dev. Growth Differ. 22:385-402.
Bavister, B. D., 1986, Animal in vitro fertilization and embryonic
    development, in: Manipulation of Mammalian Development
    (R. B. L. Gwatkin, ed.), Plenum Press, New York, pp. 81-148.
Bavister, B. D., 1987, Studies on the developmental blocks in
    cultured hamster embryos, in: The Mammalian Preimplantation
    Embryo: Regulation of Growth and Differentiation In Vitro
    (B. D. Bavister, ed.), Plenum Press, New York, pp. 219-249.

Bavister, B. D., Oocyte maturation and in vitro fertilization (IVF) in the rhesus monkey, in: The Primate Ovary (R. M. Brenner and C. H. Phoenix, series eds.; R. L. Stouffer, topic ed.), Plenum Press, New York, in press.

Bavister, B. D., A consistently successful procedure for in vitro fertilization of golden hamster eggs, Gamete Res., submitted.

Bavister, B. D. and Andrews, J. C., A rapid sperm motility bioassay procedure for quality-control testing of water and culture media, J. In Vitro Fert. Embryo Transfer, in press.

Bavister, B. D. and Yanagimachi, R., 1977, The effects of sperm extracts and energy sources on the motility and acrosome reaction of hamster spermatozoa in vitro, Biol. Reprod. 16:228-237.

Bavister, B. D., Edwards, R. G., and Steptoe, P. C., 1969, Identification of the midpiece and tail of the spermatozoon during fertilization of human eggs in vitro, J. Reprod. Fertil. 20:159-160.

Bavister, B. D., Boatman, D. E., Leibfried, L., Loose, M., and Vernon, M. W., 1983, Fertilization and cleavage of rhesus monkey oocytes in vitro, Biol. Reprod. 28:983-999.

Bavister, B. D., Dees, C., and Schultz, R. D., 1986, Refractoriness of rhesus monkeys to repeated ovarian stimulation by exogenous gonadotropins is caused by nonprecipitating antibodies, Am. J. Reprod. Immunol. Microbiol. 11:11-16.

Biggers, J. D., 1987, Pioneering mammalian embryo culture, in: The Mammalian Preimplantation Embryo: Regulation of Growth and Differentiation In Vitro (B. D. Bavister, ed.), Plenum Press, New York, pp. 1-22.

Bindon, B. M., Piper, L. R., Cahill, L. P., Driancourt, M. A., and O'Shea, T., 1986, Genetic and hormonal factors affecting superovulation, Theriogenology 25:53-70.

Boatman, D. E., 1987, In vitro growth of non-human primate pre- and peri-implantation embryos, in: The Mammalian Preimplantation Embryo: Regulation of Growth and Differentiation In Vitro (B. D. Bavister, ed.), Plenum Press, New York, pp. 273-308.

Boatman, D. E. and Bavister, B. D., 1984, Stimulation of rhesus monkey sperm capacitation by cyclic nucleotide mediators, J. Reprod. Fertil. 71:357-366.

Boatman, D. E., Morgan, P. M., and Bavister, B. D., 1986, Variables affecting the yield and developmental potential of embryos following superstimulation and in vitro fertilization in rhesus monkeys, Gamete Res. 13:327-338.

Brackett, B. G. and Oliphant, G., 1975, Capacitation of rabbit spermatozoa in vitro, Biol. Reprod. 12:260-274.

Brackett, B. G., Bousquet, D., and Dressel, M. A., 1982, In vitro sperm capacitation and in vitro fertilization with normal development in the rabbit, J. Androl. 3:402-411.

Camous, S., Heyman, Y., Méziou, W., and Ménézo, Y., 1984, Cleavage
    beyond the block stage and survival after transfer of early
    bovine embryos cultured with trophoblastic vesicles, J. Reprod.
    Fertil. 72:479-485.
Caro, C. M. and Trounson, A., 1986, Successful fertilization, embryo
    development, and pregnancy in human in vitro fertilization (IVF)
    using a chemically defined culture medium containing no protein,
    J. In Vitro Fert. Embryo Transfer 3:215-217.
Carson, R. S., Trounson, A. O., and Findlay, J. K., 1982, Successful
    fertilization of human oocytes in vitro: concentration of
    estradiol-17β, progesterone and androstenedione in the antral
    fluid of donor follicles, J. Clin. Endocrinol. Metab. 55:798-
    800.
Chang, M. C., 1959, Fertilization of rabbit ova in vitro, Nature
    (London) 184:466-467.
Critser, E. S., Leibfried-Rutledge, M. L., Eyestone, W. H., Northey,
    D. L., and First, N. L., 1986a, Acquisition of developmental
    competence during maturation in vitro, Theriogenology 25:150
    (abstract).
Critser, E. S., Leibfried-Rutledge, M. L., and First, N. L., 1986b,
    Influence of cumulus cell association during in vitro maturation
    of bovine oocytes on embryonic development, Biol. Reprod.
    34(Suppl. 1):192 (abstract).
Downs, S. M., Coleman, D. L., Ward-Bailey, P. F., and Eppig, J. J.,
    1985, Hypoxanthine is the principal inhibitor of murine oocyte
    maturation in a low molecular weight fraction of porcine
    follicular fluid, Proc. Natl. Acad. Sci. U.S.A. 82:454-458.
Edwards, R. G., 1980, Conception in the Human Female, Academic Press,
    London.
Edwards, R. G. and Steptoe, P. C., 1973, Current status of in vitro
    fertilization and implantation of human embryos, Lancet 2:1265-
    1270.
Edwards, R. G., Bavister, B. D., and Steptoe, P. C., 1969, Early
    stages of fertilization in vitro of human oocytes matured in
    vitro, Nature (London) 221:632-635.
Eppig, J. J. and Schroeder, A. C., 1986, Culture systems for
    mammalian oocyte development: progress and prospects,
    Theriogenology 25:97-106.
Eyestone, W. H., Northey, D. L., and Leibfried-Rutledge, M. L., 1985,
    Culture of 1-cell bovine embryos in the sheep oviduct, Biol.
    Reprod. 32(Suppl. 1):100 (abstract).
Fishel, S. B., Edwards, R. G., Purdy, J. M., Steptoe, P. C., Webster,
    J., Walters, E., Cohen, J., Fehilly, C., Hewitt, J., and
    Rowland, G., 1985, Implantation, abortion, and birth after in
    vitro fertilization using the natural menstrual cycle or fol-
    licular stimulation with clomiphene citrate and human menopausal
    gonadotropin, J. In Vitro Fert. Embryo Transfer 2:123-131.

Fukuda, A., Noda, Y., Tsukui, S., Matsumoto, H., Yano, J., and Mori, T., 1987, Influence of water quality on in vitro fertilization and embryo development for the mouse, J. In Vitro Fert. Embryo Transfer 4:40-45.

Gwatkin, R. B. L., 1977, Fertilization Mechanisms in Man and Mammals, Plenum Press, New York.

Hartmann, J. F., 1983, Mammalian fertilization: Gamete surface interactions in vitro, in: Mechanisms and Control of Animal Fertilization (J. F. Hartmann, ed.), Academic Press, New York, pp. 325-364.

Heyman, Y. and Ménézo, Y., 1987, Interaction of trophoblastic vesicles with bovine embryos developing in vitro, in: The Mammalian Preimplantation Embryo: Regulation of Growth and Differentiation In Vitro (B. D. Bavister, ed.), Plenum Press, New York, pp. 175-191.

Hoppe, R. W. and Bavister, B. D., 1984, Evaluation of the fluorescein diacetate (FDA) vital dye viability test with hamster and bovine embryos, Anim. Reprod. Sci. 6:323-335.

Kane, M. T., 1980, Variability in different lots of commercial bovine serum albumin affects cell multiplication and hatching of rabbit blastocysts in culture, J. Reprod. Fertil. 69:555-558.

Kane, M. T., 1987, In vitro growth of preimplantation rabbit embryos, in: The Mammalian Preimplantation Embryo: Regulation of Growth and Differentiation In Vitro (B. D. Bavister, ed.), Plenum Press, New York, pp. 193-217.

Kane, M. T. and Foote, R. H., 1970, Culture of two- and four-cell rabbit embryos to the expanding blastocyst stage in synthetic media, Proc. Soc. Exp. Biol. Med. 133:921-925.

Kuehl, T. J. and Dukelow, W. R., 1979, Maturation and in vitro fertilization of follicular oocytes of the squirrel monkey (Saimiri sciureus), Biol. Reprod. 21:545-556.

Leibfried, M. L. and Bavister, B. D., 1982, Effects of epinephrine and hypotaurine on in-vitro fertilization in the golden hamster, J. Reprod. Fertil. 66:87-93.

Leibfried, M. L. and Bavister, B. D., 1983, Fertilizability of in vitro matured oocytes from golden hamsters, J. Exp. Zool. 226:481-485.

Leibfried, M. L. and First, N. L., 1979, Characterization of bovine follicular oocytes and their ability to mature in vitro, J. Anim. Sci. 48:76-86.

Marston, J. H., Yanagimachi, R., Chang, M. C., and Hunt, D. M., 1964, The morphology of the first cleavage division in the mouse, Mongolian gerbil, golden hamster and rabbit, Anat. Rec. 148:417 (abstract).

Minhas, B. S., Capehart, J. S., Bower, M. J., Womack, J. E., McCrady, J. D., Harms, P. G., Wagner, T. E., and Kraemer, D. C., 1984, Visualization of pronuclei in living bovine oocytes, Biol. Reprod. 30:687-691.

Moor, R. M. and Crosby, I. M., 1985, Temperature-induced abnormali-
    ties in sheep oocytes during maturation, J. Reprod. Fertil.
    75:467–473.
Moor, R. M. and Trounson, A. O., 1977, Hormonal and follicular
    factors affecting maturation of sheep oocytes in vitro and their
    subsequent developmental capacity, J. Reprod. Fertil. 49:101–
    109.
Morgan, P. M., Boatman, D. E., and Kraus, E. M., 1986, Relationship
    between follicular fluid steroid hormone concentrations and in
    vitro development of rhesus monkey embryos, Biol. Reprod.
    34(Suppl. 1):132 (abstract).
Morgan, P. M., Hutz, R. J., Kraus, E. M., Cormie, J. A., Dierschke,
    D. J., and Bavister, B. D., 1987a, Evaluation of ultrasonography
    for monitoring follicular growth in rhesus monkeys,
    Theriogenology 27:769–781.
Morgan, P. M., Warikoo, P. K., Erwin, M. J., and Kraus, E. M., 1987b,
    Recovery of oocytes from spontaneously cycling rhesus monkeys:
    timing, follicular steroids and in vitro fertilization, Biol.
    Reprod. 36(Suppl. 1):90 (abstract).
Morgan, P. M., Warikoo, P. K., and Bavister, B. D., In vitro matura-
    tion and fertilization of rhesus monkey oocytes, in: The
    Primate Ovary (R. M. Brenner and C. H. Phoenix, series eds.;
    R. L. Stouffer, topic ed.), Plenum Press, New York, in press.
Pope, V. Z., Pope, C. E., and Beck, L. R., 1982, Gonadotropin produc-
    tion by the baboon embryo in vitro, in: In Vitro Fertilization
    and Embryo Transfer (E. S. E. Hafez and K. Semm, eds.), MTP
    Press, Lancaster, England, pp. 129–134.
Purdy, J. M., 1982, Methods for fertilization and embryo culture in
    vitro, in: Human Conception In Vitro (R. G. Edwards and J. M.
    Purdy, eds.), Academic Press, London, pp. 135–148.
Quinn, P., Warnes, G. M., Kerin, J. F., and Kirby, C., 1985, Culture
    factors affecting the success rate of in vitro fertilization and
    embryo transfer, Ann. N.Y. Acad. Sci. 442:195–204.
Rogers, B. J., 1978, Mammalian sperm capacitation and fertilization
    in vitro: a critique of methodology, Gamete Res. 1:165–223.
Schroeder, A. C. and Eppig, J. J., 1984, The developmental capacity
    of mouse oocytes that matured spontaneously in vitro is normal,
    Dev. Biol. 102:493–497.
Shalgi, R., 1984, Developmental capacity of rat embryos produced by
    in vivo or in vitro fertilization, Gamete Res. 10:77–82.
Silverman, I. H., Cook, C. L., Sanfilippo, J. S., Yussman, M. A.,
    Schultz, G. S., and Hilton, F. K., 1987, Ham's F-10 constituted
    with tap water supports mouse conceptus development in vitro, J.
    In Vitro Fert. Embryo Transfer 4:185–187.
Soupart, P. and Strong, P. A., 1974, Ultrastructural observations on
    human oocytes fertilized in vitro, Fertil. Steril. 25:11–44.
Spielmann, H., Eibs, H.-G., and Jacob-Miller, V., 1980, In vitro
    methods for the study of the effects of teratogens on preimplan-
    tation embryos, Acta Morphol. Acad. Sci. Hung. 28:105–115.

Staigmiller, R. B. and Moor, R. M., 1984, Effect of follicle cells on the maturation and developmental competence of ovine oocytes matured outside the follicle, Gamete Res. 9:221–229.

Steptoe, P. C. and Edwards, R. G., 1978, Birth after reimplantation of a human embryo, Lancet 2:366.

Thibault, C. and Gerard, M., 1970, Facteur cytoplasmic nécessaire a la formation du pronucleus male dans l'ovocyte de Lapine, C. R. Acad. Sci. (Paris) 270:2025–2026.

Thibault, C., Dauzier, L., and Wintenberger, S., 1954, La Fecondation in vitro de l'oeuf de la lapine, C. R. Soc. Biol. (Paris) 148:789.

Trounson, A., 1983, Factors controlling normal embryo development and implantation of human oocytes fertilized in vitro, in: Fertilization of the Human Egg In Vitro (H. M. Beier and H. R. Lindner, eds.), Springer-Verlag, Berlin, pp. 235–250.

Yanagimachi, R. and Chang, M. C., 1964, In vitro fertilization of golden hamster ova, J. Exp. Zool. 156:361–376.

13

DATA COLLECTION AND MANAGEMENT

Gregory S. Kopf

1.   INTRODUCTION

The volumes of data and reports generated through the normal
operation of in vitro fertilization and embryo transfer (IVF-ET)
programs has necessitated the development of computer-assisted modes
of information storage and retrieval. Collection, storage, retrieval
and statistical analyses of data are essential for the successful
operation of both the clinical and laboratory aspects of IVF-ET
programs, regardless of whether such programs are designed solely as
a clinical service or operate in conjunction with basic or clinical
research efforts. It has probably become apparent to any person
involved in IVF-ET that the variables contributing ultimately to
success or failure are incalculable, and that a day-to-day (or week-
to-week) assessment of pertinent laboratory and clinical parameters
is an important part of an overall quality control system in an IVF-
ET program. In addition, proper record keeping obviates the need for
long and drawn out data reduction required for prospective and
retrospective studies. The large number of records generated during
a patient's treatment cycle makes computer-assisted record and data
management a time-saving, cost-effective aid, and the current
availability and low cost of microcomputers makes such an approach
feasible for virtually any size IVF program.

The increased activity in IVF-ET over the past three to four
years has resulted in the design and marketing of specific software
programs. Examples of such prepackaged programs include "In Vitro
Clinic Plus" (Bailey and Associates, Tulsa, OK; $22,000) and IVF
Recordkeeper (Medical Research International, Boston, MA) for IVF and
"Infertility Clinic Plus" (Bailey and Associates, Tulsa, OK; $22,000)
for use in the male semen analysis work-up. Computer-based motion
analysis systems (CellSoft, Cryo Resources, Ltd., New York, NY;
CellTrack/S, Motion Analysis Corp., Santa Rosa, CA) have now been
designed specifically to handle the parameters of routine semen
analyses and are compatible with IBM PC/AT microcomputer systems.
Such systems will even prepare patient reports. However, their price
tags are considerable (in the $50,000 range) and are, therefore,
limiting to most programs. The advantages to such systems are that
the record and data analysis systems are already present as part of
the software package. However, a disadvantage is that they aren't
designed to meet the needs of a particular program and, thus, may not
grow with the demands that you might wish to place on it over the
years. In this regard, it is advantageous (and much cheaper) to have
a software program developed specifically to meet the particular
needs of the individual program. This could be accomplished in a
university setting by biostatisticians and computer programmers that
may be associated with specific academic departments. An alternative
(especially in the private sector) is to contract with computer
systems analysts and programmers; such personnel could be retained
for these purposes on a consultative basis. Still a third alternative is to use a prepackaged data base system (e.g., Data Star^{TM} and
Report Star^{TM}) for data collection and storage, and then create
individualized formats for the particular needs of the IVF program.
The latter would obviate the need for custom programming as specific
formatting could be done using the prepackaged software; this system
has been employed in the IVF-ET program at the University of Texas at
Houston (Quigley, 1984).

The degree of sophistication of data management systems will
reflect, in part, the particular needs of an individual program. For
example, the needs of a program in which only a limited number of
patients are being treated are probably going to be completely
different from a large program that has a significant clinical and
basic research component associated with it. At the Hospital of the
University of Pennsylvania, the current patient load is approximately
200 cases/year, and the computer software system that is employed has
been fashioned to incorporate patient data from patient histories,
clinical work-ups, nursing data, endocrine work-ups, results of
psychological testing, and andrological work-ups. Also included are
retrieval systems for specific laboratory parameters and quality
control procedures. Since the IVF program also serves as a central
source of biological materials for a variety of ongoing basic and
clinical research programs, the software system also incorporates an
indexing and retrieval system. The program, therefore, is specifi-

cally tailored for the needs of our program and is designed to grow
with the program.

2. DEVELOPMENT AND USE OF SOFTWARE

The IVF-ET program at the Hospital of the University of
Pennsylvania utilizes an IBM PC with a dBASE III^{TM} (Ashton and Tate,
Culver City, CA) software system that has been programmed to meet our
particular needs. The computer is located in a centralized area so
as to maximize access by the secretaries, researchers, physicians,
nursing staff, psychologists and laboratory personnel involved in the
program. Most microcomputers available on the market today (IBM,
Apple, Digital) are sufficiently sophisticated to handle the software
needs of most programs. The choice of the type of computer may,
therefore, be dictated to a great extent by types of computers
already present within the department (or facility).

As stated earlier, the development of a computer-assisted data
and record management system should be based on both the immediate
and long-term needs of a particular program. Since the purchase of a
microcomputer with its accessories, software packages, and
programming time can run into significant expense, it is important
that all personnel who will ultimately be using the computer be
involved in the decision-making processes regarding the structure and
design of the software package. These personnel (i.e., laboratory,
researchers, physicians, nurses, and secretaries) should have a clear
idea as to what they want the computer to do for them both presently
and in the future. If these aspects are not carefully considered, it
is certain that the software program ultimately developed will soon
be obsolete in terms of usefulness. It is absolutely essential,
therefore, that there be close interaction between all personnel who
intend to use the computer during the development of the system of
choice. Table I demonstrates the general outline of the computer
file systems used at the Hospital of the University of Pennsylvania.

Decisions concerning the structure of such programs should take
into account the number of people working with the computer and the
number of laboratories and/or disciplines involved. As shown in
Table I, the files which comprise our software program are divided so
that the personnel from different laboratories (disciplines) can
access a specific file that fulfills their needs. For example, all
testing of the male before a couple enters a cycle for egg retrieval
and in vitro insemination is carried out by a separate Andrology
Laboratory. All test results from this laboratory are put into the
Semen Analysis File, so it is unnecessary to access other files to
enter data. Since all of these files can be cross-accessed, data
parameters present in any of the individual files can be compared
with information in all of the other files. For example, information
in the clinical files could be accessed and compared to data listed

Table I. Outline of computer file systems used in the human in vitro
          fertilization and embryo transfer program at the Hospital
          of the University of Pennsylvania

File title	File content
1. General Clinical File 2. Initial Studies File	Patient interviews and histories; screening and testing of selected patients
3. Semen Analysis File	Screening and testing of the male
4. Treatment Files 1,2 5. Ultrasound File 6. IVF Laboratory Information File	Enter program and begin treatment; oocyte/egg retrieval and in vitro fertilization
7. Retrieval File	Oocyte/egg retrieval
8. Oocyte/Egg File	Oocyte/egg morphology; insemination information; embryo development
9. Embryo Transfer File	Embryo transfer
10. Stock Media File	General laboratory information

in the laboratory files; such flexibility would enable the user to
determine whether a particular laboratory observation (i.e., oocyte
morphology at the time of retrieval) could be related to the hormonal
profile of the patient at the time of hCG administration (i.e.,
estradiol:progesterone ratio). In addition, such files can be
interfaced with particular statistical and graphical software
packages that will enable the user to perform functions without
transposing the data to a different form. This is an important
consideration for the preparation of scientific manuscripts, as well
as for routine data analysis.

     An additional consideration in the development of a software
program is to keep the data entry format simple. This is important
for two reasons. First, the simpler the entry format the less the
chance that mistakes are made during data entry on the original forms
as well as when these forms are transcribed for computer entry.
Second, utilization of available disk space can be kept at a minimum

by avoiding, as much as possible, the entry of narrative descriptions. To allow for the use of such narrative descriptions in our
software program, each file has an area in which key words summa-
rizing a narrative can be entered. More detailed descriptions can be
listed on additional forms which are not entered into the computer.

Once a program format has been decided upon and constructed, it
is very common to find that you have overlooked some important
parameters (or deficiencies in software design) that will have to be
incorporated into the program to make it more complete and efficient.
For this reason, it is important to start using the system as soon as
possible so that these potential problems can be identified and
corrected. Designing the program(s) on paper is one thing but the
efficiency with which these programs are implemented on a daily basis
is a totally different matter, and using the program is the only way
in which these design flaws can be recognized.

The files that are used in the IVF program at the Hospital of
the University of Pennsylvania are shown in Figures 1-10. These are
intended to give the reader a general idea of the type of information
that might be considered when setting up software programs. The
reader is also advised to examine another publication dealing with
this particular subject (Quigley, 1984).

3.   REFERENCES

Quigley, M. M., 1984, Data management in an in vitro fertilization
        and embryo transfer program, in: Human In Vitro Fertilization
        and Embryo Transfer (D. P. Wolf and M. M. Quigley, eds.), Plenum
        Press, New York, pp. 383-401.

Figs. 1-10 follow on pp. 240-264.

GENERAL CLINICAL FILE New ☐ Changes ☐   PAGE 1 OF 4

SOCIAL SECURITY NUMBERS: WIFE ___ __ __ __ ___   HUSBAND ___ __ __ __ ___
PATIENT NAME: LAST   _____
                 FIRST & INITIALS _____
PATIENT DATE OF BIRTH: __ / __ / __
HUSBAND'S NAME: LAST   _____
                 FIRST & INITIALS _____
HUSBAND'S DATE OF BIRTH: __ / __ / __
ADDRESS: STREET   _____
                 CITY, STATE, ZIP _____
PHONE NUMBER: HOME   ___ - ___ - ____
                 WORK   ___ - ___ - ____
REFERRED BY: NAME   _____
                 ADDRESS: STREET   _____
                          CITY, STATE, ZIP _____
                 PHONE   ___ - ___ - ____
DATE OF IVF CONSULTATION: __ / __ / __

RACE: WIFE: White ☐1 Black ☐2 Asian ☐3 American Indian ☐4 Hispanic ☐5 Other ☐6
      HUSBAND: White ☐1 Black ☐2 Asian ☐3 American Indian ☐4 Hispanic ☐5 Other ☐6

MARRIED: Yes ☐1   No ☐2

FEMALE PREVIOUSLY FERTILE: Yes ☐1   No ☐2

OBSTETRICAL DATA
GRAVIDA: __
PARA: __
ABORTIONS: INDUCED   __   SPONTANEOUS   __
ECTOPIC PREGNANCIES: __

MEDICAL HISTORY

I.V.F. ELSEWHERE (Most Recent Results):   Yes ☐1 No ☐2
DONE AT: _____
DATE: __ / __ / __
NUMBER OF CYCLES: __
NUMBER OF EGGS: __
NUMBER OF FERTILIZED EGGS: __
NUMBER OF EMBRYOS TRANSFERRED: __
METHOD OF INDUCTION: Spontaneous ☐1 Clomid ☐2 Pergonal ☐3
          Combination(Clom. & Perg.) ☐4 GNRH ☐5   FSH ☐6
PROGESTERONE THERAPY: No ☐1 IM Progesterone ☐2 IM Delaluton ☐3 Suppositories ☐4

GENERAL	YES	NO	DON'T KNOW
CONGENITAL ANOMOLY OF REPRODUCTIVE SYSTEM	☐1	☐2	☐3
ENDOMETRIOSIS	☐1	☐2	☐3
PELVIC INFLAMMATORY DISEASE	☐1	☐2	☐3
PREVIOUS TUBOPLASTY	☐1	☐2	☐3
PREVIOUS UNILATERAL SALPINGECTOMY	☐1	☐2	☐3

University of Pennsylvania   Patient Name _____

Fig. 1(A).   Computer system for IVF-ET; details of general
             clinical file.

GENERAL CLINICAL FILE New ☐ Changes ☐   PAGE 2 OF 4

MEDICAL HISTORY
GENERAL cont'd   YES   NO   DON'T KNOW
PREVIOUS BILATERAL SALPINGECTOMY   ☐1   ☐2   ☐3
OTHER PREVIOUS PELVIC SURGERY   ☐1   ☐2   ☐3
PREVIOUS UNILATERAL OOPHORECTOMY   ☐1   ☐2   ☐3
PERITUBAL ADHESIONS   ☐1   ☐2   ☐3
HISTORY OF GENETIC DISORDER: Husband   ☐1   ☐2   ☐3
                              Wife   ☐1   ☐2   ☐3
FERTILITY DATA
HISTORY OF OVULATORY DISORDERS   ☐1   ☐2   ☐3
IF YES GIVE DIAGNOSIS: PCO ☐1 Hypothalamic ☐2 Pituitary Tumor ☐3 Endocrine ☐4
              SLP ☐5 Don't Know ☐6
OVULATION CONFIRMED BY :
        BBT ☐1 Endometrial Biopsy ☐2 Progesterone ☐3 Don"t Know ☐4
ETIOLOGY OF INFERTILITY :   PRIMARY   SECONDARY   TERTIARY
                    Tubal   ☐1   ☐1   ☐1
                    Oligospermic   ☐2   ☐2   ☐2
                    Endometriosis   ☐3   ☐3   ☐3
                    Cervical   ☐4   ☐4   ☐4
                    Immunologic   ☐5   ☐5   ☐5
                    Other - Male   ☐6   ☐6   ☐6
                    Other - Female   ☐7   ☐7   ☐7
                    Unexplained   ☐8   ☐8   ☐8
IF ENDOMETRIOSIS (CODE 3):
                    Mild   ☐1
                    Moderate   ☐2
                    Severe   ☐3
                    Extensive   ☐4
TYPE OF INFERTILITY: Primary ☐1
Secondary following: Induced Abortion ☐2 Spontaneous Abortion ☐3 Unilateral Ectopic ☐4
              Bilateral Ectopic ☐5 Term Delivery ☐6 Tubal Ligation ☐7 Other ☐8
LATEST LAPAROSCOPY: Done: At HUP ☐1 Elsewhere ☐2   DATE: _ _ / _ _ / _ _
ACCESS TO   LEFT OVARY   RIGHT OVARY
Excellent (100% of ovary- no adhesions) ☐1   ☐1
Good (> 50 % of ovary)   ☐2   ☐2
Fair (20-50% of ovary)   ☐3   ☐3
Poor (<20 % of ovary)   ☐4   ☐4
Absent   ☐5   ☐6

University of Pennsylvania   Patient Name _____

Fig. 1(B). Computer system for IVF-ET; details of general
            clinical file.

GENERAL CLINICAL FILE New ☐ Changes ☐   PAGE 3 OF 4

LATEST LAPAROSCOPY cont'd
  TUBES: LEFT   RIGHT
    Absent   ☐1   ☐1
    Normal   ☐2   ☐2
    Hydrosalphinx   ☐3   ☐3
    Phimosis   ☐4   ☐4
    Peritubal Adhesions   ☐5   ☐5
    Endometriosis   ☐6   ☐6
  OVARIES:
    Absent   ☐1   ☐1
    Normal   ☐2   ☐2
    Surgically Reduced   ☐3   ☐3
    Endometriosis   ☐4   ☐4
  DIFFICULTY OF LAPAROSCOPY: Mild ☐1 Moderate ☐2 Severe ☐3 Dangerous ☐4

MENSTRUAL HISTORY:
    NORMAL AGE OF MENARCHE (10-17): Yes ☐1 No ☐2 Don't Know ☐3
    CYCLE: Regular ☐1 Irregular ☐2 Regulated with Medication ☐3 Don't Know ☐4
    HYSTEROSALPINGOGRAM: NORMAL UTERUS Yes ☐1 No ☐2 Don't Know ☐
    ENDOMETRIAL BIOPSY SECRETORY: Inphase ☐1 Outphase ☐2

CONTRACEPTIVE HISTORY:
    EVER USED IUD: Yes ☐1 No ☐2 Don't Know ☐3
    EVER USED OC: Yes ☐1 No ☐2 Don't Know ☐3

PHYSICAL EXAMINATION:
    HEIGHT: < 5' (≤152 cm) ☐1 5'-5'7' (153-170 cm) ☐2 5'7"-6' (171-183 cm) ☐3
      > 6' (≥184 cm) ☐4
    WEIGHT: <100 lb (≤45kg) ☐1 100-125 lb (46-56kg) ☐2
    126-150 lb (57-68kg) ☐3 151-176 lb (69-79kg) ☐4 >176 lb (≥80kg) ☐5

	YES	NO	DON'T KNOW	ABS
PELVIC EXAMINATION:				
NORMAL CERVIX:	☐1	☐2	☐3	
NORMAL VAGINA:	☐1	☐2	☐3	
NORMAL SIZE FUNDUS:	☐1	☐2	☐3	☐4
NORMAL RIGHT ADNEXA:	☐1	☐2	☐3	☐4
NORMAL LEFT ADNEXA:	☐1	☐2	☐3	
FREE CUL-DE-SAC:	☐1	☐2	☐3	

POSITION OF UTERUS: Antiflexed ☐1 Retroflexed ☐2 Midline ☐3

University of Pennsylvania   Patient Name _____

Fig. 1(C). Computer system for IVF-ET; details of general
        clinical file.

GENERAL CLINICAL FILE New ☐ Changes ☐   PAGE 4 OF 4

MALE FERTILITY DATA:

	YES	NO
PREVIOUSLY FERTILE:	☐1	☐2

MEDICAL HISTORY:

	YES	NO
VENEREAL DISEASE	☐1	☐2
ALCOHOL ABUSE	☐1	☐2
TOBACCO USE	☐1	☐2
DRUG ABUSE	☐1	☐2
HYPERTENSION	☐1	☐2
DIABETIC	☐1	☐2

OTHER _

SURGERY:

VARICOCELE	☐1	☐2	If Yes DATE _ _ / _ _ / _ _
UNDESCENDED TESTES	☐1	☐2	If Yes DATE _ _ / _ _ / _ _
CONGENITAL ANOMALY OF REP. SYSTEM	☐1	☐2	

If Yes DESCRIPTION _

OTHER _

PREVIOUS DRUG TREATMENT   ☐1   ☐2
If Yes DESCRIPTION _

PREVIOUS TESTS   ☐1   ☐2
If Yes DESCRIPTION _

PREVIOUS SEMEN ANALYSIS: DATE   _ _ / _ _ / _ _
                          COUNT (mill/ml) _ _ _ . _
                          % MOTILITY _ _

University of Pennsylvania   Patient Name _ _ _ _ _ _ _ _

Fig. 1(D). Computer system for IVF-ET; details of general
            clinical file.

INITIAL STUDIES FILE New ☐1 Changes ☐2   PAGE 1 OF 2

SOCIAL SECURITY NUMBER _ _ _ _ _ _ _ _ _ _
CYCLE ATTEMPT   _ _

SCREENING LAPAROSCOPY: Yes ☐1 No ☐2
  DATE: _ _ / _ _ / _ _
  PHYSICAN'S INITIALS: _ _ _
    ACCESS TO: LEFT OVARY   RIGHT OVARY
      Excellent (100% of ovary- no adhesions)   ☐1   ☐1
      Good (> 50 % of ovary)   ☐2   ☐2
      Fair (20-50% of ovary)   ☐3   ☐3
      Poor (<20 % of ovary)   ☐4   ☐4
    TUBES:
      Absent   ☐1   ☐1
      Normal   ☐2   ☐2
      Hydrosalphinx   ☐3   ☐3
      Phimosis   ☐4   ☐4
      Peritubal Adhesions   ☐5   ☐5
      Endometriosis   ☐6   ☐6
    OVARIES:
      Absent   ☐1   ☐1
      Normal   ☐2   ☐2
      Surgically Reduced   ☐3   ☐3
      Endometriosis   ☐4   ☐4
    DIFFICULTY OF LAPAROSCOPY: Mild ☐1 Moderate ☐2 Severe ☐3 Dangerous ☐4

PRELIMINARY LABORATORY DATA

FEMALE ANTIBODIES: DATE: _ _ / _ _ / _ _
  INDIRECT ANTIBODY RESULT - FEMALE Positive ☐1 Negative ☐2 Borderline ☐3
                                     Retest ☐4 Not Done ☐5
                        RETEST   Positive ☐1 Negative ☐2 Borderline ☐3
  BLOOD TESTS:
  RUBELLA : DATE: _ _ / _ _ / _ _   Positive ☐1 Negative ☐2
  IF NEGATIVE, VACCINATED: Yes ☐1 No ☐2
     IF YES: DATE: _ _ / _ _ / _ _
  HEPATITIS: DATE: _ _ / _ _ / _ _   Positive ☐1 Negative ☐2
  CBC :   DATE: _ _ / _ _ / _ _   Normal ☐1 Abnormal ☐2

University of Pennsylvania   Patient Name _____

Fig. 2(A).   Computer system for IVF-ET; details of initial studies
              file.

INITIAL STUDIES FILE New ☐ Changes ☐   PAGE 2 OF 2

URINALYSIS: DATE: __/__/__   Normal ☐1 Abnormal ☐2 Not Done ☐3
URINE CULTURE: DATE: __/__/__   Normal ☐1 Abnormal ☐2 Not Done ☐3
PAP SMEAR: DATE: __/__/__   Normal ☐1 Abnormal ☐2 Not Done ☐3

SIX MONTH PRIOR LMP:
  1. DATE STARTED: __/__/__
  2. DATE STARTED: __/__/__
  3. DATE STARTED: __/__/__
  4. DATE STARTED: __/__/__
  5. DATE STARTED: __/__/__
  6. DATE STARTED: __/__/__

University of Pennsylvania   Patient Name _____

Fig. 2(B).   Computer system for IVF-ET; details of initial studies
              file.

SEMEN ANALYSIS FILE   PAGE 1 OF 1

WIFE'S SOCIAL SECURITY NUMBER   --- -- ----
CYCLE ATTEMPT   --
HUSBAND'S SOCIAL SECURITY NO.   --- -- ----

DATE OF SEMEN ANALYSIS   --/--/--
SAMPLE TYPE: Fresh ☐1 Frozen ☐2

TEST DONE: HUP ☐1 Other ☐2

DAYS ABSTINANCE: -- (99 if Don't Know)
TIME COLLECTED: ----
TIME ANALYZED: ----
TIME FROZEN: DATE   --/--/--
              HOUR   ----
VOLUME (ml)   --
VISCOSITY: Thin ☐1 Somewhat Thick ☐1 Thick ☐3 Very Thick ☐4
pH   --.-
AGGLUTINATION: Some ☐1 Most ☐2 All ☐3
SPERM COUNT (mill/ml):   ----.-
PARTICLES (mill/ml):   ----.-
% MOTILITY: --
PROGRESSION (± 1 - 4):   --
VIABILITY: --/--/--
MORPHOLOGY: --/--/--/--/--
NUMBER OF WBC HPF   --
NUMBER OF RBC HPF   --
NUMBER OF EPITHELIAL CELLS HPF   --
TURBIOMETRIC SPERM VEL. (µm/sec):   ---
        DATE   --/--/--
MEP SPERM VEL. (µm/sec):   ----
HAMSTER TEST: YES ☐1 NO ☐2
IF YES: % PENETRATION - CONTROL ---
                       PATIENT   ---
                       DATE   --/--/--
REPEAT HAMSTER: YES ☐1 NO ☐2
IF YES: % PENETRATION - CONTROL ---
                       PATIENT   ---
                       DATE   --/--/--

ANTIBODIES:
   TEST Not Done ☐1 Direct ☐2 Indirect ☐3 Both ☐4
      DIRECT ANTIBODY RESULTS   Positive ☐1 Negative ☐2 Borderline ☐3 Retest ☐4
                       RETEST   Positive ☐1 Negative ☐2 Borderline ☐3
      INDIRECT ANTIBODY RESULTS Positive ☐1 Negative ☐2 Borderline ☐3 Retest ☐4
                       RETEST   Positive ☐1 Negative ☐2 Borderline ☐3

University of Pennsylvania   Patient Name _____

Fig. 3.   Computer system for IVF-ET; details of semen analysis file.

TREATMENT FILE - 1   PAGE 1 OF 2

SOCIAL SECURITY NUMBER _ _ _ _ _ _ _ _ _
CYCLE ATTEMPT   _ _

PELVIC EXAMINATION

PHYSICAN'S INITIALS   _ _

                        YES   NO   DON'T KNOW
Normal Cervix   □1   □2   □3
Normal Pos. Fundus   □1   □2   □3
Nor. Rt. Adnexa   □1   □2   □3
Nor. Left Adnexa   □1   □2   □3
Free Cul-de-sac   □1   □2   □3
Position of Uterus: Anteflex □1 Retroflex □2 Midline □3

INDUCTION PROTOCOL: HMG □1   Clomid □2   Clomid/HMG □3
                     FSH □4   GRNH □5

METHOD OF RETRIEVAL: Laparoscopy □1 Laparotomy □2 Ultrasound □3
PHYSICAN'S INITIALS   _ _
DATE   _ _ / _ _ / _ _
TIME   _ _ _ _
DEGREE OF DIFFICULTY OF RETRIEVAL: Mild □1 Moderate □2 Severe □3 Dangerous □4

BLEEDING: Mild □1 Moderate □2 Severe □3 Dangerous □4
COMPLICATIONS: (comment) _
LENGTH OF ANAESTHESIA: 30-60 min □1 60-90 min □2 90-120 min □3
                        >120 min □4   N/A □5

SEDATION: Analgesia □1 Anaesthesia □2

LAPAROTOMY WITH OVUM RETRIEVAL: Lysis of Adhesions □1   Tuboplasty □2
                                 Endometriosis   □3   Other   □4

ULTRASOUND TECHNIQUE: Abdominal □1 Transurethral □2 Transvaginal □3
PHYSICAN'S INITIALS   _ _
DATE   _ _ / _ _ / _ _

POST TRANSFER BLEEDING: Spotting □1 Mild □2 Moderate □3 Severe □4 None □5
  CYCLE DAY STARTED   _ _
  CYCLE DAY STOPPED   _ _

POST TRANSFER CRAMPING: Mild □1 Moderate □2 Severe □3 None □4
  CYCLE DAY STARTED   _ _
  CYCLE DAY STOPPED   _ _

University of Pennsylvania   Patient Name _ _ _ _ _ _ _ _ _ _

Fig. 4(A).   Computer system for IVF-ET; details of treatment files
              1 and 2.

TREATMENT FILE - 1

PAGE 2 OF 2

RIA HCG LEVELS

#1) POSITIVE ☐1 EQUIVOCAL ☐2 NEGATIVE ☐3
   DATE __ /__ /__
   LEVEL _____ mIU/ml
#2) POSITIVE ☐1 EQUIVOCAL ☐2 NEGATIVE ☐3
   DATE __ /__ /__
   LEVEL _____ mIU/ml
#3) POSITIVE ☐1 EQUIVOCAL ☐2 NEGATIVE ☐3
   DATE __ /__ /__
   LEVEL _____ mIU/ml
#4) POSITIVE ☐1 EQUIVOCAL ☐2 NEGATIVE ☐3
   DATE __ /__ /__
   LEVEL _____ mIU/ml
#5) POSITIVE ☐1 EQUIVOCAL ☐2 NEGATIVE ☐3
   DATE __ /__ /__
   LEVEL _____ mIU/ml
#6) POSITIVE ☐1 EQUIVOCAL ☐2 NEGATIVE ☐3
   DATE __ /__ /__
   LEVEL _____ mIU/ml

PREGNANCY: Yes ☐1   No ☐2
IF YES, OUTCOME: Term ☐1 SAB ☐2 Ectopic ☐3 Stillbirth ☐4
                    Abnormality ☐5 Multiple ☐6 Premature ☐7
     TYPE OF DELIVERY: Vaginal ☐1 C-Section ☐2

IF MULTIPLE, GIVE NUMBER: _
SEX: Number of Males _   Number of Females _
DATE OF BIRTH: __ /__ /__

PROGESTERONE THERAPY: DATE STARTED __ /__ /__   DATE STOPPED __ /__ /__

   CYCLE DAY __ TO CYCLE DAY __
   DATE __ /__ /__
   DOSE ___ mg
   IM ☐1 SUPPOSITORY ☐2

   CYCLE DAY __ TO CYCLE DAY __
   DATE __ /__ /__
   DOSE ___ mg
   IM ☐1 SUPPOSITORY ☐2

   CYCLE DAY __ TO CYCLE DAY __
   DATE __ /__ /__
   DOSE ___ mg
   IM ☐1 SUPPOSITORY ☐2

University of Pennsylvania   Patient Name _____

Fig. 4(B).   Computer system for IVF-ET; details of treatment files
              1 and 2.

Fig. 4(C). Computer system for IVF-ET; details of treatment files
1 and 2.

TREATMENT FILE - 2   PAGE 2 OF 2

CYCLE DAY   --   CYCLE DAY   --

DATE   --/--/--   DATE   --/--/--
DRUG USED 1: Clomid □1   HMG □2   DRUG USED 1: Clomid □1   HMG □2
             FSH □3   HCG □4   FSH □3   HCG □4
             GNRH □5   GNRH □5
DOSE (units/day)   -----   DOSE (units/day)   -----

DRUG USED 2: Clomid □1   HMG □2   DRUG USED 2: Clomid □1   HMG □2
             FSH □3   HCG □4   FSH □3   HCG □4
             GNRH □5   GNRH □5
DOSE (units/day)   -----   DOSE (units/day)   -----

IF HCG (code 4) TIME GIVEN ----   IF HCG (code 4) TIME GIVEN ----
ESTRADIOL LEVEL (pg/ml)   ----   ESTRADIOL LEVEL (pg/ml)   ----
LH LEVEL (mIU/ml)   ---   LH LEVEL (mIU/ml)   ---
PROGESTERONE (ng/dl)   ---   PROGESTERONE (ng/dl)   ---
PROLACTIN (ng/ml)   ---   PROLACTIN (ng/ml)   ---
PHYSICAN'S INITIALS   ---   PHYSICAN'S INITIALS   ---

SONOGRAM   Yes □1   No □2   SONOGRAM   Yes □1   No □2

CYCLE DAY   --   CYCLE DAY   --

DATE   --/--/--   DATE   --/--/--
DRUG USED 1: Clomid □1   HMG □2   DRUG USED 1: Clomid □1   HMG □2
             FSH □3   HCG □4   FSH □3   HCG □4
             GNRH □5   GNRH □5
DOSE (units/day)   -----   DOSE (units/day)   -----

DRUG USED 2: Clomid □1   HMG □2   DRUG USED 2: Clomid □1   HMG □2
             FSH □3   HCG □4   FSH □3   HCG □4
             GNRH □5   GNRH □5
DOSE (units/day)   -----   DOSE (units/day)   -----

IF HCG (code 4) TIME GIVEN ----   IF HCG (code 4) TIME GIVEN ----
ESTRADIOL LEVEL (pg/ml)   ----   ESTRADIOL LEVEL (pg/ml)   ----
LH LEVEL (mIU/ml)   ---   LH LEVEL (mIU/ml)   ---
PROGESTERONE (ng/dl)   ---   PROGESTERONE (ng/dl)   ---
PROLACTIN (ng/ml)   ---   PROLACTIN (ng/ml)   ---
PHYSICAN'S INITIALS   ---   PHYSICAN'S INITIALS   ---

SONOGRAM   Yes □1   No □2   SONOGRAM   Yes □1   No □2

University of Pennsylvania   Patient Name _____

Fig. 4(D).   Computer system for IVF-ET; details of treatment files
              1 and 2.

ULTRASOUND FILE   PAGE 1 OF 2

SOCIAL SECURITY NUMBER   ___ ___ ___ ___ ___'___
CYCLE ATTEMPT   ___ ___

CYCLE DAY (Fill in 99 if not known)   ___ ___
DATE   ___ ___/___ ___/___ ___

                                        LEFT OVARY   RIGHT OVARY
FOLLICLES MEASURABLE: Yes □1 No □2   Yes □1 No □2
IF YES, FOLLICLE SIZE (cm)   #1   ___ · ___   ___ · ___
                                #2   ___ · ___   ___ · ___
                                #3   ___ · ___   ___ · ___
                                #4   ___ · ___   ___ · ___
                                #5   ___ · ___   ___ · ___

              FLUID IN CUL - DE - SAC   Yes □1 No □2

CYCLE DAY (Fill in 99 if not known)   ___ ___
DATE   ___ ___/___ ___/___ ___

                                        LEFT OVARY   RIGHT OVARY
FOLLICLES MEASURABLE: Yes □1 No □2   Yes □1 No □2
IF YES, FOLLICLE SIZE (cm)   #1   ___ · ___   ___ · ___
                                #2   ___ · ___   ___ · ___
                                #3   ___ · ___   ___ · ___
                                #4   ___ · ___   ___ · ___
                                #5   ___ · ___   ___ · ___

              FLUID IN CUL - DE - SAC   Yes □1 No □2

CYCLE DAY (Fill in 99 if not known)   ___ ___
DATE   ___ ___/___ ___/___ ___

                                        LEFT OVARY   RIGHT OVARY
FOLLICLES MEASURABLE: Yes □1 No □2   Yes □1 No □2
IF YES, FOLLICLE SIZE (cm)   #1   ___ · ___   ___ · ___
                                #2   ___ · ___   ___ · ___
                                #3   ___ · ___   ___ · ___
                                #4   ___ · ___   ___ · ___
                                #5   ___ · ___   ___ · ___

              FLUID IN CUL - DE - SAC   Yes □1 No □2

University of Pennsylvania   Patient Name_____

Fig. 5(A). Computer system for IVF-ET; details of ultrasound
            file.

ULTRASOUND FILE   PAGE 2 OF 2

___

CYCLE DAY (Fill in 99 if not known)   __
DATE   __/__/__

                                    LEFT OVARY   RIGHT OVARY
FOLLICLES MEASURABLE: Yes ☐1 No ☐2   Yes ☐1 No ☐2
IF YES, FOLLICLE SIZE (cm)   #1   __.__   __.__
                              #2   __.__   __.__
                              #3   __.__   __.__
                              #4   __.__   __.__
                              #5   __.__   __.__

            FLUID IN CUL-DE-SAC   Yes ☐1 No ☐2

___

CYCLE DAY (Fill in 99 if not known)   __
DATE   __/__/__

                                    LEFT OVARY   RIGHT OVARY
FOLLICLES MEASURABLE: Yes ☐1 No ☐2   Yes ☐1 No ☐2
IF YES, FOLLICLE SIZE (cm)   #1   __.__   __.__
                              #2   __.__   __.__
                              #3   __.__   __.__
                              #4   __.__   __.__
                              #5   __.__   __.__

            FLUID IN CUL-DE-SAC   Yes ☐1 No ☐2

___

University of Pennsylvania   Patient Name_____

Fig. 5(B). Computer system for IVF-ET; details of ultrasound
           file.

LABORATORY INFORMATION FILE   PAGE 1 OF 2

Social Security Number: _ _ _ _ _ _ _ _ _
Patient Number: _ _ _ _ _
Patient Name- Last Name, Initial: _

SERUM
Source: Patient □1   Donor □2   Pt/Donor □3   Other □4

PATIENT: DONOR: Number: _ _ _
Date Blood Drawn: _ _/_ _/_ _   Date Blood Drawn: _ _/_ _/_ _
Cycle Day: _ _   Cycle Day: _ _

Hormone Profile:
    Not Done □0 E_2 □1   Progesterone □2   HCG □3   Testosterone □4   Other □5

GAS PHASE:
    (5% CO_2, 95% Air) □1   (5% CO_2, 5% O_2, 90% Air) □2   Other □3
MEDIA:
    Media # (Date + Batch Code)   _ _ /_ _/_ _ _
    IVF Media Prep Date   _ _/_ _/_ _

SPERM PREPARATION:
    Source: Husband □1   Donor □2
    Donor Code   _ _ _ _

INSEMINATION:
    Fresh □1   Frozen □2
    Semen Collection Time: _ _ _ _
    Date of Last Intercourse / Ejaculation   _ _/_ _ _/_ _

Viscosity: Normal □1 Abnormal □2

Bacterial Content: Normal □1 Abnormal □2

Debris: Normal □1 Abnormal □2

Comments: Yes □1   No □2

PRE: POST:
  Vol (ml)   _ _ _ _
  Motility %   _ _ _   Motility %.   _ _ _
  Progression ±(1-4)   _ _   Progression ±(1-4)   _ _
  Conc (x10^6 ml)   _ _ _ . _ _   Conc (x10^6 ml)   _ _ _ . _ _
  Semen Vol Washed (ml)   _ _ . _   Dilution: Yes □1 No □2
                                          Sperm Vol : Media Vol (µl) _ _ _ : _ _ _

  Washing Procedure: Normal □1 Gradient Wash □3
                            Split □2   Other □4
  Overlay Vol (ml)   _ _ . _ _

University of Pennsylvania   Patient Name _ _ _ _ _ _ _ _ _ _

Fig. 6(A).   Computer system for IVF-ET; details of laboratory
              information file.

LABORATORY INFORMATION FILE   PAGE 2 OF 2

ADDITIONAL INSEMINATION
Fresh ☐1   Frozen ☐2   From Previous Day ☐3
Semen Collection Time: _ _ _ _
Date of Last Intercourse / Ejaculation _ _ / _ _ / _ _
Viscosity: Normal ☐1 Abnormal ☐2
Bacterial Content: Normal ☐1 Abnormal ☐2
Debris: Normal ☐1 Abnormal ☐2
Comments: Yes ☐1   No ☐2

PRE:
Vol (ml)   _ _ . _
Motility %   _ _ _
Progression +(1-4)   _ _
Conc (x10^6 ml)   _ _ _ _ . _ _
Semen Vol Washed (ml)   _ _ . _ ; _ _ . _ ; _ _ . _
Washing Procedure: Normal ☐1 Gradient Wash ☐3
                            Split ☐2   Other ☐4
Overlay Vol (ml)   _ . _ _ _
If Split: Used 1 Overlay ☐1 More than 1 Overlay ☐2

POST:
Motility %   _ _ _
Progression +(1-4)   _ _
Conc (x10^6 ml)   _ _ _ _ . _ _
Dilution: Yes ☐1 No ☐2
Sperm Vol : Media Vol (µl)   _ _ _ : _ _ _

University of Pennsylvania   Patient Name _____

Fig. 6(B). Computer system for IVF-ET; details of laboratory
            information file.

RETRIEVAL FILE   PAGE 1 OF 3

Social Security #   _ _ _ _ _ _ _ _ _
Patient # - - - -   _ _ _ _
Date - - - -   _ _ / _ _ / _ _

Foll. # & Ovary(L or R)   _ _ _   Foll. # & Ovary(L or R)   _ _ _
Time   _ _ _ _   Time   _ _ _ _
Volume(ml)   _ _ _   Volume (ml)   _ _ _
Oocyte Number   _ _   Oocyte Number   _ _
Follicle Size (cm)   _ _ _   Follicle Size (cm)   _ _ _

Fluid: Clear ☐1 Slightly Bloody ☐2   Fluid: Clear ☐1 Slightly Bloody ☐2
        Bloody ☐3 Clotted   ☐4   Bloody ☐3 Clotted   ☐4

Location of Eggs: Fluid ☐1   Flush ☐2   Location of Eggs : Fluid ☐1   Flush ☐2
              Fluid/Flush ☐3 Other ☐4   Fluid/Flush ☐3 Other ☐4

Foll. # & Ovary(L or R)   _ _ _   Foll. # & Ovary(L or R)   _ _ _
Time   _ _ _ _   Time   _ _ _ _
Volume(ml)   _ _ _   Volume (ml)   _ _ _
Oocyte Number   _ _   Oocyte Number   _ _
Follicle Size (cm)   _ _ _   Follicle Size (cm)   _ _

Fluid: Clear ☐1 Slightly Bloody ☐2   Fluid: Clear ☐1 Slightly Bloody ☐2
        Bloody ☐3 Clotted   ☐4   Bloody ☐3 Clotted   ☐4

Location of Eggs: Fluid ☐1   Flush ☐2   Location of Eggs : Fluid ☐1   Flush ☐2
              Fluid/Flush ☐3 Other ☐4   Fluid/Flush ☐3 Other ☐4

Foll. # & Ovary(L or R)   _ _ _   Foll. # & Ovary(L or R)   _ _ _
Time   _ _ _ _   Time   _ _ _ _
Volume(ml)   _ _ _   Volume (ml)   _ _ _
Oocyte Number   _ _   Oocyte Number   _ _
Follicle Size (cm)   _ _   Follicle Size (cm)   _ _ _

Fluid: Clear ☐1 Slightly Bloody ☐2   Fluid: Clear ☐1 Slightly Bloody ☐2
        Bloody ☐3 Clotted   ☐4   Bloody ☐3 Clotted   ☐4

Location of Eggs: Fluid ☐1   Flush ☐2   Location of Eggs : Fluid ☐1   Flush ☐2
              Fluid/Flush ☐3 Other ☐4   Fluid/Flush ☐3 Other ☐4

University of Pennsylvania   Patient Name _____

Fig. 7(A). Computer system for IVF-ET; details of retrieval file.

Fig. 7(B). Computer system for IVF-ET; details of retrieval file.

RETRIEVAL FILE   PAGE 3 OF 3

Pre-op Cul-de-sac Fluid:
Time   _ _ _ _
Volume   _ _ _ . _
Fluid: Clear ☐1   Slightly Bloody ☐2
        Bloody ☐3 Clotted   ☐4
Egg *   _ _ , _ _ , _ _

Intra-operative Cul-de-sac:
Time   _ _ _ _
Volume   _ _ _ . _
Fluid: Clear ☐1   Slightly Bloody ☐2
        Bloody ☐3 Clotted   ☐4
Egg *   _ _ , _ _ , _ _

Post-operative Cul-de-sac:
Time   _ _ _ _
Volume   _ _ _ . _
Fluid: Clear ☐1   Slightly Bloody ☐2
        Bloody ☐3 Clotted   ☐4
Egg *   _ _ , _ _ , _ _

**Other Aspirates:

Time   _ _ _ _   Time   _ _ _ _
Volume   _ _ _ . _   Volume   _ _ _ . _
Fluid: Clear ☐1 Slightly Bloody ☐2 Fluid: Clear ☐1 Slightly Bloody ☐2
       Bloody ☐3 Clotted   ☐4   Bloody ☐3 Clotted   ☐4
Egg * _ _ , _ _ , _ _   Egg * _ _ , _ _ , _ _

Time   _ _ _ _
Volume   _ _ _ . _
Fluid: Clear ☐1 Slightly Bloody ☐2
       Bloody ☐3 Clotted   ☐4
Egg * _ _ , _ _ , _ _

University of Pennsylvania   Patient Name _____

Fig. 7(C). Computer system for IVF-ET; details of retrieval file.

OOCYTE FILE   PAGE 1 OF 5

SOCIAL SECURITY #   _ _ _ _ _ _ _ _ _
PATIENT #   _ _ _ _
OOCYTE #   _ _

RETRIEVAL OBSERVATIONS
DATE: _ _ / _ _ / _ _
TIME: _ _ _ _
CUMULUS DESC: Light & Fluffy   □1   Clumping   □2
                   Blood Embedded   □3   Absent   □4
                   Loose Matrix   □5   Cells Dissected   □6
                   Dense   □7   Small Amount   □8

CORONA: Radiating   □1   Dense & Compact   □2
                   Clumping   □3   Absent   □4
                   Beaded   □5   C & C Removed   □6

ZONA: Visible   □2   Not Visible   □1
                   Even   □3   Uneven   □4

VITELLUS: Visible   □1   Not Visible   □2
            Not Retracted From Zona □3   Retracted from Zone □4
                   Spherical   □5   Aspherical   □6
                   One Pb   □7   Frag. Pb   □8

CYTOPLASM: Visible   □1   Not Visible   □2
                   Light   □3   Dark   □4
                   Uneven   □5

MATURITY: Mature 1°   □1   Mature 2°   □2
                   Immature   □3   Atretic   □4
       Immature/ Borderline Mature 1° □5
       Borderline Mature 1°/ Mature 2° □6   Unknown   □7
COMMENTS: Yes □1   No □2

INSEMINATION   Yes □1   No □2
Date   _ _ / _ _ / _ _
Motile Sperm Added (x10⁴)   _ _ _ . _
Volume(µl)   _ _ _ . _
Time   _ _ _ _
COMMENTS: Yes □1 No □2

University of Pennsylvania   Patient Name _____

Fig. 8(A).   Computer system for IVF-ET; details of oocyte file.

OOCYTE FILE   PAGE 2 OF 5

CORONA REMOVAL

DATE: __ /__ /__
TIME: ____

CUMULUS: POLAR BODIES: None Present ☐1
    Plated Down ☐1   Floating ☐2   One Present ☐2 Two Present ☐3
    Matted Over ☐3   Encasing Egg ☐4   More Than Two ☐4 Blebbs Present ☐5
                                                Fragmenting ☐6
                                                Cannot Determine ☐7

CORONA: Absent ☐1 Cleared Easily ☐2   COMMENTS: Atretic ☐1 Reinsemination ☐2
            Cleared with Difficulty ☐3   Cytoplasm with Inclusions ☐3
            Did Not Have to Remove ☐4   Disintegrated ☐4   Degenerated ☐5
              Could Not Remove ☐5   Arrested ☐6   Other ☐7

SPERM IN VICINITY OF EGG: Immotile ☐1   CELL STAGE: One Cell ☐1   › One Cell ☐2
        Motile ☐2   Circular ☐3
    Progressive ☐4   Shaking ☐5
          Agg ☐6   Other ☐7

OOCYTE: Floating ☐1   Plated Down ☐2

ZONA: Not Visible ☐1   Visible ☐2
          Intact ☐3   Fractured ☐4
    Attached Sperm ☐5

VITELLUS: Not Retracted From Zona ☐1
              Retracted From Zona ☐2
       Spherical ☐3   Aspherical ☐4

CYTOPLASM: Uneven Color ☐1
           Light ☐2   Dark ☐3
    Even Texture ☐4   Clumpy ☐5
    Transparent ☐6 Reddish Brown Color ☐7
        Graing ☐8
PRONUCLEI: One ☐1 2 Present at Syngamy ☐2
          2 Present not at Syngamy ☐3
     3 Present ☐4   Polyploid ☐5   None Present   ☐6

University of Pennsylvania   Patient Name _____

Fig. 8(B). Computer system for IVF-ET; details of oocyte file.

OOCYTE FILE   PAGE 3 OF 5

REINSEMINATION DATA
DATE: _ _ / _ _ / _ _
TIME: _ _ _ _

SPERM IN VICINITY OF EGG: Immotile ☐1
              Motile ☐2   Circular ☐3
         Progressive ☐4   Shaking ☐5
                Agg ☐6   Other ☐7

OOCYTE: Floating ☐1   Plated Down ☐2

ZONA: Not Visible ☐1   Visible ☐2
       Intact ☐3   Fractured ☐4
Attached Sperm ☐5

VITELLUS: Not Retracted From Zona ☐1
           Retracted From Zona ☐2
   Spherical ☐3   Aspherical ☐4

CYTOPLASM: Uneven Color ☐1
      Light ☐2   Dark ☐3
  Even Texture ☐4   Clumpy ☐5
  Transparent ☐6   Reddish Brown Color ☐7
   Grainy ☐8

PRONUCLEI: One ☐1   2 Present at Syngamy ☐2
       2 Present not at Syngamy ☐3
   3 Present ☐4   Polyploid ☐5
None Present ☐6

POLAR BODIES: None Present ☐1
   One Present ☐2   Two Present ☐3
More Than Two ☐4   Blebbs Present ☐5
  Fragmenting ☐6
Cannot Determine ☐7

COMMENTS:
    Atretic ☐1   Reinsemination ☐2
     Cytoplasm with Inclusions ☐3
  Disintegrated ☐4   Degenerated ☐5
   Arrested ☐6   Other ☐7

CELL STAGE:
    One Cell ☐1   > One Cell ☐2

University of Pennsylvania   Patient Name _____

Fig. 8(C). Computer system for IVF-ET; details of oocyte file.

OOCYTE FILE   PAGE 4 OF 5

EMBRYO OBSERVATIONS
DATE: _ _ / _ _ / _ _
TIME: _ _ _ _

OOCYTE: Floating ☐1   Plated Down ☐2

ZONA: Not Visible ☐1   Visible ☐2
         Intact ☐3   Fractured ☐4

SINGLE CELL: Yes ☐1   No ☐2
       Spherical ☐1   Aspherical ☐2   Light ☐3
           Dark ☐4   Cytoplasm Even ☐5
Cytoplasm Clumpy ☐6   Grainy ☐7
Blebbs Present ☐8   Arrested ☐9

BLASTOMERES: Number   _ _
        Even in Size ☐1   Uneven in Size ☐2
   Uneven in Color ☐3   Light ☐4
            Dark ☐5 Cytoplasm Even ☐6
Cytoplasm Clumpy ☐7   Grainy ☐8
Blebbs Present ☐9
PRONUCLEI: One ☐1 2 Present at Syngamy ☐2
         2 Present not at Syngamy ☐3
         3 Present ☐4   Polyploid ☐5
    None Present ☐6
POLAR BODIES: None Present ☐1
    One Present ☐2   Two Present ☐3
   More Than Two ☐4 Blebbs Present ☐5
    Fragmenting ☐6
Cannot Determine ☐7

University of Pennsylvania   Patient Name _____

Fig. 8(D). Computer system for IVF-ET; details of oocyte file.

OOCYTE FILE   PAGE 5 OF 5

SECOND OBSERVATION: Yes ☐1 No ☐2
DATE: __/__/__
TIME: ____

ZONA: Not Visible ☐1   Visible ☐2
           Intact ☐3   Fractured ☐4

SINGLE CELL: Yes ☐1   No ☐2
        Spherical ☐1   Aspherical ☐2   Light ☐3
             Dark ☐4   Cytoplasm Even ☐5
Cytoplasm Clumpy ☐6   Grainy ☐7
   Blebbs Present ☐8   Arrested ☐9

BLASTOMERES: Number __
     Even in Size ☐1 Uneven in Size ☐2 Uneven in Color ☐3
          Light ☐4   Dark ☐5 Cytoplasm Even ☐6
Cytoplasm Clumpy ☐7   Grainy ☐8 Blebbs Present ☐9
PRONUCLEI: One ☐1 2 Present at Syngamy ☐2
          2 Present not at Syngamy ☐3
       3 Present ☐4   Polyploid ☐5
     None Present ☐6
POLAR BODIES: None Present ☐1
    One Present ☐2   Two Present ☐3
  More Than Two ☐4 Blebbs Present ☐5
    Fragmenting ☐6
Cannot Determine ☐7

FINAL OBSERVATION OF CELLS: Date   __/__/__
                         Number __
                           Time   ____

COMMENTS: Yes ☐1   No ☐2

OUTCOME: Transferred ☐1   Not Transferred ☐2

OTHER OBSERVATIONS: Yes ☐1   No ☐2

University of Pennsylvania   Patient Name _____

Fig. 8(E). Computer system for IVF-ET; details of oocyte file.

SOCIAL SECURITY #   __ __ __ __ __ __
PATIENT #   __ __ __ __
TIME   __ __ __ __
DATE   __ __ / __ __ / __ __

CATHETER USED: Virginia ☐1 HUP ☐2 Other ☐3
CATHETER LENGTH INSERTED(cm)   __ __ . __

BLEEDING: None ☐1 Minimal ☐2 Extensive ☐3
CRAMPING: None ☐1 Minimal ☐2 Extensive ☐3
PATIENT POSITION for TRANSFER: Lithotomy ☐1 Knee-Chest ☐2
LENGTH of BED REST: (0-24) hr ☐1   Major ☐99

BLOOD DRAWN: Pre-Progesterone ☐1 Post ☐2

SAMPLE FROZEN: Yes ☐1   No ☐2

HORMONE PROFILE: Not Done ☐0 E_2 ☐1   Prog ☐2
                       HCG ☐3 Testosterone ☐4 Other ☐5

# of EMBRYOS TRANSFERRED   __ __

CELL STAGE OF EMBRYOS TRANSFERRED:
                         OOCYTE #1   __ __
                                 #2   __ __
                                 #3   __ __
                                 #4   __ __
                                 #5   __ __
                                 #6   __ __
                                 #7   __ __
                                 #8   __ __
                                 #9   __ __
                                 #10   __ __

ONE STAGE TRANS. ☐1 TWO STAGE TRANS ☐2
METHOD OF RETRIEVAL: LAPAROSCOPY ☐1 LAPAROTOMY ☐2 ULTRASOUND ☐3
INITIALS OF PHYSICIAN: RETRIEVAL __ __ , __ __
                       EMBRYO TRANSFER __ __

INITIALS OF LAB PERSONNEL:
               OVUM RETRIEVAL __ __ , __ __ , __ __
               INSEMINATION   __ __
               CORONA REMOVAL __ __
               EMBRYO TRANSFER __ __

COMMENTS: Yes ☐1 No ☐2
PREGNANCY: Yes ☐1 No ☐2

University of Pennsylvania   Patient Name _____

Fig. 9.   Computer system for IVF-ET; details of ET file.

STOCK MEDIA FILE   PAGE 1 OF 1

Stock Media #   _ _ /_ _ /_ _ _

Batch #   _ _ _ _ _ _

Prepared By: (initials) _ _ _

pH Before Gassing   _ . _ _

pH After Gassing   _ . _ _

Initial Osmolarity   _ _ _

Adjusted Osmolarity   _ _ _

Amount (liter)   _ _

Mouse Embryo Development
  Date Performed   _ _ /_ _ /_ _
  % to Hatching Blastocyst _ _

University of Pennsylvania   Patient Name _____

Fig. 10.   Computer system for IVF-ET; details of stock media file.

14

GAMETE INTRAFALLOPIAN TRANSFER (GIFT)

William Byrd

1. INTRODUCTION

With the unifying goal of pregnancy, several protocols have been
developed for the treatment of the infertile couple. Steptoe and
Edwards (1978) first reported a live birth following human in vitro
fertilization and embryo transfer (IVF-ET). Since that initial
report, IVF-ET has undergone exponential growth as a treatment

modality for certain types of infertility. Several other techniques
have been utilized that manipulate gametes or fertilized ova in vitro
for the treatment of infertility. Such techniques include intrauterine insemination (Kerin et al., 1984; Sher et al., 1984; Toffle
et al., 1985; Byrd et al., 1986; Confino et al., 1986; Hoing et al.,
1986) and the transfer of fertilized embryos (Buster et al., 1983) or
sperm and oocytes directly into the uterine cavity (Craft et al.,
1982).

An alternative procedure has been developed for couples with
long standing infertility in which the wife has at least one patent
fallopian tube. This procedure is known as gamete intrafallopian
transfer or GIFT. This procedure involves the placement of both
washed sperm and preovulatory oocytes into the ampullary portion of
the fallopian tube which is the natural site of fertilization (Figure
1). The first pregnancy and subsequent live birth using this
procedure was reported by Asch et al. (1984, 1985a). The rationale
for this procedure, as opposed to IVF-ET, is that it attempts to
mimic the physiological conditions that normally occur during
fertilization and early development while maximizing the number of
gametes in the fallopian tube. Recent evidence indicates that
utilizing culture medium similar in composition to tubal fluid yields
better pregnancy rates than does standard culture media for IVF-ET
(Quinn et al., 1985).

Following the protocols developed for human IVF-ET, ovarian
stimulation is induced using clomiphene citrate, human menopausal
gonadotropin, or a combination of these agents administered sequentially or simultaneously. The patient then receives human chorionic
gonadotropin and the aspiration of preovulatory oocytes is accomplished by laparoscopy or minilaparotomy. Prior to surgery, motile
spermatozoa are obtained from fresh ejaculates following washing and
sperm swim-up. If there is at least one patent fallopian tube,

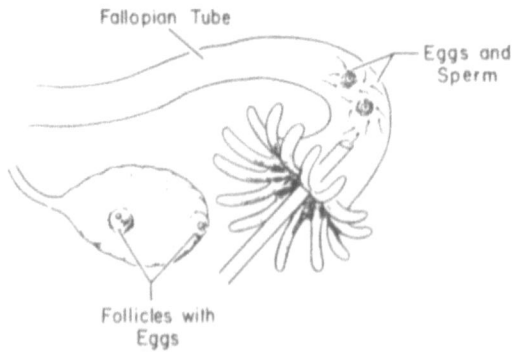

Fig. 1.   Laparoscopic transfer of oocytes and spermatozoa to the
           fallopian tube. (Reprinted from Madden et al., 1986.)

gametes are loaded into a catheter and then transferred at the time of surgery into the ampulla of the fallopian tube through the fimbrial ostium. In contrast to IVF-ET, there is no maturation of the oocytes in vitro prior to fertilization and the fertilization process takes place in vivo.

Infertile couples with endometriosis, cervical/immune factors, male factor, ovulation dysfunction or unexplained infertility do not have an absolute barrier to fertilization. In this review, the efficacy of using GIFT in the treatment of these infertile couples will be compared to IVF-ET and intrauterine insemination. The complete details of this procedure will be described as well as its success in the treatment of long standing infertility.

2. GIFT PROCEDURE

The protocol listed below is based on the original work by Asch et al. (1984, 1985a,b, 1986) and Balmaceda et al. (1986).

2.1. Ovarian Stimulation

Patients undergoing GIFT are first treated to induce follicular development using protocols similar to those used for IVF-ET. While the optimal number of oocytes for transfer has yet to be determined, it has been demonstrated in IVF-ET that the major determinant for successful outcome is the number of embryos replaced (Edwards et al., 1984; Seppälä, 1985).

Patients undergoing ovarian stimulation are given two ampules of human menopausal gonadotropin (hMG; 150 IU of follicle-stimulating hormone [FSH] and 150 IU of luteinizing hormone [LH]; Pergonal, Serono Laboratories, Inc., Randolph, MA) intramuscularly beginning on day three of the menstrual cycle. Ultrasound examinations begin on day three to rule out adnexal masses and then daily after day six. Serial estradiol levels are obtained daily from day six onwards. Human chorionic gonadotropin (hCG; Profasi, Serono Laboratories, Inc., Randolf, MA; 10,000 IU) is given intramuscularly when two or more follicles reach 16 mm in diameter on ultrasound and serum estradiol levels reach 300-500 pg/ml for each follicle greater than 16 mm in diameter. Following hCG injection, laparoscopy or minilaparotomy is performed 36 hr later to recover preovulatory oocytes.

As with human IVF-ET, there have been several different protocols utilized for ovarian stimulation in GIFT. These include the use of human FSH (Metrodin, Serono Laboratories, Inc., Randolf, MA) and hMG (Guastella et al., 1985, 1986), hMG alone (Guastella et al., 1985, 1986; Nemiro and McGaughey, 1986), clomiphene citrate alone (Guastella et al., 1985, 1986; Balmaceda et al., 1986; Corson

et al., 1986; Nemiro and McGaughey, 1986), clomiphene citrate
followed by hMG (Molloy et al., 1985; Corson et al., 1986; Nemiro and
McGaughey, 1986; Grunert et al., personal communication;) and the
simultaneous use of clomiphene citrate and hMG (Gordts, 1986; Madden
et al., 1986; Molloy et al., 1987). See Chapter 19 for further
discussion.

2.2. Supplies and Materials

Since the GIFT protocol is similar to IVF-ET to the point
of tubal transfer, the reader is referred to the review by Sokoloski
and Wolf (1984) for the laboratory details of gamete preparation.
The following information provides a basic list of materials and
supplies needed for GIFT.

a. Sterile working environment (e.g., laminar flow hood)

b. Stereo dissecting microscope (optional video camera
    attachment) and an inverted phase microscope for high
    resolution evaluation of oocyte quality (optional)

c. Incubator: tri-gas 5% CO_2:5% O_2:90% N_2 or 5% CO_2/air

d. Media: The most widely used medium for GIFT is Ham's F-10
    medium supplemented with heat inactivated maternal serum
    (HIMS).

    (1) Basic Ham's F-10 stock: Ham's F-10 dry powder is
         mixed with highly purified water and supplemented with
         40 mM calcium lactate, 25 mM sodium bicarbonate and
         penicillin-streptomycin. The final stock solution is
         adjusted to 280 mOsmol final concentration and stored
         at 4°C until used.

    (2) Sperm washing medium: Ham's F-10 with 7.5% (HIMS)

    (3) Medium for flushing follicles: Ham's F-10 with 7.5%
         HIMS or phosphate buffered saline supplemented with
         100 IU of heparin per ml.

    (4) Medium for embryo transfer: Ham's F-10 with 50% HIMS

e. Plasticware and glassware: Follicular fluid collection
    tubes (example: Falcon 3033), organ culture dishes (Falcon
    3037), petri plates (Falcon 3001 and 1007), 15 ml conical
    tubes (Falcon 2095), fire polished, tissue culture cleaned
    Pasteur pipets. See also Chapter 8.

f. Laparoscopic equipment.   Reviewed by Beauchamp (1984)

g.   Transfer catheter: See Gamete Transfer (2.5) for a
discussion of the catheters currently used for GIFT.

2.3. Sperm Preparation

While the number of sperm at the site of fertilization in
the fallopian tube is probably less than 1000 (Ahlgren, 1975), these
sperm represent a highly motile, selected population of cells. Asch
et al. (1984, 1985a,b, 1986) empirically chose 100,000 motile sperm
as the concentration to be used per oviduct based on the numbers of
motile cells used in human IVF-ET (Trounson et al., 1980).

Sperm specimens are collected by masturbation approximately 2.5
hr prior to surgery and allowed to liquefy. Following liquefaction,
specimens are suspended in Ham's F-10 medium (1 volume semen:3
volumes medium) supplemented with 7.5% HIMS and centrifuged for 6-10
min at 200-400 g. The supernatant is removed and the pellet is
resuspended in 1 ml of fresh medium and centrifuged again. A highly
motile sperm fraction is recovered using a swim-up isolation technique in which the final pellet after centrifugation is overlaid with
0.5 to 1.0 ml of medium and the sperm are allowed to swim into the
overlaying layer for 45-90 min at 37°C in an atmosphere of 5% CO_2:5%
O_2:90% N_2. The supernatant is then recovered and sperm numbers are
adjusted to 100,000 motile sperm in 25 μl of medium.

If motile sperm preparations are to be recovered from severely
oligospermic and/or asthenozoospermic males, the initial incubation
time should be increased to obtain more motile sperm. Alternatively,
sperm can be recovered by the following technique. Following
centrifugation, the final sperm pellet is resuspended in 1 ml of
medium and incubated for 45-120 min which allows the nonmotile cells
and debris to settle. The motile fraction of sperm remaining in
suspension is then recovered, analyzed and adjusted to 100,000 motile
cells in 25 μl of medium as above.

While most groups report the transfer of 100,000 motile cells
per tube, when dealing with male factor patients many centers use
200,000-400,000 motile sperm per fallopian tube (Grunert et al.,
personal communication; Madden et al., personal communication; Nemiro
and McGaughey, 1986).

2.4. Oocyte Recovery

For a review of laparoscopic follicular aspiration, the
reader is referred to Beauchamp (1984). There has been some concern
over the gas composition used for the pneumoperitoneum. Exposure to
100% CO_2 in the peritoneum could alter the pH of the bicarbonate-
buffered Ham's F-10 medium during prolonged transfers. This could be
dealt with either by the addition of Hepes buffer to the transfer
medium or the utilization of a 5% CO_2:5% O_2:90% N_2 mixture in the

peritoneum (Grunert et al., personal communication). Following follicular aspiration using conventional IVF-ET oocyte retrieval methods, preovulatory oocytes are graded for maturity based on the expansion of cumulus and corona cells and the presence of the first polar body if it can be visualized (see Chapter 8) and then incubated from the time of recovery to transfer in Ham's F-10 medium with 50% serum in an atmosphere of 5% CO_2:5% O_2:90% N_2 at 37°C before replacement in the ampulla of the fallopian tube.

2.5. Gamete Transfer

Upon completion of oocyte recovery, gametes are loaded into a catheter for transfer. Several catheters have been used for GIFT. These include the Deseret Intracath (#3132; Deseret Medical Inc., Sandy, UT; Asch et al., 1986; Madden et al., personal communication), Semm Catheter (Wisap, W. Germany; Guastella et al., 1985), a 50 cm, 16-gauge Teflon catheter (William A. Cook Pty. Ltd., Australia; Yovich et al., 1986), and a number 5 French side port transfer catheter with the tip cut off (Nemiro and McGaughey, 1986). The catheter is attached to a disposable tuberculin or a Hamilton syringe for injection of the gametes. Prior to loading, the catheter is flushed with Ham's F-10 medium containing 50% serum and checked to insure that air leaks are absent. The catheter is first loaded with 25 μl of sperm suspension containing 100,000 sperm. A 25 μl sample of sperm is placed on an inverted culture dish and then drawn into the catheter. Following aspiration of the sperm, 5 μl of air is drawn into the catheter to prevent mixing of the gametes followed by one to four mature oocytes in 25 μl of medium. This is followed by another 5 μl air bubble and finally 10 μl of medium. Alternately, it is possible to mix the sperm suspension and oocytes together before transfer which would allow a much smaller volume of fluid to be transferred. While most groups report the use of a pneumatic (air) column in the transfer catheter, Molloy et al. (1985) and Nemiro and McGaughey (1986) report the use of a catheter filled with medium. The use of a fluid-filled catheter has been suggested to be superior to the air-filled catheter in transcervical insertion of fertilized embryos in IVF-ET (Poindexter et al., 1986).

While initial reports included the use of both minilaparotomy and laparoscopy for oocyte pick up and transfer (Asch et al., 1985b), most groups are now comfortable with laparoscopy exclusively. A blunt metal sleeve that fits snugly in the aspirating needle cannula is inserted into the abdominal cavity. This sleeve guides the catheter to the fimbria. To insert the catheter, the fimbrial end of the fallopian tube is gently grasped with forceps either by the serosa of the tube or by holding the whole thickness of the tube behind the fimbria. Once the ostium is visualized, the ostium and ampulla are aligned with the metal sleeve. The distal tip of the transfer catheter is then threaded approximately 1.5 to 2.0 cm into the fimbriated end of the fallopian tube. To determine the degree of penetration by the catheter, the catheter can be marked prior to use. The contents of the catheter are gently discharged and the catheter returned to the embryologist and examined to ensure that no oocytes have been retained. The same procedure is then repeated for the second fallopian tube, if present and patent. It is appropriate to attempt the first transfer in the tube that appears to be in the worst shape since this allows, in the event of a failed transfer, the embryos to be transferred to the remaining tube. Depending upon the number of oocytes recovered and the availability of patent fallopian tubes, between one and six oocytes are then transferred to a tube. If excess oocytes or immature oocytes are recovered, they can be fertilized and either transferred transcervically into the uterus or cryopreserved for future transfer.

2.6. Postoperative Treatment

The patient is given 25 mg of progesterone (I.M.) daily from day 4 after transfer until 8 weeks after transfer. Tropho-blastic development is then monitored by serial serum beta hCG levels. Other groups report the use of progesterone supplementation either by injection of progesterone (12.5 mg I.M.; Guastella et al., 1985) or the use of vaginal suppositories (25-50 mg progesterone suppositories/twice daily; Corson et al., 1986, Nemiro and McGaughey, 1986) or 1500 IU of hCG on days 6, 9 and 12 following transfer (Molloy et al., 1987). While most groups support the luteal phase with progesterone or hCG, the usefulness of progesterone in luteal phase support in IVF-ET has been questioned (Trounson et al., 1986). There is also evidence that an imbalance of progesterone-estrogen levels may influence tubal transport of embryos (Edwards, 1980) which should be examined.

3.   DISCUSSION

3.1. Incidence of Pregnancy, Pregnancy Loss and Multiple Pregnancies Following GIFT

Ultimately the selection of GIFT, compared to other available treatment modalities such as IVF-ET and intrauterine insemination, requires an evaluation of pregnancy results. While GIFT is a relatively new procedure, there are a sufficient number of cases reported to allow preliminary evaluation of its potential. The range of pregnancies achieved per transfer is from 20-49% of all reported pregnancies and 15-39% of clinical pregnancies (n = 899 transfers, Table I). The mean combined biochemical or clinical pregnancy rate per transfer is 34.8% while the clinical pregnancy rate per transfer is 28.8%. Comparisons between pregnancies per treatment cycle in GIFT, IVF-ET and intrauterine insemination (IUI) are presented in Figure 2, but they must be interpreted with caution. The type and severity of infertility may exclude certain patients

Table I.   Summary of pregnancy rates following GIFT

Reference	Number of patient cycles with transfer	Total number of pregnancies (percent)[a]	Number of clinical pregnancies (percent)
Asch and Cittadini[b]	311	--[c]	103 (33.1)
Corson et al. (1986)	20	4 (20.0)	3 (15.0)
Grunert et al.[b]			
a) GIFT	120	--[c]	25 (20.8)
b) GIFT + IVF-ET	16	--[c]	4 (25.0)
Guastella et al. (1986)	44	17 (38.6)	17 (38.6)
Madden et al.[b]	40	14 (49.2)	12 (30.0)
Molloy et al. (1987)	71	23 (32.4)	19 (26.7)
Nemiro and McGaughey[b,d]	223	83 (37.2)	63 (28.2)
Yee (1986)	36	--[c]	10 (27.7)
Yovich et al. (1986)	18	4 (22.2)	4 (22.2)
Total	899	145 (34.8)	259 (28.8)

[a]Both biochemical and clinical pregnancies.
[b]Personal communication.   Refer to acknowledgements.
[c]Not reported.
[d]Nemiro and McGaughey (1986) refer to the laparoscopic aspiration of oocytes and transfer of gametes to the fimbria of the fallopian tube as tubal transfer (TT).

from GIFT or IUI which may lower the overall pregnancy success in IVF-ET.   While IUI has a much lower pregnancy rate per treatment cycle, it can be repeated with little risk and minimal cost to the patient.   Most patients undergoing IUI do not require ovarian stimulation and the overall risk of multiple pregnancies is much lower.   When more information is available, a comparison of cumulative pregnancy rates in GIFT, IVF-ET and IUI would be of interest. Comparisons of cumulative data from established IVF-ET centers can be misleading, since the overall success of IVF-ET has increased in

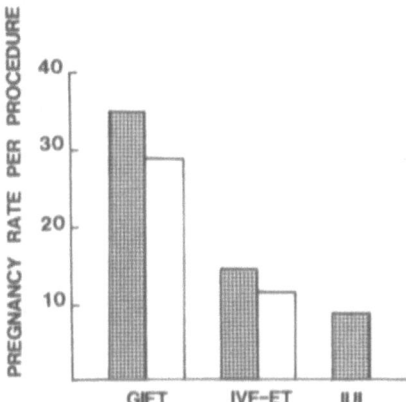

Fig. 2.   Pregnancy rate of GIFT, IVF-ET and IUI based on the number
           of laparoscopies or inseminations completed. The GIFT data
           is taken from Table I. IVF-ET data is taken from Seppälä
           (1985). IUI data is based on the results of Toffle et al.
           (1985), Wiltbank et al. (1985), Byrd et al. (1986), Confino
           et al. (1986), DiMarzo and Rakoff (1986) and Hoing et al.
           (1986). Biochemical and clinical pregnancies (hatched
           bar), clinical pregnancies (open bar).

recent years. If one considers the recent data reported by Jones
(1986), the success rate per treatment cycle has improved from 12.7%
in 1981 to 26.7% in 1985.

     The overall risk of multiple pregnancies following GIFT in all
reporting groups is 20.5% of clinical pregnancies (Table II). The
number of ectopic pregnancies reported is 3.7% of the clinical
pregnancies with an abortion/miscarriage rate of 26.5%. Seppälä
(1985), in his collaborative survey of IVF-ET, reported a multiple
pregnancy rate of 12.6%, approximately 1.8% of which were ectopic and
29.9% of which aborted. While there is little difference in the
percentage of ectopic and aborted pregnancies in GIFT (Table II) and
IVF-ET (Seppälä, 1985), the multiple pregnancy rate is much higher in
GIFT. Further information is needed to establish the optimal number
of oocytes transferred in GIFT.

     3.2. Pregnancy Success Based on the Number of Oocytes or Embryos
          Transferred

     In IVF-ET the pregnancy rate per transfer is dependent upon
the number of embryos replaced in utero (Seppälä, 1985). A
comparison of pregnancies per treatment cycle based on the number of
oocytes (GIFT) or embryos (IVF-ET) transferred can be seen in Table
III. The average number of oocytes transferred per GIFT cycle is
3.4. There have been no reported pregnancies with the transfer of

Table II. Incidence of multiple, ectopic and aborted pregnancies
           following GIFT

Reference	Number of multiple pregnancies (percent)[a]	Number of ectopic pregnancies (percent)[a]	Number of aborted pregnancies (percent)[a]
Asch and Cittadini[b]	c	3 (2.9)	27 (26.2)
Corson et al. (1986)	0	0	1 (25.0)
Grunert et al.[b]			
a) GIFT	2 (8.0)	1 (4.0)	5 (20.0)
b) GIFT + IVF-ET	0	0	2 (50.5)
Guastella et al. (1986)	3 (17.6)	1 (5.8)	2 (11.7)
Madden et al.[b]	0	0	7 (50.0)
Molloy et al. (1987)	3 (15.7)	0	3 (15.8)
Nemiro and McGaughey[b,d]	21 (25.3)	4 (4.8)	20 (24.1)
Yee (1986)	c	c	c
Yovich et al. (1986)	1 (25.0)	0	0
Total	30 (20.5)	9 (3.7)	66 (26.5)

[a]Based on clinical pregnancies.
[b]Personal communication. Refer to acknowledgements.
[c]Not reported.
[d]Tubal transfer.

immature oocytes. Surprisingly, when the pregnancy success rate
based on the total number of oocytes transferred is examined (Figure
3), there is an initial plateau with the transfer of one, two, three
or four oocytes which increases with the transfer of five oocytes.
In contrast, in IVF-ET there is a linear increase in the percentage
of patients pregnant with increasing numbers of embryos transferred
(Figure 3). However, since data from only 342 GIFT transfers are
available, it is premature to draw final conclusions concerning
pregnancy outcome in GIFT and IVF-ET based on the number of oocytes
or embryos transferred. Further information is needed to evaluate

Table III. Pregnancy rate based on number of oocytes or embryos transferred[a]

| Reference | Number of oocytes transferred per treatment cycle | | | | |
	1	2	3	4	5
	Number of cycles (percent pregnant)				
GIFT					
Corson et al. (1986)	0	7 (14.2)	4 (0)	2 (0)	4 (50.0)
Guastella et al. (1986)	1 (0)	13 (30.7)	19 (52.6)	5 (20.0)	6 (33.3)
Molloy et al. (1987)	1 (0)	3 (0)	4 (25.0)	63 (28.5)	0
Grunert et al.[b]	12 (25.0)	9 (22.2)	36 (22.2)	50 (16.0)	10 (30.0)
Madden et al.[b]	1 (0)	4 (25)	7 (0)	16 (50.0)	12 (41.7)
Nemiro and McGaughey (1986)[c]	5 (40.0)	17 (29.4)	14 (21.4)	13 (46.1)	4 (0)
	Number of embryos transferred				
IVF-ET					
Seppälä (1985)[d]	3321 (9.5)	2514 (14.6)	1340 (19.3)	818 (24.1)	---

[a]All pregnancies (clinical and biochemical).
[b]Personal communication. Refer to acknowledgements.
[c]Tubal transfer.
[d]Clinical pregnancies per transfer cycle.

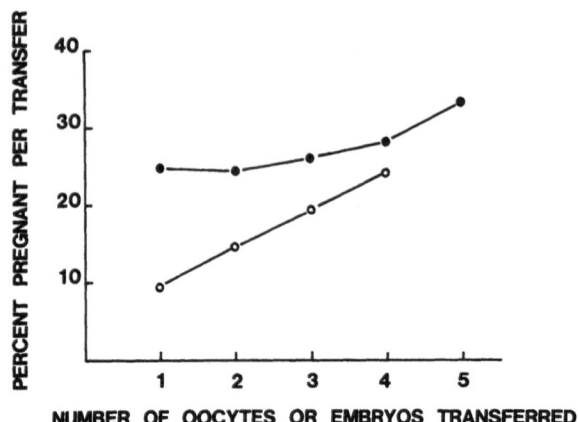

Fig. 3.   Pregnancy rate based on the number of oocytes (GIFT) or
           embryos (IVF-ET) transferred. Data taken from Table IV.
           (o) Seppälä (1985) clinical pregnancies only, (●) GIFT –
           all pregnancies reported.

the optimal number of oocytes to be transferred, both total and per
tube. Other questions to be answered are: 1) Is there an increased
incidence of multiple gestations or ectopic pregnancies with multiple
transfers in one tube?; 2) What is the success rate of patients with
one tube verses two tubes following transfer of the same number of
oocytes?; 3) What is the transfer efficiency and retention of gametes
in the ampulla following transfer?; 4) What is the cumulative
probability of pregnancy in women undergoing multiple transfers?; and
5) What is the relationship between estradiol levels, the number of
large follicles and pregnancy outcome?

   3.3. Patient Selection for GIFT

        Selection of GIFT over IVF-ET or IUI will ultimately be
predicated upon the pregnancy rate for a particular etiology. Table
IV presents accumulated data for pregnancy success based on etiology.
While the etiology of infertility is often multifactorial, the
patient populations in most studies are divided into different
infertility categories based on their primary etiology. The pregnancy rate per transfer ranges from 6.7 to 47.0%. Based solely on
etiology, couples with mixed male and female factors or male factor
alone have the poorest outcome in GIFT. The overall pregnancy rate
per transfer in male factor patients (16.1%) or patients with a
combination male and female factor (6.7%) is significantly lower (p <
0.05, adjusted Chi-square) than couples with unexplained infertility
(29.5%).

Table IV. Success of GIFT based on etiology

Reference	Male factor	Un-explained	Endo-metriosis	Cervical/immune	Ovulatory dysfunction	Pelvic adhesions	Multiple factors male & female	Multiple factors female	Other
				Number of transfers/(percent pregnant)					
Corson et al. (1986)	0	3 (33.3)	2 (0)	4 (25)	2 (0)	0	0	0	2 (100)
Guastella et al. (1986)	1 (0)	27 (25.9)	3 (66.6)	1 (0)	0	3 (66.6)	0	0	9 (66.6)
Yovich et al. (1986)	9 (0)	2 (100)	2 (100)	0	0	0	0	0	5 (0)
Asch and Cittadini[a]	30 (23.3)	180 (30.0)	35 (40.0)	9 (11.1)	0	15 (46.6)	0	0	42 (47.6)
Grunert et al.[a]	1 (0)	7 (0)	25 (12.0)	0	17 (23.5)	8 (37.5)	27 (7.4)	35 (43.3)	0
Madden et al.[a]	8 (12.5)	10 (30.3)	7 (42.8)	3 (33.3)	0	1 (0)	3 (0)	6 (50.0)	4 (25.0)
Nemiro and McGaughey (1986)	13 (15.3)	11 (36.3)	10 (0)	7 (28.5)	3 (66.7)	10 (50.0)	0	0	6 (50.0)
Totals	62 (16.1)	240 (29.5)	84 (29.6)	24 (20.8)	22 (27.2)	37 (44.9)	30 (6.7)	41 (29.2)	68 (47.0)

[a]Personal communication. Refer to acknowledgements.

3.4. Male Factor Patients

The major disadvantage of GIFT versus IVF-ET is the inability to determine whether or not fertilization has occurred in the absence of an established pregnancy. This is particularly important in the male factor category. The collaborative study on IVF-ET by Seppälä (1985) reported an overall pregnancy rate of 6.9% per treatment cycle in male factors (n = 872) while patients with other factors experienced an overall rate of 9.7% (n = 4136; p < 0.025). Similar results have been reported by Guzick et al. (1986) with an overall pregnancy rate in IVF-ET for male factor of 7.14% (n = 70) and 14.27% in other patient groups (n = 987). The lowered success rate can be attributed to poor fertilization since the pregnancy rates per embryo transfer are similar in both groups.

To solve this dilemma patients with known male factors may elect to undergo at least one cycle of IVF-ET before undergoing GIFT to determine if oocytes can be fertilized. Alternately, GIFT and IVF-ET can be performed in the same cycle. This would allow evaluation of male gamete function in vitro. Or following GIFT, excess oocytes may be fertilized and cryopreserved for later transfer. Is testing of male gamete function using the traditional sperm penetration assay (SPA) of value in some patients electing to undergo GIFT? Many groups have found a limited correlation of the SPA with IVF-ET outcome in male factor patients (Foreman et al., 1984; Ausmanas et al., 1985; Belkien et al., 1985; Cohen et al., 1985; Margalioth et al., 1986). Further investigation is needed to determine if the SPA will be of use in predicting outcome in GIFT.

While the number of sperm used in GIFT is generally 100,000 motile cells, the optimal number needed to fertilize eggs has yet to be established. Some groups are empirically using 200,000–400,000 motile cells per tubal transfer when a male factor is present (Nemiro and McGaughey, 1986; Grunert et al., personal communication; Madden et al., personal communication). To enhance pregnancy success with male factor patients, it may be necessary to institute prospective randomized studies using different sperm concentrations to determine if pregnancy results can be improved. Analysis of the original semen parameters as well as the motile sperm fraction used for insemination compared to outcome would be of interest in determining which, if any, semen parameters are predictive of success. Earlier results for IVF-ET demonstrated that male factor patients with oligospermia required at least a ten-fold higher motile sperm concentration to achieve comparable fertilization rates as patients with normal semen parameters (Byrd and Wolf, 1984).

3.5. Reproductive Success Using GIFT

Increased reproductive success in certain women undergoing GIFT as opposed to IVF-ET or IUI may be due in part to the mechanical

placement of isolated motile sperm fractions and oocytes directly
into the ampulla of the fallopian tube. Infertility may result from
abnormal transport of either spermatozoa or oocytes to the site of
fertilization. In the case of certain types of immune infertility,
women with poor or hostile cervical mucus, or anatomical problems
such as cervical stenosis, GIFT may enhance fertilization and
subsequent pregnancy by circumventing the lower portion of the
reproductive tract. While intrauterine insemination is an alternative treatment option for some of these patients, in couples with a
male factor the percentage of pregnancies following treatment is
highly variable. Some investigators (Kerin et al., 1984; Byrd et
al., 1986; DiMarzo and Rakoff, 1986, Hoing et al., 1986) report a
high success rate while others do not find intrauterine insemination
efficacious in treatment of male factor patients (Confino et al.,
1986; Hull et al., 1986). The discrepancies between these groups may
result from differences in the definition of male factor, the
severity of the male factor and selection of patient populations for
study.

Some couples with unexplained infertility or endometriosis may
suffer from abnormal follicular rupture or expulsion of the egg. It
has been suggested that some women exhibit a luteinized unruptured
follicle syndrome (Koninckx et al., 1978; Marik and Hulka, 1978).
While the existence of this syndrome has been questioned (Dhont et
al., 1984), it is possible that some women have abnormal oocyte
release which can be overcome using GIFT. Oocyte entrapment has been
noted in patients undergoing IVF-ET that were stimulated with
clomiphene citrate or combinations of clomiphene citrate and human
menopausal gonadotropin (Stanger and Yovich, 1984). At laparoscopy,
they found at least 60% of the ruptured follicles still had ova
entrapped in the follicles.

Alternately, infertility may result from abnormal egg transport.
It is believed that the ciliary action of the mucosal cells of the
fimbriae play an important role in the transport of the ovulated egg-
cumulus mass over the surface of the fimbria to the ampulla (Blandau
et al., 1977). While the absence of the fimbriae does not prevent
conception (Beyth, 1986), its presence is probably conducive to ovum
transport. Any damage to the fimbriae, tubes or the presence of
adhesions may impair fertility by decreasing the efficiency of ovum
transport which can be overcome by GIFT.

Our knowledge of the role of the tubal environment, hormonal
cues, muscular activity, and cilial action on human reproduction is
limited and the influence these factors have on the success of GIFT
is unknown. The tubal environment may contribute to the success of
fertilization, embryonic growth or to the synchronization of embryo
arrival into the uterus. All of these factors may represent
advantages of GIFT over IVF-ET.

4.   CONCLUSION

GIFT combines the advantages of natural physiology and IVF-ET
technology to offer a high expectation of pregnancy success in
infertile women with patent fallopian tubes. GIFT should not be
viewed as a challenge to IVF-ET but rather as a complementary
technique. Obviously, GIFT cannot be used in women who have diseased
or absent tubes or inaccessible ovaries unless ultrasound techniques
compatible with GIFT are developed. Unexpected surgical findings
such as extensive adhesions, poor semen parameters or immature
oocytes recovered by aspiration may necessitate using IVF-ET. With
the widespread use of ultrasound-guided follicular aspiration, the
cost-risk factor of IVF-ET compared to GIFT must be considered.
Patients may elect to undergo ultrasound-guided aspiration and embryo
transfer rather than GIFT with its higher pregnancy rate. IVF-ET and
GIFT can be combined such that excess oocytes that are not transferred during GIFT can be fertilized and transferred during the
treatment cycle or cryopreserved for replacement during future non-
stimulated cycles. This combined approach would also yield more
information about fertilization and subsequent development of the
patient's oocytes. Thus the combination of GIFT and IVF-ET will
provide a potent tool in the treatment of certain infertile couples.

5.   ACKNOWLEDGEMENTS

I gratefully acknowledge the following groups who have provided
me with their most recent results on GIFT. Without their cooperation
it would have been much more difficult to assemble this chapter on
GIFT.

R. Asch[1] and E. Cittadini[2]; [1]Department of Obstetrics and
Gynecology, University of California at Irving, Orange, CA;
[2]Instituto di Endocrinologae Fisiopatologia della Reproduzione Umana,
Universita di Palermo, Palermo, Italia.

G. Grunert, W. Gibbons, E. Lotze, L. Rodriguez-Rigau, E.
Steinberger and R. Woodward; Womans Hospital of Texas, Houston, TX.

J. Madden, D. M. Bookout, D. Silverstein, M. Spelts, D. W.
Stein, S. Stephens and A. Toofanian; Presbyterian Hospital, Dallas,
TX.

J. S. Nemiro[1] and R. W. McGaughey[2]; [1]Department of Obstetrics
and Gynecology, Good Samaritan Medical Center, Phoenix; [2]Department
of Zoology, Arizona State University, Tempe, AZ.

6.   REFERENCES

Ahlgren, M., 1975, Sperm transport to and survival in the human
    fallopian tube, Gynecol. Invest. 6:206-214.
Asch, R. H., Ellsworth, L. R., Balmaceda, J. P., and Wong, P. C.,
    1984, Pregnancy after translaparoscopic gamete intrafallopian
    transfer, Lancet 2:1034-1035.
Asch, R. H., Ellsworth, L. R., Balmaceda, J. P., and Wong, P. C.,
    1985a, Birth following gamete intrafallopian transfer, Lancet
    2:163.
Asch, R. H., Balmaceda, J. P., Ellsworth, L. R., and Wong, P. C.,
    1985b, Gamete intra-fallopian transfer (GIFT): a new treatment
    for infertility, Int. J. Fertil. 30:41-45.
Asch, R. H., Balmaceda, J. P., Ellsworth, L. R., and Wong, P. C.,
    1986, Preliminary experiences with gamete intrafallopian
    transfer (GIFT), Fertil. Steril. 45:366-371.
Ausmanas, M., Tureck, R. W., Blasco, L., Kopf, G. S., Ribas, J., and
    Mastroianni, L., Jr., 1985, The zona-free hamster egg penetra-
    tion assay as a prognostic indicator in a human in vitro
    fertilization program, Fertil. Steril. 43:433-437.
Balmaceda, J. P., Wong, P. C., Ellsworth, L., and Asch, R. H., 1986,
    Results on one hundred consecutive cases of infertility treated
    with gamete intra-fallopian transfer (GIFT), Fertil. Steril.
    [Suppl.]:26 (abstract).
Beauchamp, P. J., 1984, Laparoscopic follicular aspiration, in:
    Human In Vitro Fertilization and Embryo Transfer (D. P. Wolf and
    M. M. Quigley, eds.), Plenum Press, New York, pp. 149-169.
Belkien, L., Bordt, J., Freischem, C. W., Hano, R., Knuth, U. A., and
    Nieschlag, E., 1985, Prognostic value of the heterologous ovum
    penetration test for human in vitro fertilization, Int. J.
    Androl. 8:275-284.
Beyth, Y., 1986, Ovum pick-up mechanism: a reappraisal, Curr. Probl.
    Obstet. Gynecol. Fertil. 9:81-122.
Blandau, R. J., Brackett, B., Brenner, R. M., Boling, J. L.,
    Broderson, S. H., Hamner, C., and Mastroianni, L., Jr., 1977,
    The oviduct, in: Frontiers in Reproduction and Fertility Control
    (R. O. Greep, ed.), The MIT Press, Cambridge, MA, pp. 132-145.
Buster, J. E., Bustillo, M., Thorneycroft, I., Simon, J. A., Boyers,
    S. P., Marshall, J. R., Seed, R. G., and Louw, J. A., 1983, Non-
    surgical transfer of an in-vivo fertilised donated ovum to an
    infertility patient, Lancet 1:816-817.
Byrd, W. and Wolf, D. P., 1984, Oogenesis, fertilization and early
    development, in: Human In Vitro Fertilization and Embryo
    Transfer (D. P. Wolf and M. M. Quigley, eds.), Plenum Press,
    New York, pp. 213-273.
Byrd, W., Ackerman, G., Carr, B., Edman, C., Grun, B., Guzick, D.,
    Johnson, N., and McConnell, J., 1986, Results of intrauterine
    insemination in couples with male factor, idiopathic, and
    cervical factor infertility, Fertil. Steril. [Suppl.]:105
    (abstract).

Cohen, J., Fehilly, C. B., and Walters, D. E., 1985, Prolonged
    storage of human spermatozoa at room temperature or in a
    refrigerator, Fertil. Steril. 44:254-262.
Confino, E., Friberg, J., Dudkiewicz, A. B., and Gleicher, N., 1986,
    Intrauterine inseminations with washed human spermatozoa,
    Fertil. Steril. 46:55-60.
Corson, S. L., Batzer, F., Eisenberg, E., English, M. E., White,
    S. M., Laberge, Y., and Go, K. J., 1986, Early experience with
    the GIFT procedure, J. Reprod. Med. 31:219-223.
Craft, I., Djahanbakhch, O., McLeod, F., Bernard, A., Green, S.,
    Twigg, H., Smith, W., Lindsay, K., and Edmonds, K., 1982, Human
    pregnancy following oocyte and sperm transfer to the uterus,
    Lancet 1:1031-1033.
Dhont, M., Serreyn, R., Duvivier, P., Vanluchene, E., De Boever, J.,
    and Vandekerckhove, D., 1984, Ovulation stigma and concentration
    of progesterone and estradiol in peritoneal fluid: relation
    with fertility and endometriosis, Fertil. Steril. 41:872-877.
DiMarzo, S. J. and Rakoff, J. S., 1986, Intrauterine insemination
    with husband's washed sperm, Fertil. Steril. 46:470-475.
Edwards, R. G., 1980, Conception in the Human Female, Academic Press,
    New York.
Edwards, R. G., Fishel, S. B., Cohen, J., Fehilly, C. B., Purdy,
    J. M., Slater, J. M., Steptoe, P. C., and Webster, J. M., 1984,
    Factors influencing the success of in vitro fertilization for
    alleviating human infertility, J. In Vitro Fert. Embryo Transfer
    1:3-23.
Foreman, R., Cohen, J., Fehilly, C. B., Fishel, S. B., and Edwards,
    R. G., 1984, The application of the zona-free hamster egg test
    for the prognosis of human in vitro fertilization, J. In Vitro
    Fert. Embryo Transfer 1:166-171.
Gordts, S., 1986, Pregnancy following gamete intrafallopian transfer:
    a case report, J. In Vitro Fert. Embryo Transfer 3:260-261.
Guastella, G., Comparetto, G., Gullo, D., Palermo, R., Venezia, R.,
    Cefalu, E., Ciriminna, R., Salerno, P., and Cittadini, E., 1985,
    Gamete intra-fallopian transfer (GIFT):   a new technique for the
    treatment of unexplained infertility, Acta Eur. Fertil. 16:311-
    316.
Guastella, G., Comparetto, G., Palermo, R., Cefalu, E., Ciriminna,
    R., and Cittadini, E., 1986, Gamete intrafallopian transfer in
    the treatment of infertility: the first series at the
    University of Palermo, Fertil. Steril. 46:417-423.
Guzick, D. S., Wilkes, C., and Jones, H. W., Jr., 1986, Cumulative
    pregnancy rates for in vitro fertilization, Fertil. Steril.
    [Suppl.]:20 (abstract).
Hoing, L. M., Devroey, P., and Van Steirteghem, A. C., 1986, Treat-
    ment of infertility because of oligoasthenoteratospermia by
    transcervical intrauterine insemination of motile spermatozoa,
    Fertil. Steril. 45:388-391.

Hull, M. E., Magyar, D. M., Vasquez, J. M., Haynes, M. F., and
    Moghissi, K. S., 1986, Experience with intrauterine insemination
    for cervical factor and oligospermia, Am. J. Obstet. Gynecol.
    154:1333-1338.
Jones, H. W., Jr., 1986, The impact of in vitro fertilization on the
    practice of gynecology and obstetrics, Int. J. Fertil. 31:99-
    111.
Kerin, J. F. P., Kirby, C., Peek, J., Jeffery, R., Warnes, G. M.,
    Matthews, C. D., and Cox, L. W., 1984, Improved conception rate
    after intrauterine insemination of washed spermatozoa from men
    with poor quality semen, Lancet 1:533-534.
Koninckx, P. R., Heyns, W. J., Corvelyn, P. A., and Brosens, I. A.,
    1978, Delayed onset of luteinization as a cause of infertility,
    Fertil. Steril. 29:266-269.
Madden, J. D., Bookout, D. M., Silverstein, E. H., Stephens, S. R.,
    Spelts, M. A., Toofanian, A., and Stein, D. W., 1986, GIFT-
    pregnancy by translaparoscopic gamete intra-fallopian transfer,
    Dallas Med. J. 72:15-17.
Margalioth, E. J., Navot, D., Laufer, N., Lewin, A., Rabinowitz, R.,
    and Schenker, J. G., 1986, Correlation between the zona-free
    hamster egg sperm penetration assay and human in vitro
    fertilization, Fertil. Steril. 45:665-670.
Marik, J. and Hulka, J., 1978, Luteinized unruptured follicle
    syndrome: a subtle cause of infertility, Fertil. Steril.
    29:270-275.
Molloy, D., Speirs, A., du Plessis, Y., Gellert, S., Bourne, H., and
    Johnston, W. I. H., 1985, Gamete intra-fallopian transfer, Med.
    J. Aust. 143:428.
Molloy, D., Speirs, A., du Plessis, Y., McBain, J., and Johnston, I.,
    1987, A laparoscopic approach to a program of gamete intrafal-
    lopian transfer, Fertil. Steril. 47:289-294.
Nemiro, J. S. and McGaughey, R. W., 1986, An alternative to in vitro
    fertilization-embryo transfer: the successful transfer of human
    oocytes and spermatozoa to the distal oviduct, Fertil. Steril.
    46:644-652.
Poindexter, A. N., III, Thompson, D. J., Gibbons, W. E., Findley,
    W. E., Dodson, M. G., and Young, R. L., 1986, Residual embryos
    in failed embryo transfer, Fertil. Steril. 46:262-267.
Quinn, P., Kerin, J. F., and Warnes, G. M., 1985, Improved pregnancy
    rate in human in vitro fertilization with the use of a medium
    based on the composition of human tubal fluid, Fertil. Steril.
    44:493-498.
Seppälä, M., 1985, The world collaborative report on in vitro ferti-
    lization and embryo replacement: current state of the art in
    January 1984, Ann. N.Y. Acad. Sci. 442:558-563.
Sher, G., Knutzen, V. K., Stratton, C. J., Montakhab, M. M., and
    Allenson, S. G., 1984, In vitro sperm capacitation and
    transcervical intrauterine insemination for the treatment of
    refractory infertility: phase I, Fertil. Steril. 41:260-264.

Sokoloski, J. E. and Wolf, D. P., 1984, Laboratory details in an in
    vitro fertilization and embryo transfer program, in: Human In
    Vitro Fertilization and Embryo Transfer (D. P. Wolf and M. M.
    Quigley, eds.), Plenum Press, New York, pp. 275-296.
Stanger, J. D. and Yovich, J. L., 1984, Failure of human oocyte
    release at ovulation, Fertil. Steril. 41:827-832.
Steptoe, P. C. and Edwards, R. G., 1978, Birth after the reimplan-
    tation of a human embryo, Lancet 2:366.
Toffle, R. C., Nagel, T. C., Tagatz, G. E., Phansey, S. A., Okagaki,
    T., and Wavrin, C. A., 1985, Intrauterine insemination: the
    University of Minnesota experience, Fertil. Steril. 43:743-747.
Trounson, A., Howlett, D., Rogers, P., and Hoppen, H.-O., 1986, The
    effect of progesterone supplementation around the time of oocyte
    recovery in patients superovulated for in vitro fertilization,
    Fertil. Steril. 45:532-535.
Trounson, A. O., Leeton, J. F., Wood, C., Webb, J., and Kovacs, G.,
    1980, The investigation of idiopathic infertility by in vitro
    fertilization, Fertil. Steril. 34:431-438.
Wiltbank, M. C., Kosasa, S., and Rogers, B., 1985, Treatment of
    infertile patients by intrauterine insemination of washed
    spermatozoa, Andrologia 17:22-30.
Yee, B., 1986, Update: Gamete transfer and embryo cryopreservation,
    Fertil. News 20(#1):3-5.
Yovich, J. L., Matson, P. L., Turner, S. R., Ricardson, P., and
    Yovich, J. M., 1986, Limitation of gamete intrafallopian
    transfer in the treatment of male infertility, Med. J. Aust.
    144:444.

15

ENDOCRINE MEASUREMENTS

Albert S. Berkowitz

1. INTRODUCTION

    The endocrinology laboratory participating in an in vitro
fertilization-embryo transfer (IVF-ET) program must provide timely
and accurate data throughout the period of follicular recruitment.
In addition, the laboratory may also be responsible for confirming
pregnancy and contributing to the understanding of luteal phase
function during successful and unsuccessful treatment cycles. The
correct interpretation of daily hormone assay results will enhance
the chance of obtaining a pregnancy during any given cycle since

these results directly influence the timing and amount of medication
provided and, therefore, the length of the "follicular" phase.
Accurate timing for the administration of human chorionic gonadotropin (hCG) which induces completion of oocyte maturation can be
assisted by measurement of gonadotropins, primarily luteinizing
hormone (LH), during the latter part of the recruitment portion of
the treatment cycle. Even though distinct and different problems may
be encountered during the measurement of steroid and protein
hormones, certain techniques may be applied to all hormone assays to
increase validity and clinical relevance.

The results of hormone assays, usually in tandem with estimation
of follicular diameter obtained via ultrasound, can be used by the
clinician to determine the progress of follicular recruitment. The
result of each day's hormone measurement is an independent entity,
but it is particularly important when considered in light of the
previous assays. The endocrine laboratory must be able to guarantee
(within statistical guidelines) that results are subject to as few
perturbations as possible. In fact, most assay systems are affected
by both human and machine error, as well as variations in reagents.
The successful laboratory recognizes the potential errors and insti-
tutes procedures to identify them when and as they occur. It is
naive and counterproductive to believe that problems will not and do
not arise; only stringent quality control methods will permit
accurate assessment of a patient's hormonal milieu.

The answers to three questions are of paramount importance to
the success of an endocrinology laboratory. Coupled with this
information is the fact that the laboratory director should have a
broad background in reproductive physiology and endocrinology, as
well as a history of successful interaction with both clinicians and
basic scientists. In short, no contributor to the IVF-ET program can
exist in a vacuum; information from each segment of the program must
be shared. The three key questions (in order of complexity) are:
1) What information is important to the clinician and other team
members?; 2) What impact do results from the endocrinology laboratory
have on follicular recruitment?; and 3) How does the endocrinology
laboratory maintain efficient and meaningful operation?

2.   LABORATORY RESULTS

It is perhaps odd to discuss the information provided to the
clinician before documenting the methods by which the information is
obtained, but most laboratories will be judged solely by their timely
reporting of endocrine values. The hormone data becomes part of the
patient's continuing record which also includes daily follicular
measurements and medication doses. Our program depends on rapid
turnover of samples, and the laboratory attempts to provide hormone
data within 4 hr of the time the blood reaches the lab. This

schedule allows the clinician time to interpret or question the data,
and schedule a meeting with the patient if necessary. It also allows
the nurse coordinator time to administer medication either early or
late in the afternoon, depending on patient constraints. The earlier
the decision to administer hCG can be reached (and therefore whether
or not to collect extra blood samples), the more likely an endogenous
LH surge (if present) will be detected. Since we reschedule
follicular aspiration if a surge of LH is detected several hours
prior to the administration of hCG, the rapid reporting of results
may spare a patient surgery if her ovulation has already fallen
outside the surgical "window."

3.   IMPACT OF LABORATORY RESULTS

      Hormone measurements generally complement and augment informa-
tion obtained via ultrasound measurement of follicular diameter.
Occasionally, follicles will appear to increase in diameter without a
concomitant, detectable rise in estradiol. The patient who has
developed ovarian cysts may present such a picture, and it is
incumbent upon the endocrinology laboratory to substantiate its
values in this situation. A patient exhibiting this response will
often be allowed "one more day" to see if the situation resolves
itself, but the cycle is often terminated. Another (more common)
problem is the patient who shows an early plateau or a decrease in
peripheral estradiol levels during follicular recruitment. Foremost
is the ability to discern a statistical rise, fall, or plateau
(versus apparent changes). The demarcation of relevant changes in
serum LH is also critical when attempting to discern the actual onset
of an endogenous LH surge. The definition of each patient's baseline
is critical in the recognition of the surge. Levels of serum
estradiol and/or LH generally reinforce other data indicating
successful recruitment; this data may, however, signal the end of an
unsuccessful recruitment cycle even if sonograms indicate follicular
growth.

4.   OPERATION OF THE LABORATORY

      4.1. Choice of Kits

         In order to provide consistent, meaningful results, a
laboratory must maintain a rigorous quality control program that
extends to the choice of reagents and/or kits.

      All of our results are generated via radioimmunoassay (RIA) or
immunoradiometric assay (IRMA) technology; we maintain this methodology because of an investment in equipment (gamma counter, centrifuges, etc.) and a belief that these systems are sensitive and
biologically relevant. A large number of commercial kits are

available to measure reproductive hormones, and we have made choices
based on comparisons with assays used and/or developed over the past
15 years in this laboratory. We formerly relied heavily on reagents
supplied by the National Pituitary Agency of the NIH to accomplish
research studies involving humans, and have chosen commercial kits
for LH and human chorionic gonadotropin (hCG) based on comparisons
with those reagents. Protein hormone assays are relatively simple
and straight forward as compared to steroid systems, which rely on
tedious extraction techniques, inefficient and time consuming
quantitation of β-radiation, correction factors for extraction
efficiency, and the exposure of personnel to potentially dangerous
solvents and fluor systems. Our development and use of these steroid
assays has, however, given us the experience necessary to evaluate
critically commercial steroid RIA kits in light of our admittedly
time consuming but highly accurate in-house systems.

4.2. LH Assay

        In choosing a system to measure LH, speed, accuracy, and
specificity must be considered. Accuracy can be defined as the
ability to measure correctly the LH content of a sample, and specificity as the ability to discern LH from other hormones; the latter
criterion is mainly an attribute conferred on the assay system by the
specific antibody or antibodies used. For years we depended on a
double-antibody RIA which required incubation for 96 hr at 22°C; the
RIA was specific, but obviously time-consuming and unsuitable for an
IVF-ET program. In an effort to modify this system, samples were
incubated with the specific antibody at 37°C prior to the addition of
the tracer; after addition of the tracer, incubation continued at
37°C and the reaction was terminated and precipitated with goat anti-
rabbit gammaglobulin (GARGG; "2nd antibody") and polyethylene glycol
(PEG). While this procedure produced results in only 5 hr, the assay
was 5-fold less sensitive than the 96-hr system (minimum reliable
potency estimates for the 96-hr and 2-hr assays were 1.5 and 7.5
mIU/ml, respectively). However, since all samples obtained from any
given patient were quantitated in the same assay, this system was
rigorous enough to detect a relative surge of LH. The detection of
the surge, not the accuracy of the individual potency estimates, was
the objective. Hormone measurements reported by us are all based on
the 96-hr system (Quigley et al., 1984a,b).

        Recent advances in antibody production via hybridoma culture
techniques have resulted in the availability of kits that are
accurate, sensitive, and provide results in a fraction of the time
previously required. This laboratory presently relies on Serono
Diagnostics (Norwell, MA) LH MAIAclone system. This kit's virtues
include minimal pipetting steps, color-coded reagents, magnetic
sedimentation (which replaces centrifugation), and an incubation
period of only 15 min at 37°C. The minimum reliable potency estimate

of the LH MAIAclone system is 2.0 mIU/ml as compared to 1.5 mIU/ml
obtained in the sensitive 96-hr assay.

4.3. Estradiol Assay

This laboratory resisted the use of commercial kits for the
quantitation of estradiol (and other steroids) for many years. Cross
reactivity of supposedly specific antibodies was a very real problem,
as was the (lack of) range of the dose-response curve. Ovarian
stimulation caused by such medications as human menopausal gonadotropins results in the recruitment of multiple follicles and hyper-
physiological levels of estradiol. Stimulated cycles may result in a
circulating level of estradiol which reaches 10 times that normally
observed. We required one system that accurately measured estradiol
during stimulated cycles and could also detect the low level of
estradiol associated with various aspects of endocrine-related
infertility. A direct double antibody RIA for estradiol is presently
employed (Pantex, Santa Monica, CA) that exhibits low cross reactivity with other steroids and can accurately quantitate estradiol
between 35 and 3000 pg/ml. There is a total of 75 min incubation
time, and a 15-min centrifugation. This system permits results to be
reported within 3.5 to 4 hr of receipt of the blood sample. A
comparison of the direct Pantex system with a (subsequently abandoned) system involving extraction and chromatography (Haning et al.,
1984) convinced us to evaluate the Pantex direct assay kit.

The evaluation of the Pantex direct estradiol kit involved
several steps. A large backlog of stored (and quantitated) samples
enabled us to run entire patient cycles in the new system. We
specifically chose cycles that exhibited the spectrum of changes in
estradiol, a rapid increase, slow increase, no increase, or a
statistical decrease after an increase. We also reran extremely low
($\leq$ 30 pg/ml) and high ($\geq$ 2000 pg/ml) samples. Aliquots of patient
samples were also spiked with estrone and estriol (up to 500 pg/ml),
testosterone (up to 5000 pg/ml), and progesterone (up to 50,000
pg/ml). Purified estradiol was also added to aliquots of patient
samples and steroid-stripped serum to ascertain the "recovery" of the
added estradiol. Table I is a summary of estradiol recovery and
estrone cross reactivity during evaluation of the Pantex direct
estradiol RIA kit.

4.4. Choice of Quality Control Samples

Most RIA or IRMA kits include one or more samples to be run
in each assay as a control for interassay variability. The samples
are in actuality, often relabeled standards and may not adequately
represent or approximate the milieu (proteins, hormones, etc.) of
human serum samples. The laboratory also has no control over the
potency of these samples, e.g., where they fall on their dose-
response curve. There are several commercially available immunoassay

Table I. Recovery of estradiol and estrone from stripped serum and
patient serum during evaluation of the Pantex direct
estradiol RIA Kit[a]

Treatment	pg/ml	% Recovery	% Recovery of added hormone
Stripped Serum (SS)	<1	0	
SS + 250 pg estradiol	258	103	
SS + 500 pg estradiol	476	93.4	
2016-3	65	100	
2016-3 + 250 pg estrone	67	103	0.8
2016-3 + 500 pg estrone	74	114	1.8
2016-3 + 250 pg estriol	70	108	2.0
2016-3 + 500 pg estriol	71	109	1.2

[a]Results are expressed as percent of the unaugmented sample and/or
as percent recovery of the added hormone.

control preparations, such as American Dade's Tri-Rac R Tri-level
Immunoassay Control (American Dade, Miami, FL). Multi-level commercial controls are accompanied by potency estimates derived from
several methodologies, but not all methodologies (especially new
ones) are always represented in statistically meaningful numbers.
For this reason, as well as cost containment, our laboratory has
developed in-house control serum pools. These pools are derived from
pooled patient samples and are often augmented by the addition of
purified hormone. We use the same control pools for 8 months to 1
year; these pools are especially critical when lots of reagents
change. Figure 1 indicates the potency of our 2 estradiol pools in
relation to the estradiol dose-response curve. The development of
in-house pools also allows a laboratory to tailor results that are
meaningful for its procedures. Even a cursory examination of potency
estimates provided with commercial products indicates the divergence
of results obtained from a single sample run with different reagents.

We find the inclusion of in-house control samples especially
critical in eliminating false-positive results in pregnancy testing.
The entire staff of our IVF program endures the same anxious 2 weeks
as the patient waiting for a post-transfer pregnancy test. Pregnancy
testing is accomplished via a quantitative βhCG serum IRMA (Serono
Diagnostics, Norwell, MA) that is up to 50 times more sensitive than
qualitative urine tests frequently used in the physician's office; we
can confidently detect 0.4 mIU/100 µl serum, i.e., 4 mIU/ml. This
laboratory has quantitated samples of up to 60 mIU/ml in serum from

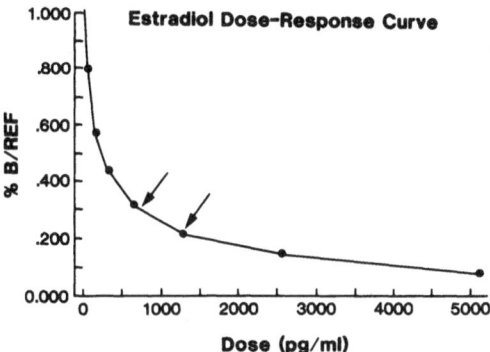

Fig. 1. Potency of two estradiol quality control pools superimposed
         upon the estradiol dose-response curve. These pools repre-
         sent values commonly observed in many patients stimulated
         with Serophene and Pergonal.

patients who only a few hours earlier had a negative urine pregnancy
test.   Most (if not all) serum RIA and IRMA tests have an inherent
level of background "noise," and hCG is clearly no exception.   We
routinely can detect 2-3 mIU/ml βhCG in nonpregnant female and male
samples when kits are fresh.   For this reason, nonpregnant and
pregnant female serum controls are routinely included in every assay
to account for background levels of βhCG and help to avoid misre-
porting a pregnancy.

     An additional facet of quality control testing is membership in
an external proficiency testing program, such as that provided by the
College of American Pathologists (CAP).   Through a proficiency
testing program, a laboratory may compare its results with other labs
using the same or different technology.

     4.5. Data Reduction

          After properly delimiting appropriate assay methodology,
choosing a method for determining potency estimates of patient
samples from raw counts is of paramount importance.   Data reduction
has been (and can still be) done by hand by plotting on graph paper,
but in my opinion, this method is at best "semi-quantitative."
Subtle day-to-day changes can be missed, to say nothing of the
procedure being inefficient and tedious.   Most laboratories use some
type of computerized data reduction, be it via a dedicated computer
on the counting instrument or a free-standing unit.   The choice of
type of data reduction is more important than what computer runs the
program.   Data processing has progressed through point-to-point
linear interpolation, linear-log relationships, polynomial curve-
fitting, spline-fitting and logit transformation to the now common

four-parameter logistic curve-fitting routines. Four-parameter curve-fitting routines give linearization of results in about 99% of all cases. Use of a 4-parameter program also provides another type of quality control for the laboratory, namely, an analysis of the dose-response curve. If curve parameters shift beyond preset limits, assay results can be severely affected. Dose-response curve analysis can be used to identify deleterious changes in kit reagents and technical errors. An entire lot of assay kits was once rejected because dose-response curve analysis (the statistical comparison of individual and/or groups of dose-response curves in order to detect changes in assay performance) indicated the center of the curve had shifted more than 2 standard deviations from previous kit lots. Table II indicates a computer-generated analysis of 20 estradiol assays. Parameter A is the theoretical limit of binding with no cold mass ("B_0"), while parameter D is the limit of the curve at an infinite dose of cold mass ("nonspecific" binding). Parameter B is the slope of curve at parameter C, the midpoint. This laboratory accepts a coefficient of variation (cv) of no more than 10 percent for any parameter (Becan-McBride, 1982) during our continuous dose-response curve analysis. The computer automatically batches results in groups of 20, which means we fully analyze our estradiol results at a minimum of every three weeks. Often reagent lots change during a series of 20 assays, but we endeavor not to change reagent lots during any given patient's follicular recruitment period. We have conducted studies in which samples from patients were repeated in subsequent assays to determine if there was an advantage to running all samples in the same assay. The results indicated that patient progress, treatment, outcome, and daily values were the same when samples were assayed one at a time versus in a batch. Statistical analysis becomes more complicated when reagent lots change during the accumulation of assay statistics, but the program allows specific assays to be "flagged" and "unflagged" to account for reagent change. In this way, several analyses from the partitioned data can be obtained.

4.6. Running Assays

4.6.1.   Blood Collection/Sample Logging

In our IVF program, hormone and ultrasonographic monitoring begins on day 8 of the menstrual cycle. The initial blood sample (and each subsequent morning sample) is obtained between 0730 and 0830. All blood is collected in Serum Separation Tubes ("Red/Gray"; Becton-Dickinson, Rutherford, NJ) and delivered to the laboratory for centrifugation and logging into our data files. Each patient is assigned a unique identification number by the nurse-coordinator, and the Endocrinology Laboratory then assigns a "dash number" to each sample. The dash number is not unique to a particular date, but only to a particular sample. In this way, every sample carries an identification number consisting of the patient

Table II.   Computer-generated analysis of the estradiol assay
            dose-response curve[a]

QC DATA FOR PROTOCOL # 1,   ASSAY TYPE: 4 PARAMETER
PANTEX E_2

Assay #	PARM A	PARM B	PARM C	PARM D	CNTRL 1	CNTRL 2
43	17211	0.7610	223.6	887.6	707.0	1211.0
44	16631	0.7742	225.1	835.1	728.0	1237.0
45	16913	0.7876	236.9	1036.0	757.0	1421.0
46	16433	0.7752	215.8	879.4	697.2	1285.0
47	16097	0.7716	218.6	916.0	703.7	1284.0
48	15715	0.7633	220.0	848.5	710.0	1278.0
49	15076	0.7751	238.0	836.3	713.0	1247.0
50	15547	0.7989	224.4	881.3	639.2	1138.0
51	14920	0.7947	231.9	918.7	662.4	1180.0
52	15112	0.7818	219.9	915.8	738.2	1265.0
53	14542	0.7827	218.5	837.2	659.6	1213.0
54	14045	0.7868	241.0	831.3	690.9	1288.0
55	13995	0.8283	249.3	822.7	630.0	1170.0
56	14263	0.7973	226.3	826.0	671.2	1201.0
57	14338	0.7981	233.1	867.9	713.0	1419.0
58	14204	0.7957	231.9	792.9	733.9	1339.0
59	13332	0.7722	240.7	710.1	687.5	1351.0
60[b]	13948	0.7856	220.8	796.6	1061.0	1005.0
61	13598	0.7989	228.1	800.8	709.1	1374.0
62	18089	0.7853	219.1	985.5	674.6	1226.0
MEAN	15266	0.7857	228.5	864.7	696.1	1270.0
+2SD	17933	0.8173	247.4	1009.0	763.3	1432.0
−2SD	12600	0.7541	209.7	720.6	628.9	1108.0
CV%	8.7	2.0	4.1	8.3	4.8	6.4

[a]This data is used to evaluate the daily performance of the assay
as well as indicate subtle long-term trends not noticeable on a
day-to-day basis.
[b]Data not used in calculations.

identification number <u>and</u> a chronologically assigned dash number.
The initial sample is "-0." A chronological master list of all
patients is also maintained in the laboratory. Figure 2 illustrates
the type of card system utilized to maintain our laboratory's "first
line" of data management. After assignment of the identification
number, all pertinent information is entered onto the patient data
card. This information is important in order to properly assess the
daily endocrine and ultrasonographic results. Once all data is
recorded, the blood is decanted in a storage vial (Kimble Glass,
Toledo, OH) and a separate aliquot (in a 12 x 75 mm culture tube) is
prepared for use in the daily assay. A second aliquot is prepared,
labeled, and frozen for use in the LH assay. By aliquoting the
sample in this manner repeated freezing and thawing is avoided. The
major portion of the sample is also safeguarded from mishandling and
cross-contamination; this may prove to be especially serendipitous
when future studies requiring serum samples are suggested. A data
("flow") sheet is completed for each assay. The dose-response curve
and quality control samples are printed on each flow sheet along with
a section used to identify each assay. In case troubleshooting
becomes necessary, the assay date and technician, as well as the
reagent lot and expiration date are readily available. Complete
instructions for each assay are also printed on each flow sheet.
Each sample is placed on the flow sheet according to patient name,
identification number, and date drawn. This information is cross-
checked with the information on the data card before results are
reported.

LAST	FIRST	M.I.			ENDOCRINE DATA				M.D.	M __ F __	IVF Rx# __	ID#
Date Drawn	Time	Dash No.	Days Since LMP	Age	Diag	Test Req't	E2	Prog	T	DHEA-S	LH	FSH	Prl	BhCG

Fig. 2.   A representation of the card used to record patient data.
          The unique ID # and Dash # are used to track samples
          throughout the laboratory.

4.6.2.   Laboratory Equipment and Supplies

a.   Blood Collection

Centrifuge (blood): Fisher Centrific Model 228 (Pittsburgh, PA)

Blood Collection Tubes: Becton Dickinson SST Vacutainer #6510 (Rutherford, NJ)

Blood Storage Vials: Kimble 19 x 65 mm 3 dram #60910-L (Toledo, OH)

b.   RIA/IRMA Laboratory

Waterbath: Precision Scientific Model 86 (Chicago, IL)

Centrifuge (RIA): Damon/IEC DPR6000 (Needham Heights, MA)

Gamma Counter: Rohm and Haas Micromedic 4/200 Plus with Micromedic Assay Compucenter (MACC) (Horsham, PA)

Repetitive Pipet: Brinkmann Instruments Eppendorf Repeator 4780 (Fisher #21-380-8, Pittsburgh, PA)

Multivolume Pipets: Gilson Electronics Pipetman P20, P200, P1000, P5000 (Middleton, WI)

LH IRMA Kit: Serono Diagnostics LH MIAclone 13203 (Norwell, MA)

βhCG IRMA Kit: Serono Diagnostics βhCG MAIAclone 12304 (Norwell, MA)

Estradiol RIA Kit: Pantex Direct 174 (Santa Monica, CA)

c.   Troubleshooting

Geiger Counter: Ludlum Measurements Model 2 (Sweetwater, TX)

4.6.3.   Assay Procedures

Some general considerations will be mentioned as
far as the actual running of assays. Each manufacturer should
provide a complete list of reagents, instructions, suggested pattern
of tube setup, performance characteristics, and control results. It
is strenuously suggested that instructions concerning sample and
reagent volumes, incubation timing and temperatures, and centrifugation (for other precipitation techniques) be closely followed. Until
you become thoroughly familiar with the performance of the kit in
your laboratory, change nothing. If more than one technician is
likely to run any given assay, strict adherence to protocol is not
only a matter of good laboratory technique, but of survival of
confidence in the laboratory. In this laboratory, when a technician
begins work on a system unfamiliar to him/her, the individual
initially re-runs previously reported samples and controls. The
individual also will frequently include the previous days' samples in
their current assay in order to assess interassay variability and
technical competence. The new technician's dose-response curve
parameters are also compared to the accepted laboratory norms. If
all performance facets check out, the probationary period (at least
5-6 independent assays) is complete.

The first step in reporting results is recording the assay dose-
response curve statistics in a quality control notebook. Figure 3
depicts the form "shell" used for the estradiol assay. The recorded
data is a twin to that automatically stored in the gamma counter's
memory, and, in fact, could be generated by the computer after each
assay. However, by transferring the data by hand into the quality
control notebook, the individual running the assay is forced to
mentally review the current assay in light of the previously
completed assays. The entries at the top of each quality control
page represent a summary of the previous assays to date. The entry
of the dose-response curve data is the initial "review" of the
acceptability of the assay. Any one parameter on any given day may
be no more than 3 standard deviations from the mean; all other values
must be within 2 standard deviations, or the assay is repeated. Once
the dose-response curve parameters and quality control sample values
have met statistical requirements, the results are transferred to the
patient data card and reported to the nurse-coordinator.

A final facet involved in the reporting of daily results is
substantiating the difference between 2 or more consecutive samples.
Referring to the summary data in Table II, confidence limits may be
placed on each control value (utilizing the cv). The limits of
control 1 are 662.6-729.5 (696.1 ± 4.8%) and the limits of control 2
are 1188.7-1351.3 (1270.0 ± 6.4%).

Increases (or decreases) must exceed the confidence limits
(e.g., no overlap) to be considered statistically significant, and

4 PARM RIA QC for: Date (#)	$\frac{Bo}{T}$%	A Bo calc.	B Slope	C mid-point ( )	D NS cpm calc.	20% Curve ( )	80% Curve ( )	QC 1 ( )	QC 2 ( )	QC 3 ( )	Lot #	Assay by:	QC Updated: __/__/__ MACC disk ID#
$\bar{X}$ (n=)													
-2 Std. Dev.													
+2 Std. Dev.													
C.V. %													
21													
22													
23													
24													

Fig. 3.   A representation of the estradiol assay page in the quality control notebook. The 20 assays on each page represent the current data stored in the gamma counter's computer. The quality control data is updated after every 20 assay runs.

more importantly, biologically relevant. We generally assume the cv will slowly increase as the limits of each assay are reached (the "tails" of the dose-response curve) and will compute sample confidence limits based upon a sample's position on the dose-response curve **relative** to the controls.

4.6.4.   Preventative Maintenance/Troubleshooting

Many equipment and procedural problems can be prevented (or at least delayed) by adhering to a sensible schedule of preventative maintenance. The key to our operation is the gamma counter, a Micromedic 4/200 Plus AGC with a "MACC" Compucenter, a dedicated computer for data reduction and storage (Rohm and Haas, Horsham, PA). Each day begins with the following routine: 1) Five empty assay tubes are counted by each of the four detectors (channels), and the mean machine background count/detector is computed. This value is automatically stored and subtracted from the raw counts; 2) A source containing an isotope of extremely long half-life (e.g., ^{129}I; half-life = 1.7×10^7 years) is counted in each detector. We know the number of μCi of ^{129}I in the source, and from that have determined our "potential" detection, namely 197,580 dpm. Dividing the number of counts obtained/detector by 197,580 results in the percent efficiency of each detector. Detector efficiency may not vary by more than 2%; and 3) A special built-in program tests for each detector's "peaking ability." This is a check of the counter's electronics; each detector can vary no more than 2% from the computed "peak mean"; excess variability requires recalibrating the counter's electronic gain.

Our normal machine background averages 50 cpm. This value can vary from 30-75 cpm, but if any detector's background doubles to 100 cpm the tube transport and balance rods, as well as the inside of the detector housing is decontaminated with a product such as Radiacwash (Atomic Products Corp., Center Moriches, Long Island, NY). Even though high background counts would continue to be automatically subtracted from raw counts, the contamination might be spread from tube to tube, detector to detector, or counter to operator. Widely varying day-to-day background counts are inevitably due to sloppy technique in assay preparation, centrifugation, aspiration, or decanting. In order to prevent contaminating areas with radiation, we observe the following safety procedures: 1) radioactive use areas are restricted and clearly labeled; 2) disposable gloves are always worn when handling radioactive materials; 3) distinct disposal barrels are maintained in the laboratory for solid and liquid waste; be sure these barrels are graphically or pictorially labeled to avoid emptying by the cleaning staff; 4) plastic-backed disposable absorbent pads are used on laboratory benches and in hoods, are changed after each assay, and sometimes between steps within an assay; 5) the laboratory is frequently surveyed with a Geiger counter; we complete a multi-point wipe test monthly; 6) kits are stored in a secluded and well-marked area outside the main procedures room; only the kit being used to conduct an assay may be in the laboratory; and 7) quarterly thyroid scans are scheduled and must be completed to maintain our institutional license.

Additional procedures that seem on the surface to be only time-consuming and pedestrian are actually critical to assay precision. They include: 1) maintaining daily records of refrigerator and/or freezer temperature. This is especially critical for long-term storage of patient samples. We suggest purchasing units with temperature alarms or adding the alarms to on-site units. Be sure personnel understand your accepted temperature range for each unit (we use 0 to 6°C for refrigerators, -12 to -25°C for freezers, and -70 to -80°C for ultra-low freezers); 2) maintaining daily records of water bath performance. Several assays require incubation at 37°C with a tolerance of only ± 1°C. Aberrant temperatures will cause major shifts in the dose-response curve. We suggest using thermometers that are calibrated and traceable to National Bureau of Standards specifications, such as the Fisher 15-043 series (Fisher Scientific, Pittsburgh, PA). Use only distilled water in your water bath. As it evaporates, normal tap water may leave heavy mineral deposits on the heating element, resulting in uneven heating and temperature regulation; 3) calibration of pipets. We attempt to maintain pipets at no more than ± 2% of the maximum volume. A calibration kit such as Medical Laboratory Automation's (MLA) Pipet Volume Calibration Kit (Mount Vernon, NY) will help standardize pipets in the 10 to 1000 μl range. It does, however, require a discrete-sample or flow-through spectrometer. One can also simply weigh various volumes of distilled water on a balance such as the Mettler H54 (Mettler Instrument Corp., Hightstown, NJ). Due to the specific gravity of distilled water at 20°C (0.99862 g/ml), 1 μl for our purposes weighs 1 mg; 4) use of electronic timers. We threw out mechanical timers years ago when we discovered assay timing varied up to 9% depending upon which mechanical timer was used. We now use electronic timers (such as Fisher's 06-662-5 and 06-662-6); it is well worth changing the batteries of the portable units on a regular basis rather than risking timer failure during an incubation; and 5) use of a scheduling board. This laboratory is responsible for a general clinic population and acts as a reference laboratory besides working with IVF patients. Every IVF and long-term infertility patient is posted for testing as far in advance as possible. Patient scheduling is reviewed every afternoon in order to efficiently assign the next day's assays and other jobs. The afternoon review also permits critiquing the day's results.

5. THE FUTURE: LESS LAB WORK?

Several new products have been introduced that may help manage the treatment of IVF patients. One, the OvuSTICK[TM] Self-Test (Monoclonal Antibodies, Inc., Mountain View, CA) uses solid-phase mouse monoclonal antibody for human LH bonded to plastic sticks. The antibody binds LH, which then binds to more mouse monoclonal anti-LH linked to alkaline phosphatase. When placed in the proper substrate, a color change occurs. The OvuSTICK is marketed as a product that helps predict ovulation by urinary detection of the "mid-cycle" surge of LH. OvuSTICKS are used in our clinic as an adjunct to the treatment of female infertility, and have been used in several IVF cycles to assess women with long and continued estrogen increase and slow follicular growth. Serum and ultrasonographic monitoring begins on day 8 and Serophene and Pergonal (Serono Laboratories, Randolph, MA) are given starting on day 3; most patients will be ready for hCG after 3 to 4 days of monitoring. Past this point, OvuSTICKS can help ascertain if a surge has begun or is in progress due to the long exposure to a high and increasing estrogen level. The urinary level of LH, however, reflects serum changes with about a 6-hr lag. We, therefore, can use OvuSTICKS to help pinpoint a series of previously drawn blood samples to be studied via our LH IRMA. In all cases examined so far, OvuSTICKS and LH IRMAs have provided complimentary results. It is unclear whether a product measuring urinary LH can in

all cases provide the precise information needed to reschedule oocyte
aspiration, but this methodology does provide, in a noninvasive
manner, information that benefits the IVF patient and endocrinology
laboratory.

Another new technique that may prove useful in predicting the
correct time for oocyte aspiration ("ovulation") is the Cue^{TM}
Ovulation Predictor (Zetek, Aurora, CO). This instrument monitors
electrical readings in saliva and vaginal mucus and claims to predict
ovulation 5 to 6 days prior to the peak of LH. The proposed mechanism for changes in salivary and vaginal mucus conductivity involves
estrogen effects on aldosterone via increasing renin substrate and
activation of the renin-angiotensin system. It is also proposed that
the Cue Ovulation Predictor is accurate during drug-induced ovulation
induction as well as normal cycles (Fernando et al., in press). Only
extensive experience will determine if the Cue Ovulation Predictor
provides the detailed information necessary to determine if and when
an endogenous LH surge has occurred, or whether this instrument will
serve as an adjunct to the familiar quantitative serum LH test.

6.   REFERENCES

Becan-McBride, K. (ed.), 1982, Textbook of Clinical Laboratory
    Supervision, Appleton-Century-Crofts, New York.
Fernando, R. S., Regas, J., and Betz, G., Physiological mechanisms
    associated with ovulation prediction using the Cue^{TM} Fertility
    Monitor, Arch. Gynecol. Obstet., in press.
Haning, R. V., Jr., Boehnlein, L. M., Carlson, I. H., Kuzma, D. L.,
    and Zweibel, W. J., 1984, Diagnosis-specific serum 17β-estradiol
    (E_2) upper limits for treatment with menotropins using a ^{125}I
    direct E_2 assay, Fertil. Steril. 42:882-889.
Quigley, M. M., Berkowitz, A. S., Gilbert, S. A., and Wolf, D. P.,
    1984a, Clomiphene citrate in an in vitro fertilization program:
    hormonal comparisons between 50- and 150-mg daily dosages,
    Fertil. Steril. 41:809-815.
Quigley, M. M., Schmidt, C. L., Beauchamp, P. J., Pace-Owens, S.,
    Berkowitz, A. S., and Wolf, D. P., 1984b, Enhanced follicular
    recruitment in an in vitro fertilization program: clomiphene
    alone versus a clomiphene/human menopausal gonadotropin
    combination, Fertil. Steril. 42:25-33.

16

CRYOPRESERVATION OF MAMMALIAN EMBRYOS

Frank B. Kuzan and Patrick Quinn

1.   INTRODUCTION AND BASIC PRINCIPLES

1.1. History

    Early investigations of the effects of cooling and rewarming cells employed both spermatozoa and ova as experimental material. Spermatozoa were selected because of the ease of collection, inherent motility, and small size. Oocytes were employed mainly for their large size which allowed direct morphological observations. In 1776, Spallanzani investigated the effects of "winter snow and cold" on stallion semen and silkworm eggs. Upon warming, the spermatozoa were "reactivated" and became motile. This was the first report of the successful cold storage of sperm cells. It was not until 1938 that Jahnel, working on the problem of syphilis, found that sperm cells cooled to $-79°C$ retained some motility after 40 days of storage. Luyet and Hodapp (1938) used the vitrification technique of Stiles (1930) to preserve frog spermatozoa successfully. In 1940, Shettles found that survival rates varied among semen samples; however, no greater that 10% viability was maintained in any sample. Hoagland and Pincus (1942) extended Shettles' findings that some males did not produce semen samples suitable for cryopreservation and applied vitrification to human semen with favorable results. Parkes (1945) reported that spermatozoa survived at higher rates when large compared to small volumes were cooled. Parkes disregarded the physical law that cooling rate is proportional to surface area to volume ratio, but suggested some mechanism involving the semen surface area exposed to air as being critical to survival rate. Finally, Polge et al. (1949) discovered the cryoprotectant nature of glycerol. This discovery led to semen storage in farm animals and in 1953 to the first attempt to store human sperm (Sherman, 1973).

Based on the work of Parkes and others, Chang (1947, 1954) investigated low temperature storage of rabbit ova. Chang's contributions included: the recognition of the importance of cooling rate in maintenance of viability; the artificial activation of oocytes by rapid cooling; and the production of litters from embryos previously stored at 0°C. Smith (1952) found that stepwise addition of cryoprotectant to 1-cell rabbit embryos improved viability. Finally, in 1972 two groups working independently (Whittingham et al., 1972; Wilmut, 1972) reported the successful cryopreservation of 8-cell mouse embryos. This success was based on the following principles of cryobiology: stepwise cryoprotectant addition/removal; controlled slow-cooling; initiation of ice crystal formation (seeding); final storage temperature; and warming rate. Since 1972 several thousand mouse, rabbit, and cattle embryos have been successfully stored by cryoprotective techniques which rely on the basic principles Wilmut and Whittingham et al. described in 1972.

1.2. Cooling Rate

Parkes (1945) discovered, quite by accident, that cooling rate is associated with survival rate. This was implied by the various container sizes and volumes of semen he employed for cooling. The large containers with relatively large sample volumes gave the best postthaw motilities; hence, slower cooling rates were associated with higher cell viabilities. This was true because of the physical principle of osmotic dehydration. As ice crystals formed in the suspending solution, the osmolality of this solution increased and the cells responded osmotically by losing water. Thus, slower (relative) cooling rates were associated with greater amounts of cellular dehydration and greater postthaw viabilities. Mazur (1963) was the first to observe that different cells possessed different optimal cooling rates. Based on the hydraulic conductivity (the rate at which a cell responds to an osmotic challenge) and the temperature coefficient of water permeability (the effect of temperature on water movement) for each cell type, Mazur was able to construct a mathematical equation which rendered a quantitative explanation of optimal cooling rate. Leibo et al. (1978) constructed a graph of cooling rate against survival rate based on data from Thorpe et al. (1976, lymphocytes) and Morris and Farrant (1972, erythrocytes). Over a 1000-fold difference in optimal cooling rate was found between ova (0.3°C/min) and erythrocytes (1000°C/min). Mazur's equation would predict this, since ova possess a low hydraulic conductivity and high temperature coefficient of water loss and would respond at a slow rate compared to erythrocytes which have a very high hydraulic conductivity and low temperature coefficient. Slow rates of water loss indicate that comparatively slow cooling rates are required to maintain cellular integrity.

1.3. Seeding

The discussion of osmotic behavior implies that some change has occurred in the suspending medium relative to the cell. The change which occurs as the solution is cooled is the solidification of pure water (ice crystals), leaving behind a hypertonic suspending solution compared to cellular cytoplasm. However, in cases where small volumes are cooled at slow rates, the phenomenon of super-cooling can occur. Supercooled describes a solution which has been cooled to well below the freezing point without ice formation. Supercooling is detrimental because the cell is exposed to dehydrating conditions for a lesser amount of time and the cell is exposed to a rapid temperature shift when ice does form. To prevent supercooling, ice crystal formation is induced (seeding) near the freezing point of the cryosolution. Seeding causes the cryosolution to become hypertonic to the cell and the cell can respond osmotically during the cooling process. In summary, cooling rate varies among cell types and must allow for an osmotic response by the cell. Further, without seeding the cell is exposed to a large sudden temperature shift when ice forms in the supercooled solution. This temperature shift is detrimental to the cell.

1.4. Cryoprotectants

Cryoprotectants improve the survival of cells during the cooling/thawing process; however, the mechanism of action of these compounds is not fully understood. Cryoprotectants are divided into two groups, intracellular and extracellular. Intracellular cryopro-tectants include glycerol, dimethylsulfoxide (DMSO), and 1,2-propanediol (propylene glycol). These are relatively small molecules with polar and lipophilic tendencies that effect their rate of movement into cells. Extracellular cryoprotectants are usually large molecules and include sucrose, polyvinyl pyrrolidone, hydroxyethyl starch, and dextran.

One of the main actions of cryoprotectants is to lower the freezing point of a solution. With intracellular cryoprotectants such as DMSO (1 M), intracellular ice formation may not occur until −40 to −50°C (Whittingham, 1980). Since extracellular ice formation is induced in the cryosolution at seeding, an increased extracellular osmotic pressure results and water flows from the cell to the cryosolution. Additionally, partial cellular dehydration can be achieved by using a cryosolution containing an extracellular cryoprotectant (Renard et al., 1984).

Cryoprotectants may also protect cells by interacting with cell membranes during transitions from fluid to rigid to fluid states during the cooling/thawing cycle. Protection may be afforded by lessening the effects of high concentrations of molecules within the cells caused by dehydration during the cooling process.

Intracellular cryoprotectants diffuse into cells at rates much slower than water due to membrane permeabilities. Thus, when cells are placed into cryoprotectant solutions, they tend to dehydrate and shrink due to the hypertonicity induced by the cryoprotectant. With time osmotic equilibrium is reached and the cell returns to normal size. If, however, the osmotic shock is too great, the cell will not recover. Schneider and Mazur (1984) calculated that 0.4 shrinkage of osmotic volume was the most bovine blastocysts could tolerate. This is why cryoprotectants are added in a stepwise manner. Temperature, cryoprotectant, and cell type must be considered when constructing cryoprotectant addition schedules because this equilibration is an osmotic response.

Removal of cryoprotectant also requires stepwise decreases, since cells loaded with cryoprotectant will swell when placed in medium without cryoprotectant. With stepwise cryoprotectant removal, the cells still swell but not to a lethal extent. Alternatively, cells loaded with intracellular cryoprotectant can be placed into a solution containing a nonpenetrating molecule, such as sucrose. Under these conditions the sucrose counteracts the intracellular cryoprotectant, the cell does not swell but shrinks when placed into the sucrose solution, and the cryoprotectant tends to leave the cell. The cell then reaches normal size when removed from sucrose. In general, the cells of embryos tend to survive shrinkage better than swelling. This is why the sucrose dilution method is successful.

Cryoprotectants can be toxic to cells; thus, strict protocols for addition and removal of cryoprotectants have been developed. In one study propanediol was shown to be less toxic to 8-cell mouse embryos than DMSO (Renard, 1985). To overcome toxicity a mixture of several cryoprotectants may be used at lower individual concentrations. Reduced exposure times at ambient temperatures while possible may result in insufficient equilibration of the cell with cryoprotectant and therefore inadequate protection during freezing and thawing.

In summary, the cell type, rate of addition, rate of removal, toxicity, and whether or not the cryoprotectant penetrates the cell must be considered when selecting a cryoprotectant.

1.5. Final Cooling Temperature

The final cooling temperature before storage in liquid nitrogen is also important. Final cooling temperature determines the hydration state of the cell before intracellular ice actually forms. In general, large cells, which tend to lose water at a slow rate, require lower final temperatures. Two- to 8-cell embryos are cooled to -80°C while blastocyst stage embryos are cooled to a final temperature of -36°C before storage in liquid nitrogen. Based on the length of time required to run the cooling program, these procedures have been termed the slow (-80°C, Whittingham et al., 1972) and fast

(-36°C, Willadsen et al., 1978) programs. Generally, the slow
program takes 3.5 to 4 hr and the fast program 2 to 2.5 hr to
complete; these times include cryoprotectant addition.

1.6. Warming Rate

        Finally, warming rate is dependent on cooling rate and the
final hydration state of the cell. The most profound interaction is
said to be between cooling and warming rates. Leibo et al. (1974)
investigated the effect of differing warming rates when two cooling
rates were employed with 8-cell mouse embryos. When embryos were
cooled at 0.2°C per min, a single warming rate (3°C/min) was
associated with embryo survival. By contrast, embryos cooled at
1.7°C/min could survive a wide range of warming rates (10°C to
100°C/min). Leibo (1981) explained these data based on the hydraulic
conductivity and the final hydration state of the cells. During the
warming process, the cells were hypertonic compared to the suspending
medium, water entered the cell to reach equilibrium, and cell lysis
resulted when the warming rate was too rapid. However, when the
warming rate was within a moderate range, equilibrium could be
maintained and the cells survived.

        Leibo et al. (1974) concluded that the mechanism of cell damage
when slow cooled cells were very slowly warmed (0.3°C/min) was not
immediately apparent. Rall et al. (1984) using a cryomicroscope
addressed the problem of cell death associated with slow-cooling and
slow-warming (0.3°C/min). The results suggested that damage was not
the direct effect of formation, decrystallization, or melting of ice.
Rall suggested that the presence of intracellular ice, although
innocuous, might lead to changes in other critical cellular components. Alternatively, this warming injury may be related in some way
to the water content of the blastomeres. Thus, embryo viability is
best preserved when slow-warming is used in conjunction with slow-
cooling and fast-warming with fast-cooling.

        In conclusion, the basic principles of cryopreservation include:
cell type; cryoprotectant selection; cryoprotectant addition;
seeding; cooling rate; final cooling temperature; hydration state of
the cells; warming rate; cryoprotectant removal; and measure of cell
viability.

2.   APPLICATION OF CRYOPRESERVATION TO HUMAN EMBRYOS

        The replacement of more than three embryos after an in vitro
fertilization stimulated cycle does not improve pregnancy rate (Wood
et al., 1985). Cryopreservation of human embryos alleviates the
problem of multiple gestations due to an excessive number of embryos
transferred in the stimulation cycle (Kerin et al., 1983; Trounson

and Mohr, 1983; Zeilmaker et al., 1984; Fehilly et al., 1985; Quinn and Kerin, 1986).

Cryopreservation can also be used when problems arise due to adverse maternal conditions at the time of embryo transfer (Trounson and Mohr, 1983) and allows for more flexibility in the synchronization of cycles in an embryo/oocyte donation program (Lutjen et al., 1984). Ovarian hyperstimulation has been associated with a decreased uterine receptivity for the embryo; this may be caused by the antiestrogenic effect of clomiphene citrate (Rogers et al., 1984). Thus, it may be more efficient to cryopreserve embryos for subsequent natural ovulatory cycles when the uterine environment may be more conducive for implantation (Zeilmaker et al., 1984). A marked degree of asynchrony occurs in the maturity of oocytes collected in cycles where hyperstimulation has been induced with human menopausal gonadotropin (hMG). Although immature oocytes at the germinal vesicle stage can be matured in vitro, fertilized and viable pregnancies result after transfer (Veeck, 1985); it is likely that the embryos resulting from such oocytes would be out of synchrony with the development of the uterus at the time of transfer. It may, therefore, be more appropriate to freeze such embryos for subsequent transfer in a natural ovulatory cycle.

2.1. Problems of Cell Size and Number

The preimplantation embryo at the early cleavage stages (1- to 8-cell) is composed of some of the largest cells in the body. Freezing techniques, as discussed above, are based on the removal of water from cells prior to the formation of intracellular ice, which is lethal to most mammalian cells. Increased solute concentration that results as pure water freezes both inside and outside the cell also reduces the viability of cells. One of the factors determining the rate at which water can leave the cell during cooling is the ratio of volume to surface area. Since this ratio is very high in the large spherical cells of early embryos, cooling rates 10 to 100 times slower than those required to cryopreserve cell lines, spermatozoa or red blood cells have been used for preimplantation embryos.

Successful methods for cryopreservation of mammalian embryos require the survival of at least 50% of the frozen-thawed blastomeres (Mohr et al., 1985). This is unlike the situation when cell lines comprising many millions of cells are frozen and survival of only 1-5% may be adequate to reestablish the line and thus constitute acceptable cryosurvival. Consequently, freezing methods for mammalian embryos have to be very effective to justify the effort on the part of the patient and medical scientist.

2.2. Factors Affecting Cryosurvival

The main factors affecting the survival of frozen-thawed embryos are the stage of development of the embryos, the cryoprotectant used and the rates of cooling and thawing.

The human embryo appears to be similar to that of the mouse in that all stages from the 1-cell fertilized zygote to expanded blastocyst can be successfully frozen and thawed. There appears to be an interaction with the type of cryoprotectant used, however, which may reflect differences in permeability of the different stages to the different cryoprotectants. Glycerol has been used successfully at the blastocyst stage whereas success with 4- to 8-cell stage embryos is lower (Cohen et al., 1985). DMSO is more permeable than glycerol in 4- to 8-cell embryos and is used by the Monash group for these stages (Trounson and Mohr, 1983). More recently, French workers have shown that propanediol can be used for 1- to 4-cell stage embryos, which survive better under these conditions than 8-cell and later stages (Lassalle et al., 1985).

The rates of cooling and thawing are related to the amount of intracellular ice remaining in the embryos when slow-cooling is terminated at sub-zero temperatures by rapid plunging into liquid nitrogen (Whittingham, 1980). In general, if slow-cooling is terminated at relatively high sub-zero temperatures (-30°C to -40°C), enough water remains in the cells so that the warming rate has to be rapid (> 500°C/min) to allow for rapid dispersal of these ice crystals and prevention of their recrystallization into damaging larger crystals. If slow-cooling continues until -80°C before plunging into liquid nitrogen, the degree of dehydration within the cell is much greater and warming rates have to be considerably lower (4-25°C/min) to allow for adequate rehydration. Thus, optimal warming rates are critically dependent on the cooling procedures used.

2.3. Length of Culture Prior to Freezing

In the two most successful protocols reported in the literature, embryos are frozen at different times of development after insemination. Mohr et al. (1985) culture embryos to the 4- or 8-cell stage (35-48 hr and 60-72 hr after insemination, respectively) whereas Cohen et al. (1985) recommend freezing blastocysts at 116-145 hr after insemination. Because embryos show a decreased viability with increased periods of culture in vitro, the longer period of culture required for embryos to reach the blastocyst stage may act as a quality control to select only those embryos capable of continued growth. Whichever protocol is adopted, the end result in terms of the number of pregnancies initiated as a proportion of the embryos available for freezing at the time of transfer of the fresh embryos is similar (Table I). It would seem more appropriate to freeze

Table I. Comparison of embryo stage at cryopreservation with the proportion of embryos transferred and resulting pregnancy rates

Method	Extra 4- to 8-cell	Percent frozen	Percent transferred	Pregnancy rate/transfer	Pregnancy rate/ extra 4- to 8-cell
			Proportion of embryos at each stage calculated with respect to previous stage		
Mohr et al. (1985)	100	100	50	16.2	8.1
Fehilly et al. (1985)					
5-10 cell	100	100	35	15.4	5.4
Blastocyst	100	31	66	32	6.5

embryos as soon as possible after insemination to minimize the
adverse effects of in vitro culture. Therefore, the method of
Lassalle et al. (1985) using propanediol and day 1 or day 2 embryos
might prove superior to either of the aforementioned procedures.

2.4. Duration of Storage

No known biological activity occurs in cells frozen in
liquid nitrogen and they should be able to exist indefinitely under
these conditions. Only photophysical reactions take place at −196°C,
such as ionizing radiation. This type of radiation could cause
genetic damage during storage at −196°C because the normal enzymatic
repair mechanisms do not function. Experiments with mouse embryos
showed that background radiation over 200 to 1000 years would be
required to cause enough genetic damage to appreciably reduce the
survival of embryos stored in liquid nitrogen (Whittingham, 1980;
Lyon et al., 1981).

2.5. Individual Variation in Success Rates

It has been suggested that there are variations in uterine
receptivity with different hyperstimulation regimes (Rogers et al.,
1984). Trounson (1985) recently presented data showing that the
pregnancy rate obtained with cryopreserved embryos was related to the
outcomes obtained with the transfer of fresh embryos from the same
hyperstimulation treatment cycles. When a pregnancy had not been
initiated with the fresh embryos, only 6% of the women became
pregnant on subsequent transfer of their cryopreserved embryos,
whereas 38% became pregnant after transfer of their frozen-thawed
embryos if they had achieved a clinical pregnancy from the transfer
of fresh embryos. Whether this difference is due to differences in
uterine receptivity or the quality of embryos amongst individuals has
yet to be determined. It does highlight, however, the difficulties
inherent in working in the field of early human development and the
need to apply caution in extrapolating from one group of patients to
another. It is essential, therefore, that any group embarking on the
cryopreservation of human embryos optimize their system with an
animal model such as the mouse.

3.   MATERIALS FOR CRYOPRESERVATION

3.1. Chemicals/Solutions

Two basic types of media are usually employed during the
process of human in vitro fertilization (IVF) and embryo transfer
(ET). Phosphate or Hepes buffered solutions are utilized for
manipulations in room air and bicarbonate buffered medium for 5 to 6%
CO_2 in air applications. Phosphate or Hepes buffered solutions are
usually employed during oocyte retrieval to "flush" lines and dilute

follicular fluid samples. Since Kane (1975) found an absolute requirement for CO_2 by rabbit embryos, bicarbonate based media have been employed for long term culture. The formulae for 2 phosphate buffered saline solutions are presented in Table II. Both have been applied to human IVF and cryopreservation successfully. Dulbecco's phosphate buffered saline, dPBS, is used at the University of Washington for cryopreservation at the blastocyst stage. PBI is used at Cedars Sinai Medical Center for cryopreservation from 1-cell unfertilized to 10-cell embryos. As seen in Table II, the composition of these media are quite similar. Whichever media is used, the chemicals must be of reagent grade with water of the highest quality. At the University of Washington, high pressure liquid chromatography water is employed since the city water supply is inconsistent resulting in variable degrees of "pure" water.

In the United States, the basic IVF growth medium has been Ham's F-10 (HF-10) with a serum supplement. At the University of Washington HF-10 is made on a weekly basis, tested with 2-cell mouse embryos, and discarded after 10 days storage. Medium which does not support the development of more that 85% 2-cell mouse embryos to blastocyst and more than 45% through the hatching process is discarded. The serum supplement has been human fetal cord serum

Table II. Formulation of phosphate-buffered media used in human IVF-ET and for embryo cryopreservation

Component[a]	dPBS mg/100 ml	PBI mg/100 ml
NaCl	800	800
KCl	20	20
Na_2HPO_4	115	114
KH_2PO_4	20	20
Glucose	100	100
Na Pyruvate	3.6	0.07 ml Stock A[b]
$CaCl_2 \cdot 2H_2O$	13.2	0.51 ml Stock B[c]
$MgSO_4 \cdot 7H_2O$	12.1	18
Pen/Strep		6
Phenol Red	---	1

[a]Add all components but $CaCl_2$ to 80 ml dH_2O. Stir until in solution. Dissolve $CaCl_2$ in 5 ml dH_2O; add slowly to above solution. Bring to 100 ml. Add pen/strep solution. Store at 4°C up to 14 days.
[b]Stock A Na Pyruvate .051 g/10 ml
[c]Stock B $CaCl_2 \cdot 2H_2O$ .262 g/10 ml

(hFCS), which is heat inactivated, aliquoted, and frozen within 36 hr of collection. Cord sera is used for up to 3 months and must meet the same embryo culture criteria set for HF-10. (See Chapter 5 for details of quality control testing of medium.)

Synthetic human tubal fluid (HTF) medium (Quinn et al., 1985) has been used successfully by both the Adelaide and the Cedars Sinai Medical Center IVF groups. The composition of HTF is based on human oviductal fluid (Table III) and does not contain the co-factors, vitamins, lipids, and amino acids which are present in HF-10. As with other media, chemicals and water must be of the highest quality and HTF must be tested with mouse embryos prior to use in IVF

Bicarbonate buffered media have been used mainly for embryo growth in a CO_2 environment. Although these media can be used as

Table III.   Composition of synthetic human tubal fluid
             (HTF) medium: a bicarbonate buffered medium
             used in human IVF-ET

Stock	Component	Amount	Volume	Concentration in HTF	Stability
A	NaCl	5.931 g		101.6 mM	3 mo., 4°C
	KCl	0.350 g		4.7 mM	
	MgSO$_4$·7H$_2$O	0.050 g		0.2 mM	
	KH$_2$PO$_4$	0.050 g	100 ml	0.37 mM	
	Na lactate	3.700 ml		21.4 mM	
	Glucose	0.500 g		2.78 mM	
	Penicillin	0.060 g		100.0 U/ml	
B	NaHCO$_3$	1.050 g	50 ml	25.0 mM	2 wk., 4°C
	Phenol Red	0.005 g		0.001%	
C	NaPyruvate	0.051 g	10 ml	0.33 mM	2 wk., 4°C
D	CaCl$_2$·2H$_2$O	0.262 g	10 ml	2.04 mM	3 mo., 4°C
E	Hepes – NaSalt	3.254 g	50 ml		3 mo.,. 4°C
	Phenol Red	0.005 g	Adjust pH 7.5 w HCl		

Final Medium Prep:

	Stock	HCO$_3$ Buffered	Hepes Buffered
Adjust osmolarity to 280	A	1.0	1.0
Gas: 5% O$_2$, 5% CO$_2$, 90% N$_2$	B	1.0	0.16
	C	0.071	0.071
	D	0.115	0.115
	E	----	0.84
	H$_2$O	7.81	7.81

basic cryosolutions, the interaction of pH, bicarbonate ion, and the embryo during the cooling process is poorly understood. Therefore, most IVF programs employ one of the PBS formulations for use in cryopreservation.

3.2. Glassware and Disposables

Glassware which comes into contact with any of the cryosolutions must be thoroughly cleaned and, where possible, dry heat sterilized, 156°C for 3 to 6 hr. A 10% solution of 7 X detergent (Flow Laboratories, Inc., McLean, VA) in deionized water is employed to clean media bottles, erlenmeyer flasks, volumetric flasks, and graduated cylinders. The protocol is as follows: 16 hr soak in 7 X detergent in distilled water, 3x wash in 7 X in distilled water, 10x rinse with distilled water, 5x rinse with Type I water, and place in drying oven overnight. Media bottles are dry heat sterilized; other glassware is covered and stored dust-free until use.

Culture tubes and petri dishes are purchased sterile (gamma irradiated) and, for convenience, pipets are purchased individually wrapped and sterile. Embryo storage containers can be glass ampules (Figure 1), plastic ampules, or French semen straws (Figure 2). A wash procedure similar to the one above has been recommended for glass ampules. An additional boiling step, in 0.5 M HCl, has been recommended for this container type. Also the tendency for these containers to explode upon warming has curtailed their use for cryopreservation. Plastic ampules can be used; however, the thermodynamics of the ampules' plastic walls may not be identical to glass.

Currently, 0.5 and 0.25 cc French semen straws (IMV International Corp., see Table IV) are the most popular containers for cryopreservation. Straws can be purchased in bulk, do not appear to be toxic to embryos, and can be gas sterilized with no adverse effect to the embryo. The only problem with straws is learning how to load embryos into the straw. The most straight forward method of loading straws is with a finely drawn "egg pipet" (Figure 3). A note of caution at this point: all of the procedures described in this chapter need to be thoroughly tested with mouse embryos prior to implementing them in a human IVF program.

3.3. Cryoprotectants

Glycerol (Cohen et al., 1985), DMSO (Trounson and Mohr, 1983), and propanediol (Testart et al., 1986) are three intracellular cryoprotectants which have been used successfully with human embryos. Care must be taken in the storage of these chemicals since DMSO is known to oxidize at a slow rate when stored at room temperature in containers with large air spaces. It is recommended that DMSO be

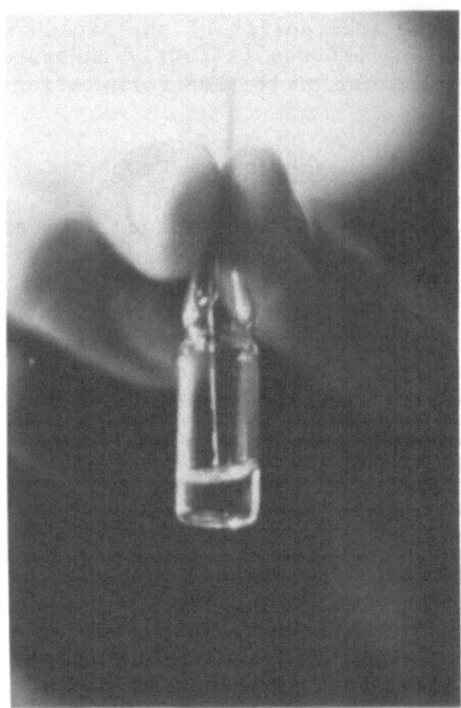

Fig. 1.   Typical glass ampule employed for cryopreservation of
          embryos.

aliquoted into small containers, tightly sealed, and stored at 4°C
prior to use.   This is good practice for all cryoprotectants because
the storage of chemicals in small volumes reduces the chance of
contamination by repeated use.   Sucrose has been used as a diluent
for both cryoprotectant removal and rapid dehydration of blastomeres
prior to cooling (Testart et al., 1986).

3.4. Embryos

        An excess number of embryos is required for cryopreserva-
tion.   Success has been achieved with cryopreservation at all
preimplantation stages of human embryo development (Cohen et al.,
1985; Trounson and Mohr, 1983; Testart et al., 1986).   Specific
techniques relating to embryo stage will be discussed in section 4 of
this chapter.   In general, a greater percentage of early cell-stage
embryos survive cryopreservation when cooled by a slow program with
DMSO as cryoprotectant (Trounson and Mohr, 1983) or by the
sucrose/propanediol dehydration method (Testart et al., 1986).   A
large proportion of blastocyst stage embryos survive cryopreservation

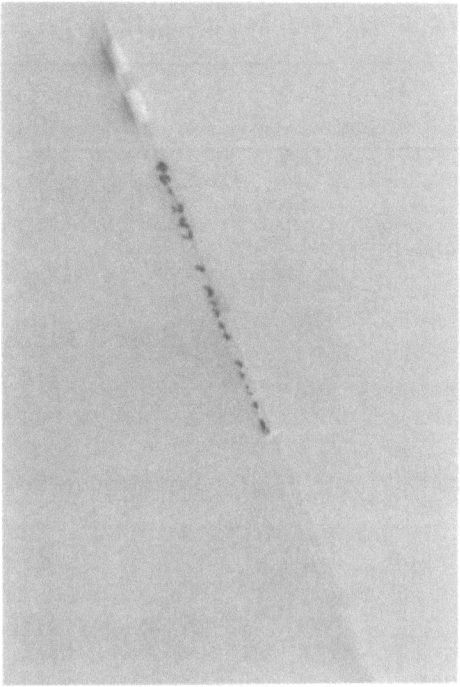

Fig. 2. A 0.5 cc French semen straw which is currently employed for
         cryopreservation of embryos.

with glycerol as cryoprotectant when cooled with the fast program
(Fehilly et al., 1985). Embryos allocated to cryopreservation should
be of fair quality as the stress of cooling, cryoprotectant, and
warming most often results in blastomere death when poor quality
embryos are employed.

3.5. Biological Cell Freezers

        Several programmable cell freezers are currently available
(Table IV). The most popular machines employ nitrogen vapor as a
cooling medium and can be programmed for several rate changes during
a cooling program (Figure 4). Prior to purchase of a machine, become
familiar with the operation of several machines. Mouse embryos must
be employed for checking the entire system, including the freezing
machine, before the cryopreservation of human embryos is begun.

Table IV. Sources of programmable cell freezers

Distributor	Machine / Type
FTS Systems, Inc. P. O. Box 158 Stone Ridge, NY 12484-0158 (914) 687-7664	FTS / Methanol Bath Programmable
Ts Scientific, Inc. P. O. Box 318 Quakertown, PA 18951 (215) 257-4756	Planer / Nitrogen Vapor Programmable
Applied Animal Genetics 625 Imperial Way Napa, CA 94558 (707) 257-7807	Freeze Control / Nitrogen Vapor Programmable
CryoMed East 49659 Leona Dr. Mt. Clemens, MI 48045 (313) 949-4507	CryoMed / Nitrogen Vapor Programmable
CryoMed West 14252 Culver Dr. Irvine, CA 92714 (714) 857-1764	
IMV International Corporation 6870 Shingle Creek Parkway Suite 100 Minneapolis, MN 55430 (612) 560-4986	Minicool / Nitrogen Vapor Programmable
Colorado Agriculture Services 11928 Williams Way Denver, CO 80233 (303) 457-3606	Hoxan / Nitrogen Vapor Programmable
Cryogenetic Technology, Inc. 400 Hoover Road Soquel, CA 95073 (408) 475-7000	CG 100 / Nitrogen Vapor Programmable

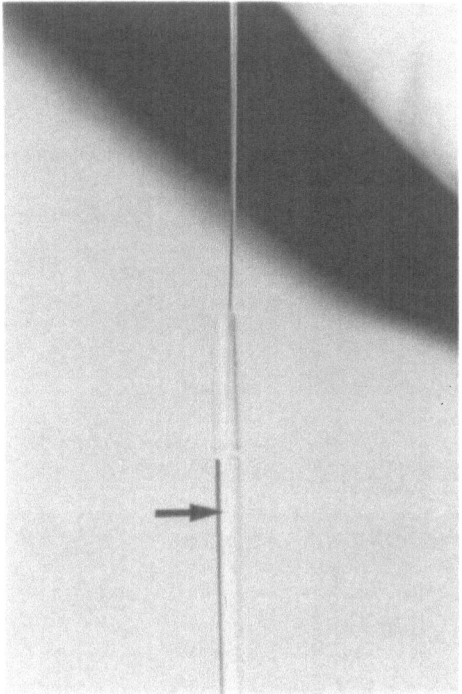

Fig. 3.   Loading embryos into a 0.5 cc French straw. Pipet tip
          (arrow) is well below fluid level within the straw.

4.   PRACTICAL APPLICATIONS

    4.1. Embryo Selection

          Perhaps the most difficult aspect of cryopreservation with
IVF is the selection of embryos for fresh transfer versus those for
cryopreservation. There is a tendency to select the "best" embryos
for transfer while relegating those which have little chance of
initiating a pregnancy to cryopreservation. Recently, our IVF team
has employed a random number table to allocate embryos for transfer
and cryopreservation. Additional criteria for cryopreservation
include: cleavage to at least the 6-cell stage by 60 hr after sperm
addition, minimal numbers of blebs/fragmentations, patient consent,
and an excess of 3 cleaving embryos at the time of transfer. Other
criteria can be found in: Trounson and Mohr, 1983; Zeilmaker et al.,
1984; and Cohen et al., 1985. As an aside, the selection of
livestock and laboratory animal embryos differs radically from that
of human embryos; the "best" embryos produced by livestock are always
reserved for cryopreservation.

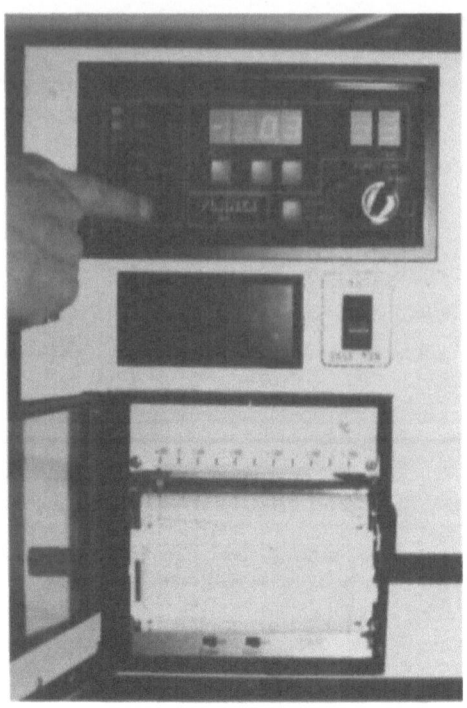

Fig. 4. Programmable biological cell freezer and strip recorder.

Currently, there are no methods available to indicate which
embryo will initiate a pregnancy. The best data concerning embryo
quality versus pregnancy rate comes from the livestock embryo
transfer industry. Data from cattle, sheep and swine point to the
fact that a minimum of 75% of fresh embryos recovered at the morula
to blastocyst stage are capable of initiating pregnancies. This, of
course, does not account for differences in recipient animals,
problem transfers, or environmental shocks given the embryo while
outside the reproductive tract. Pregnancy rates after cryopreservation may range from 0% to 65% with an average of 50%. Thus, a little
more than 25% of the best grade embryos are lost due to cryopreservation. Estimates of pregnancy rate per embryo transferred after in
vitro fertilization in the human is 10-15%, that is, no more than 15%
of cleaving embryos implant after transfer. Extrapolating from the
cattle data, then, only 5-7% of the cryopreserved "best" grade human
embryos can be expected to initiate pregnancies. World-wide data is
not available at present; however, the Monash and Bourne Hall IVF
teams report pregnancy rates approaching 15% with multi-embryo
transfers and 2-5% with single cryopreserved embryo transfers. These
rates are improving as "better" quality embryos are selected for
cryopreservation.

To answer the question of what are best grade embryos, we need to look at the bovine embryo transfer industry where Elsden et al. (1978) compared pregnancy rate and embryo quality. Embryos were graded as poor, fair, good, or excellent based on morphological appearance. Excellent embryos had no visible imperfection; good embryos contained an extruded blastomere or one other minor imperfection; fair embryos were retarded by no more than two days or possessed more than two extruded blastomeres; poor embryos were more than two days retarded and possessed several extruded blastomeres. No difference was found in pregnancy rate between good (58%) and excellent embryos (63%), fair embryos achieved a 30% pregnancy rate, and poor embryos a 12% pregnancy rate. These data indicate that morphology alone is not an indicator of an embryo's ability to implant; however, based on single embryo transfers, good or excellent embryos do have an advantage in achieving pregnancy.

In summary, the selection of cooling program and cryoprotectant depend on the cell-stage embryo to be preserved. Early cell-stage embryos tend to retain viability at higher rates when cooled slowly in DMSO, while fast cooling with glycerol preserves a large percentage of blastocyst stage embryos. New methods are required to assess embryo quality which may in turn increase embryo survivability during cryopreservation.

4.2. Classification of Embryos Postthaw

The critical test of viability for any embryo after various in vitro manipulations, including cryopreservation, is the birth of a healthy offspring following transfer. Gross damage from cryopreservation is easily recognized because the embryo appears degenerate or soon takes on a characteristic necrotic appearance after thawing. Other embryos, however, appear morphologically intact upon thawing but fail to divide in culture or do not produce a pregnancy on transfer. Very little work has been done on these aspects with human embryos and we must rely on experience gained from studies with laboratory animals and domestic species.

The general experience with cryopreservation of bovine morulae and blastocysts is that "better quality" embryos are more likely to survive the trauma of cryopreservation. Certain practices in human IVF and GIFT procedures will confound the random selection of embryos and/or oocytes for fresh transfer or cryopreservation. Since there is a loss of viability following cryopreservation, it is obvious that the better quality embryos are selected for fresh transfer leaving embryos of poorer quality for assignment to cryopreservation. To overcome this problem oocytes can be assigned to various treatments based on numerical order of collection (Table V). The selection of oocytes for the various treatments is still not totally randomized, because the larger, more visible follicles are more likely to contain better quality oocytes and tend to be aspirated first. One advantage

Table V. A scheme for assigning oocytes to treatment during human
          IVF–ET and GIFT in an attempt to randomize allocation

Oocyte no.	Normal IVF	GIFT Normal semen	GIFT Abnormal or unknown semen
1	IVF	GIFT	IVF ± freeze
2	IVF	GIFT	IVF ± freeze
3	IVF	GIFT	GIFT
4	IVF	GIFT	GIFT
5	IVF	IVF ± freeze	GIFT
6	IVF	IVF ± freeze	GIFT
7	oocyte freeze	oocyte freeze	oocyte freeze
8	oocyte freeze	oocyte freeze	oocyte freeze
9	IVF + freeze	IVF + freeze	oocyte freeze
10	IVF + freeze	IVF + freeze	oocyte freeze

of cryopreserved oocytes and embryos in the same treatment cycle is
that the oocytes can be thawed first and, if no fertilization
results, the stored embryos can be thawed for transfer on successive
days.

     One persistent problem facing both the livestock industry and
human IVF is a means of assessing embryo quality. A noninvasive
method which does not compromise the viability of the embryo is
required. Several methods have been suggested for this purpose, none
of which are satisfactory when used alone. However, a battery of
such tests may give a general impression of an embryo's viability.
Following are some representative methods currently employed for
assessing embryo viability.

     4.3. Rate of Cell Division

          One noninvasive method is the rate of cell division; this
parameter is useful for the assessment of "fresh" embryo quality
(Cummins et al., 1986), although no data support its use for
cryopreserved embryos (Fishel et al., 1985). Nevertheless,
differences in developmental rate can be detected in human embryos
maintained in culture until the blastocyst stage and long term
culture such as this may allow for the selection of the "best"
quality embryos. However, pregnancy rates after replacement of
frozen/thawed blastocysts is similar to that of earlier cell-stage
embryos (Fehilly et al., 1985).

4.4. Morphological Appearance

Embryo morphology, as previously discussed, can be correlated to pregnancy rate. Even though more pregnancies result from the transfer of human embryos graded "good," based on morphological appearance, than from those graded "poor," many "good" embryos fail to result in pregnancy. Additionally, Mohr (1984) found no correlation between embryo quality and pregnancy after transfer. In that study embryos were graded "good" when blastomeres were of even size and the embryo was not fragmented or irregular in appearance.

4.5. Noninvasive Techniques

Fluorescein diacetate is a nonfluorescent dye which passes freely through cell membranes. If the cell is viable, an intracellular esterase removes the acetate molecules; the resulting fluorescein molecule cannot cross intact cell membranes and remains intracellular. Fluorescein is then detected within the cell by fluorescence microscopy. Retention of fluorescein has been correlated to the ability of embryos to develop in vitro. Apparently exposure to fluorescein diacetate and ultraviolet light does not impair in vitro or in vivo developmental rates (Mohr and Trounson, 1980). This procedure has not been used with human embryos because morphological appearance renders just as much useful information about the embryo (Mohr et al., 1985).

Studies in both mice and humans (O'Neill, 1985; O'Neill et al., 1985) have shown that pregnancy occurs when embryos which are capable of producing significant amounts of a platelet activating factor (PAF) are present in the reproductive tract. Platelet activating factor can be measured in the culture medium of human IVF embryos and PAF secreting embryos were correlated with viable pregnancies (O'Neill et al., 1985). A simple assay for the level of PAF and other products of embryo metabolism may, therefore, provide a reliable guide to the viability of embryos and a means of assessing their pregnancy potential.

4.6. Postthaw Viability

A preliminary guide to whether embryos have survived thawing is their general morphological appearance and whether they undergo shrinkage on transfer to medium containing sucrose for dilution of the cryoprotectant.

Cracks in the zona pellucida are thought to result when fracture planes form in the extracellular ice matrix during rapid cooling (Rall and Polge, 1984). The fracturing is thought to result from mechanical stresses caused by temperature gradients within the ice crystals or by differences in the thermal expansion of adjacent materials (e.g., the wall of the container, ice crystals and the vitrified interior of embryos). There has been no systematic study of this subject but, empirically, we have observed that a) zona cracks are more numerous when there has been difficulty seeding the sample (e.g., seeding at lower than $-7°C$) and b) no cracked zonae have been observed when samples have been frozen in straws following the protocol of Lassalle et al. (1985).

A reliable guide to whether embryos have survived cryopreservation is their ability to continue development upon subsequent culture in vitro. With blastocysts, this is evidenced by re-expansion of the blastocoel cavity within several hours of culture (Fehilly et al., 1985). The rate of development of frozen-thawed embryos is slowed in comparison to their nonfrozen counterparts and must be considered when scheduling transfer of the frozen-thawed embryos back into the uterus. For example, 2-cell mouse embryos which have been cryo-preserved can take an extra 24 hr to reach the blastocyst stage in vitro. Although there is no substantial evidence on this point with human embryos, it may be preferable to transfer frozen-thawed embryos into a uterus which is 12-24 hr behind the chronological age of the embryos.

Several pregnancies have resulted from the transfer of cryopre-served human embryos at the 4- to 8-cell stage where several of the blastomeres had been destroyed following thawing (Mohr et al., 1985; Quinn and Kerin, 1986). This indicates that the blastomeres of human embryos, like those of the mouse, remain totipotent at least to the 8-cell stage. A 4- to 8-cell embryo with 50% or more of its blastomeres intact upon thawing is considered worthwhile for transfer (Mohr et al., 1985).

Lassalle et al. (1985) have reported that more 2-, 4- and 8-cell human embryos with regular sized blastomeres survive freezing and thawing than 3-, 5-, 6- and 7-cell embryos containing fragments and uneven sized blastomeres. It would appear, therefore, that the state of the nucleus and cytoplasm in relation to the stage of the cell cycle may influence survival of blastomeres during cryopreservation.

An overall pregnancy rate of 10% is found in the published results from several IVF groups using cryopreservation. Although the sample sizes are small, this pregnancy rate, in terms of the number of embryos available for freezing after transfer of the fresh embryos, is similar between different cryopreservation methods and slightly lower than the rates obtained from the transfer of fresh nonfrozen embryos. Somewhat higher survival rates (80%) for good quality 2-, 4- and 8-cell human embryos have been reported by Lassalle et al. (1985) and on transfer of these embryos to 21 patients, 9 became pregnant (Testart et al., 1985). The simplicity of the technique used by these authors, the reduced toxicity of propanediol used as the cryoprotectant and its applicability to unfertilized oocytes and day 1 and day 2 embryos (1- to 4-cell stage), indicate that this method may be superior to others currently in use and it deserves further assessment for routine application.

4.7. Cryoprotectant Solutions

Phosphate buffered saline supplemented with 20% maternal serum or human cord serum is generally employed for the dilution of cryoprotectant. Solutions of various concentration, either molar or percent, are required for the stepwise addition and removal of cryoprotectant. Following is a brief description of the preparation of molar solutions.

One mole of any chemical substance is defined by the molecular weight of that chemical. The molecular weight of glycerol is 92.1; thus, 92.1 grams (g) of glycerol equals 1 mole. A 1-molar solution of glycerol would be defined by 92.1 g of glycerol in a volume of 1 liter (1). This is because molarity is defined as the number of moles of a chemical in one liter of solution.

As discussed in section 3.3., the three main cryoprotectants used with human embryos are glycerol, DMSO, and propanediol. These chemicals are liquid; thus, specific gravity (s.g.) must be considered when formulating molar solutions. Specific gravity is a ratio of the weight of a specific volume of one substance compared to the weight of that same volume of water at 0°C. The following example may clarify the importance of specific gravity: 1 mole of glycerol is 92.1 g, glycerol is a liquid which is difficult to weigh; however, specific gravity can be used to convert weight (g) to volume (ml) by this formula:

Weight (g) = s.g. (g/ml) x Volume (ml)

Rearrange to: Volume (ml) = Weight (g) x 1/s.g. (g/ml); thus: 92.1 g/1.25 g/ml = 73.7 ml equals 1 mole of liquid glycerol. The three cryoprotectants are listed in Table VI along with weights and volumes required for 1 M solutions. For practical application 10 to 20 ml of 3 M cryoprotectant solution is prepared fresh the day of use. The working solutions, for addition or removal of cryoprotectant, are prepared by dilution of the 3 M stock. The following formula gives the parts diluent (PBS + 20% serum) needed for each part 3 M stock to make the desired concentration:

Molarity stock solution - desired molarity
desired molarity

Table VII illustrates this principle.

Cryoprotectant solutions are also formulated based on a volume percent basis. For example, a 10% glycerol solution contains 1 ml

Table VI. Conversion chart of weight to volume based on specific
           gravity

Cryoprotectant	Molecular weight	Specific gravity	g/liter 1 molar	ml/liter 1 molar
Glycerol	92.10	1.25	92.10	73.70
DMSO	78.13	1.10	78.13	71.00
Propanediol (Propyleneglycol)	76.10	1.04	76.10	73.17

glycerol and 9 ml PBS + serum, or if a solid cryoprotectant is used 1
g in 10 ml final volume. Stock solutions can be constructed and the
formula % stock - % desired/% desired can be used to prepare desired
dilutions.

In addition to formulating % or M solutions glycerol, DMSO, and
propanediol have unique physical characteristics which should be
considered. Glycerol is a viscous material; accurate measurement
requires patience, care and cleanliness. To prepare a glycerol or
propanediol solution add the cryoprotectant with a pipet, carefully
wipe the pipet with a tissue after removal from the pure chemical
before addition to PBS, thoroughly wash the inside of the pipet, mix
well, and then filter through a 0.22 micron filter.

Since DMSO will dissolve membrane filters, sterile technique
must be employed when mixing DMSO solutions. Thus, the PBS plus
serum should be filter sterilized prior to DMSO addition and
sterility must be maintained during subsequent dilution steps.

In summary, cryoprotectant solutions can be formulated on a
percent or molar basis, dilutions are made based on the concentration
of a stock cryoprotectant solution, and cryoprotectants must be
handled with care.

4.8. Cooling/Warming Programs

Three basic cooling programs will be described; each was
designed for a specific cell-stage embryo and cryoprotectant combination. Pregnancies have resulted from all three programs; only
reference to human embryos will be made. The techniques will be
described from cryoprotectant addition to warming and cryoprotectant
removal. Usually cryoprotectant is added and removed by moving the

Table VII. Sample calculations for preparation of working solutions
from a 3 M stock cryoprotectant solution

Desired molarity	3.0 M stock (ml)	Diluent (ml)	Total volume (ml)
0.25	1.0	$\dfrac{3.0 - .25}{.25} = 11.0$	12
0.50	1.0	$\dfrac{3.0 - .5}{.5} = 5.0$	6
1.00	1.0	$\dfrac{3.0 - 1.0}{1.0} = 2.0$	3
1.32	1.0	$\dfrac{3.0 - 1.32}{1.32} = 1.27$	2.27

embryo through a series of separate solutions rather than diluting
the medium containing the embryo. An overall view of the manipulations of the sample container during cooling/warming and embryo
recovery is shown in Figure 5.

(1) DMSO - Slow-cooling - Early Embryos

Trounson and Mohr (1983) described a cooling program based
on DMSO, 4- to 10-cell embryos, slow-cooling, and slow-warming.
Since that report, the following modified program has been in general
use.

a. Six-step cryoprotectant addition at room temperature
at 10 min per step. 0.25 M, 0.5 M, 0.75 M, 1.0 M,
1.25 M, 1.5 M DMSO

b. Load embryos into container at 1.5 M DMSO step.

c. Place container into programmable freezer.

d. Cool sample from room temperature to -6°C at -2°C/min.

e. Seed sample (initiate ice formation) at -6°C.

f. Hold at -6°C for 20-30 min.

g. Cool sample at -0.3°C/min to -80°C.

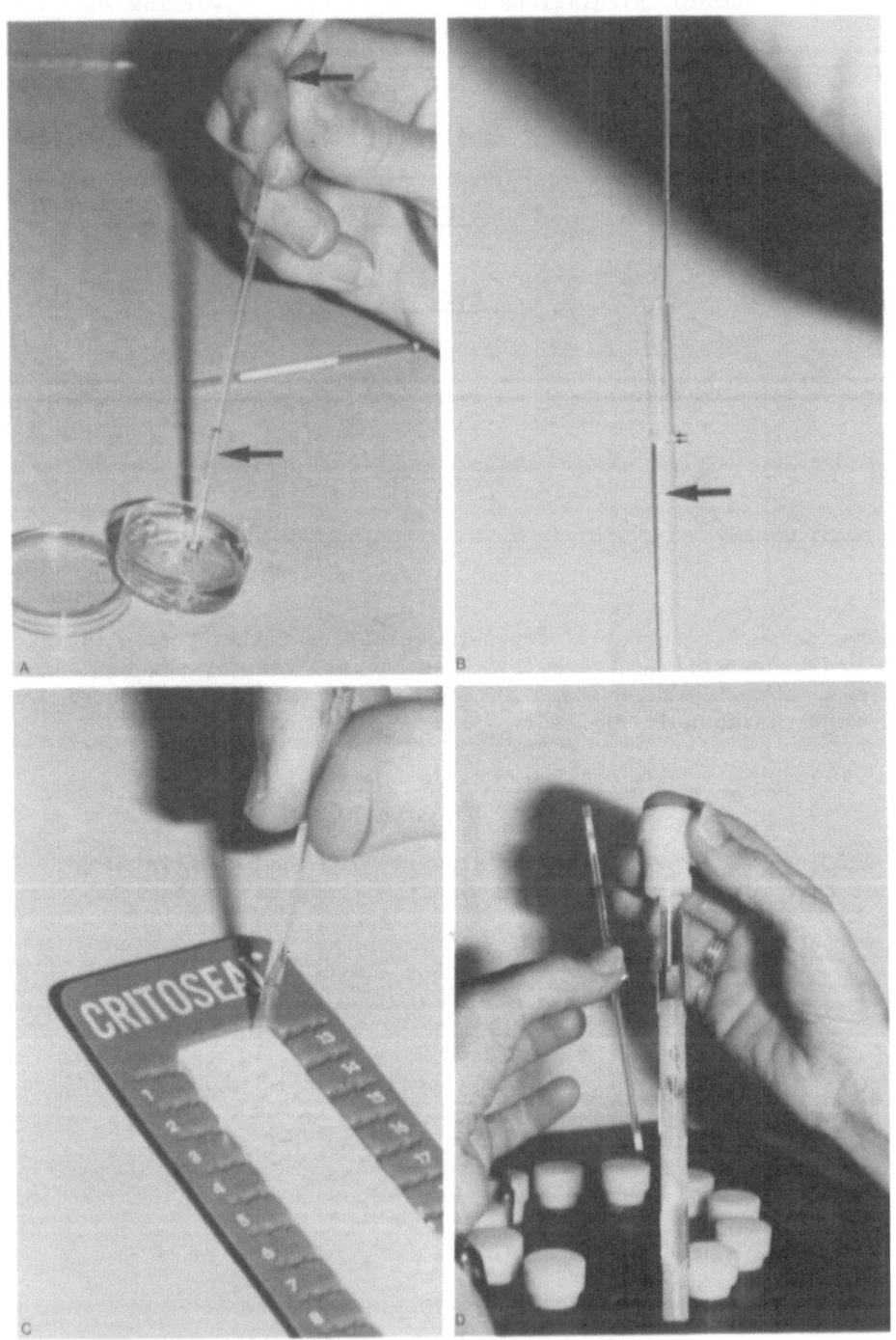

Fig. 5A. Loading container with cryosolution. An air space (arrow) is recommended to prevent the possibility of embryo adherence to the cotton plug within the straw. The cotton plug should be wetted with cryosolution (cotton plug at upper arrow).

Fig. 5B. Loading embryo into container. Pipet tip (arrow) is well below the cryosolution to avoid introduction of air into the cryosolution. After the embryo is loaded, a layer of paraffin oil is introduced at the double arrow if straws are held in the vertical position during the cooling process. The layer of paraffin oil should not be more than 1 mm thick to avoid thermal interference.

Fig. 5C. Sealing the container. The straw can be sealed with a variety of substances (see text). Here the container is sealed with critoseal; the inside of the straw must be dried thoroughly if critoseal is employed.

Fig. 5D. Loading the straw into the cooling machine.

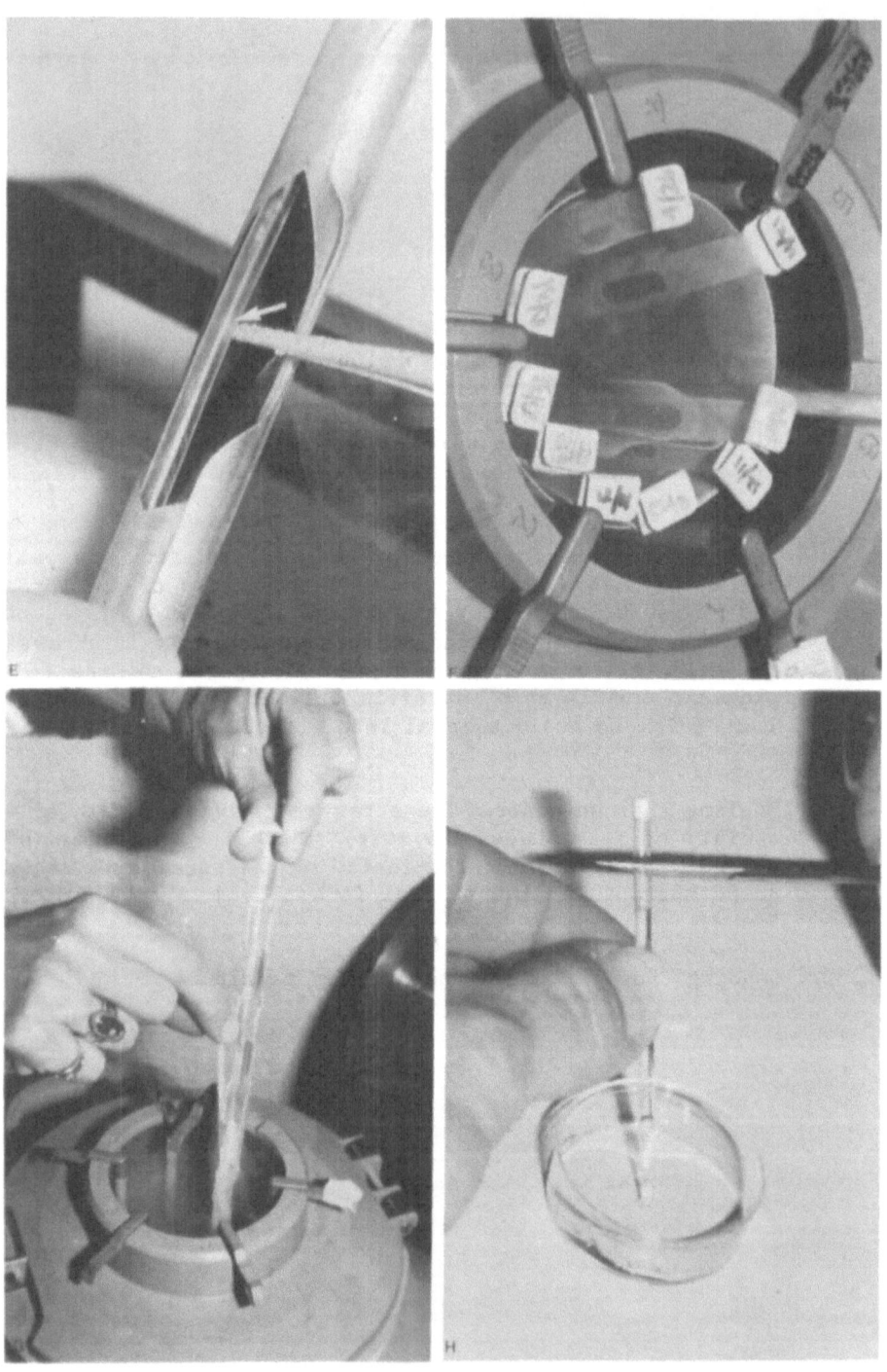

Fig. 5E.   Seeding the straw with precooled forceps.

Fig. 5F.   View through the neck of the liquid nitrogen tank showing
            the storage unit: canister, cane, goblet, and straw.

Fig. 5G.   Removing a straw from the liquid nitrogen storage tank:
            goggles, gloves or forceps are recommended to avoid
            nitrogen burns.

Fig. 5H.   Removing the embryo from the container. The cotton plug
            end is removed and then the sealed end is opened while the
            straw is held over a culture dish. The embryo is then
            located with the aid of a dissection microscope.

h.   Plunge sample into liquid nitrogen for storage.

i.   Warming - warm sample from -80°C to room temperature
       at +8°C/min.

j.   Six-step, 10 min per step, removal of cryoprotectant
       at room temperature. 1.5 M, 1.25 M, 1.0 M, 0.75 M,
       0.5 M, 0.25 M, 0.00 M DMSO

k.   Assess embryo viability.

Our experience with this program and 2-cell mouse embryos
packaged in 0.5 cc French straws has been: 95% embryo recovery and
80% developed to blastocyst after 4 days in culture. Five hundred
mouse embryos were employed in this study. Human embryos did not
survive the cooling/warming process as well; of 16 embryos cooled, 12
were warmed, 8 were recovered and 4 transferred. It must be
reiterated that thorough testing of a system with mouse embryos prior
to working with human embryos is recommended. The blastomere size of
the 2-cell mouse embryo is equivalent to those of 8-cell human
embryos.

(2) Glycerol - Fast Program - Rapid Thaw

        Cohen et al. (1985) reported success with a cooling program
based on blastocyst stage embryos, glycerol as cryoprotectant, the
fast cooling program, and rapid warming.

a.   Human embryos are maintained in culture to the
       blastocyst stage, 5 to 7 days after insemination.

b.   Glycerol is added at room temperature to morula and
       expanded blastocyst stage embryos in 6 steps at 10 min
       per step. (1, 2, 4, 6, 8, 10%)

c.   Embryos are packaged in 10% glycerol solution and
       placed into a programmable freezer.

d.   Samples are cooled from room temperature to -7°C at
       -1°C/min.

e.   Samples are seeded at -7°C and held for 10 min.

f.   Samples cooled to -36°C at -0.3°C/min.

g.   Samples plunged into liquid nitrogen for storage.

h.   Warming - sample placed into a 30°C water bath for
       approximately 1 min.

i. Six-step, 10 min/step, removal of glycerol. 10, 8, 6, 4, 2, 1, 0%

j. Slow-warm to 37°C, observe for reexpansion of blastocoel.

k. Assess viability.

To date we have cooled and stored 38 human blastocysts in 0.5 cc French straws. Seven blastocysts have been warmed and transferred; all 7 were recovered and 5 of the 7 re-expanded before transfer. Rabbit blastocyst stage embryos are comparable in size to the human blastocyst; the mouse blastocyst, although smaller in size, can be utilized for experimentation. We have cooled 150 mouse blastocysts, recovered 145 (96%) embryos after thaw and 138 (92%) hatched from the zona pellucida upon culture.

(3)   Propanediol, Dehydration, Fast Thaw

Lassalle et al. (1985) and Renard et al. (1984) reported two cooling methods employing propanediol as cryoprotectant. Renard et al. (1984) will be referred to as Method I and Lassalle et al. (1985) will be referred to as Method II.

Method I.   Unfertilized 1-cell to 4-cell human embryos.
            Propanediol = $PROH_2$

a. Solutions required

1. 4 ml PBI + 1 ml human serum + 0.856 g sucrose (20% serum, 0.5 M sucrose, PBI)

2. 16 ml PBI + 4 ml human serum (20% serum, PBI)

3. 8.37 ml #2 add 1.63 ml $PROH_2$ and 1.712 g sucrose (20% serum, 2.2 M $PROH_2$, 0.5 M sucrose)

4. 8.89 ml #2 plus 1.11 ml $PROH_2$ (20% serum, 1.5 M $PROH_2$ PBI)

b. Propanediol addition

Molar $PROH_2$	Vol #4, ml (1.5 M $PROH_2$)	Vol #2, ml (PBI)	Minutes at Room Temperature
0.5	1	2	5
1.0	2	1	5
1.5	3	0	10

Propanediol is added to embryos in three steps as indicated above.

c.   Container is filled with solution #3
     (i.e., 20% serum, 2.2 M $PROH_2$, 0.5 M sucrose)

d.   Embryos or oocytes are placed in solution #3 for 5 min,
     loaded into prefilled container (c above) and placed at
     −30°C for 30 min.

e.   Samples are plunged into liquid nitrogen.

f.   Warming, 37°C water bath for approximately 1 min with
     vigorous agitation.

g.   Remove embryos from container and place in solution #1 at
     room temperature for 5 min.

h.   Transfer embryos to PBI + 20% serum + 0.25 M sucrose, 5 min
     at room temperature.

i.   Transfer to culture medium (10 min at room temperature).

j.   Assess viability.

Method II.   Unfertilized 1-cell to 4-cell human embryos

a.   Solutions

     1.   16 ml PBI + 4 ml human serum (PBI + 20% serum)

     2.   10 ml of #1 + 1.25 ml $PROH_2$ (1.5 M $PROH_2$ in PBI + 20%
          serum)

Prepare the following:

Molar $PROH_2$	Vol #2, ml (1.5 M $PROH_2$)	Vol #1, ml (PBI)	Room Temperature (Min)
0.5	1	2	5
1.0	2	1	5
1.5	3	0	10

Propanediol is added to embryos in three steps as above.

b.   Prepare 1.5 M $PROH_2$ in PBI + 20% serum + 0.2 M sucrose
     (0.342 g sucrose in 5 ml #2). Place embryos in 1.5 M
     $PROH_2$, 0.2 M sucrose, PBI for 5 min, load into container.

c. Cool from room temperature to $-7°C$ at $-2°C/min$; hold at $-7°C$ for 60 sec before seeding.

d. Seed and hold at $-7°C$ for 5 min.

e. Cool at $-0.3°C/min$ to $-30°C$.

f. Plunge into liquid nitrogen and store.

g. Warming; vigorous agitation in $37°C$ water bath until ice disappears.

h. Remove embryos from container and place into following solutions at room temperature, 5 min each step:

   PBI + 1.0 M $PROH_2$ + 0.2 M sucrose
   PBI + 0.5 M $PROH_2$ + 0.2 M sucrose
   PBI +   0.2 M sucrose
   Fresh medium

i. Assess embryo viability.

Results from a 12-month period using these two procedures are given in Table VIII. At present, Method I is the procedure of choice for unfertilized oocytes whereas Method II gives the best results with pronucleate embryos.

Table VIII. Results of propanediol Method I and II with human oocytes and embryos[a]

	Method I[b]			Method II[c]		
	Oocytes	Pronucleate	Cleaved	Oocytes	Pronucleate	Cleaved
No. patients	5	1	14	15	20	23
No. ova	12	1	51	36	53	55
No. ova surviving (%)	7 (58)	0 (0)	13 (25)	1 (3)	27 (51)	24 (44)
No. patients receiving ET (%)	2 (40)	0 (0)	7 (50)	0 (0)	15 (71)	11 (48)
No. patients pregnant	0	–	0	–	1	2

[a]Unpublished, Quinn et al., Adelaide, Australia.
[b]Renard et al., 1984 (see section 4.8).
[c]Lassalle et al., 1985 (see section 4.8).

In summary, three cryopreservation protocols have been presented. To achieve maximum success strict compliance with cryoprotectant/embryo/cooling/warming program must be met. That is, blastocyst stage embryos must be utilized with glycerol and fast cooling/warming for maximal survival, etc.

4.9. Experience with Oocytes

4.9.1.   Hamster

Hamster oocytes have been successfully cryopreserved for later use in assays to test the penetrating ability of human spermatozoa (Fleming et al., 1979; Quinn et al., 1982). Up to 80% of the oocytes survived cryopreservation when DMSO and slow-cooling to -80°C were used and their penetration rate was no different from that of fresh oocytes. Decondensation of the sperm heads within the cytoplasm of frozen-thawed oocytes was retarded, however, indicating that the mechanism for this process had been affected in some way.

4.9.2.   Mice

Wood (1985) recently reported her results of an extensive study on the ability of cryopreserved mouse oocytes to be fertilized and their subsequent development in vitro and in vivo. Using DMSO and slow-cooling to -80°C, 36% of oocytes surviving freezing and thawing were fertilized in vitro. The incidence of polyspermy was greater than in unfrozen controls but in the embryos which developed from normally fertilized oocytes, the incidence of aneuploidy was no different from that of the fresh controls. Implantation and further development was similar with embryos derived from frozen oocytes and controls. The observations of the incidence of polyspermy and chromosomal abnormalities has important implications for human oocytes. The number of pronuclei should be checked at the end of the insemination period since it would seem likely that cryopreservation could damage the zona pellucida or cortical granules and increase the incidence of polyspermy. If the fertilized oocyte contains two normal pronuclei, concern over other chromosomal abnormalities is not warranted based on the experience with mouse oocytes. A thorough study, however, with oocytes specifically donated so that the chromosome karyotype could be determined in embryos derived from cryopreserved oocytes would be desirable.

4.9.3.   Human Studies

Several groups have initiated studies of the cryopreservation of human oocytes.

Trounson (1985) reported that oocytes survived freezing and thawing which was carried out with DMSO as cryoprotectant and utilized slow-cooling to -80°C. Some of the thawed oocytes were capable of fertilization and developed to blastocysts in culture.

Chen (1986) has obtained survival of oocytes after freezing with DMSO as cryoprotectant and a twin pregnancy was established after three oocytes were frozen for 6 hr, thawed, fertilized and transferred in the same cycle during which the oocytes were collected.

Our own experience has been with oocytes frozen using propanediol and sucrose as cryoprotectants. Bernard et al. (1985) found that human oocytes which had been equilibrated in propanediol and sucrose and had the cryoprotectants diluted out were capable of fertilization and development to the 8-cell stage, whereas fertilization was not observed in oocytes similarly treated with glycerol. Our results are given in Table VIII and the most promising procedure appears to be Method I.

Although these observations are only preliminary, they are encouraging and further work is continuing in the area. The protocol used at present is as follows:

a.   Oocytes are processed 2-4 hr after collection, which normally takes place 36 hr after the injection of human chorionic gonadotropin (hCG).

b.   The surrounding cumulus cells are removed by incubating the oocyte-cumulus complex in Hepes-buffered culture medium (Quinn et al., 1985) containing 300 IU bovine hyaluronidase/ml and 5 mg bovine serum albumin/ml; oocytes are incubated for 2-3 min which is sufficient to detach the cumulus cells and allow the oocyte with its surrounding corona cells to be handled with a finely drawn glass pipet.

c.   The oocyte is then washed in PBI medium containing 20% heat inactivated human serum and processed by the procedures given for Method I outlined previously.

4.10. Vitrification

Vitrification is the solidification of a liquid brought about not by crystallization (i.e., the formation of ice) but by extreme elevation in viscosity during cooling. During vitrification, highly concentrated aqueous solutions of cryoprotectants become so viscous that they pass from the liquid state to a nonstructured solid state called a glass (Fahy et al, 1984). Cooling has to be sufficiently rapid ($\sim$ 2500°C/min) to allow water to pass straight into a glass state without forming ice crystals.

Since it is thought that embryos vitrify intracellularly after slow-cooling to between -25°C and -45°C and then plunging into liquid nitrogen (Rall and Polge, 1984), several laboratories have been active in developing methods to achieve external vitrification. This requires an increased level of cryoprotectants over that used in conventional protocols.

The only successful vitrification procedure reported in the literature thus far is that of Rall and Fahy (1985). These authors used a mixture of both intracellular and extracellular cryoprotectants: 20.5% w/v DMSO, 15.5% w/v acetamide, 10% w/v 1,2 propanediol and 6% w/v polyethylene glycol; total concentration 13 molal.

Eight-cell mouse embryos were first incubated in a 25% dilution of the above mixture for 15 min at room temperature, then a 50% dilution at 4°C for 10 min and finally a 90-100% mixture for 10 min before direct immersion in liquid nitrogen. Placing the embryos at 4°C at the 50% dilution step helps to reduce the toxicity of the cryoprotectant mixture by preventing further permeation. Dehydration occurs at this step because of the high extracellular concentration of cryoprotectants. This dehydration concentrates both intracellular protein and previously permeated intracellular cryoprotectant and permits intracellular vitrification without additional cryoprotectant permeation.

If the embryos are slowly warmed (10°C/min), this allows for sufficient time for the devitrification (crystallization) of the glassy cytoplasm and extracellular liquid, i.e., the growth of large ice crystals which are damaging to the embryo and this leads to low survival. Rapid warming (500°C/min, e.g., agitation of straws in a water bath at 0°C) does not allow time for devitrification and leads to high rates of survival.

The next step is the removal of the cryoprotectants from the thawed embryos. Rall and Fahy (1985) diluted the thawed embryos into a 50% dilution of the cryoprotectants at 4°C for 10 min, then a 25% dilution at 4°C for 10 min and then 12.5% and 6.25% at room temperature for 10 min each followed by a final wash in cryoprotectant-free medium. Survival and development of 80-90% of the embryos was observed.

Our own experience using the vitrification process with 2- and 8-cell mouse and human embryos has been disappointing (Quinn and Kerin, 1986). The embryos appear alive morphologically when first thawed and placed in the 50% dilution of cryoprotectants but then invariably swell and burst when placed in decreasing dilutions of the cryoprotectant mixture. More work is needed to optimize the dilution gradient and the temperature at which this is carried out.

Several other groups have reported preliminary results of studies of vitrification of mouse embryos. Wood (1985), using the cryoprotectant mixture of Rall and Fahy (1985) at a concentration of 90%, obtained 86% survival of 8-cell mouse embryos and 18% of these developed into live fetuses on transfer compared to 50% of controls. Again, however, results have been variable and difficult to replicate (personal communication).

A slightly different mixture of cryoprotectants has been used by Peura et al. (1985). They tested various mixtures of DMSO, glycerol, propanediol and sucrose and reported survival and development of 60-90% of 2-cell mouse embryos vitrified in mixtures of 2-4 M DMSO containing 0.25 M sucrose. The embryos were frozen in straws, thawed rapidly at 37°C, and diluted into 0.25 M sucrose.

Obviously, vitrification has several advantages over conventional freezing methods. It does not require expensive instrumentation, it is quick and economical on liquid nitrogen usage, and it may provide superior survival rates because of the avoidance of ice crystal formation. Further studies are clearly needed. The technique is not recommended for use with human material at present.

5.   TROUBLESHOOTING

Cryopreservation technologies are so precise that problems can arise during any phase of the procedure with the result of poor embryo survival. Some problems are readily apparent, such as the explosion of an ampule or straw; others are more subtle, such as improper storage of cryoprotectant. In this section problems which effect embryo survival will be discussed.

5.1. Containers

The most common storage containers are plastic straws (Figure 2) and glass ampules (Figure 1); either container type can explode upon removal from the liquid nitrogen storage tank. Most often the cause is improper sealing of the container. The result of a leak in any container type is the influx of liquid nitrogen; the expansion coefficient of liquid nitrogen is so great that the container will explode when it is removed from the storage tank. To remedy this problem straws should be filled with fluid so that the cotton plug end is wetted. The open end of the straw can then be sealed with critoseal (used for blood capillary tubes), polyvinyl pyrrolidone (PVP), heat, or ultrasound. The distributor of French straws suggests that PVP be used as a sealant for their product because PVP forms a solid when wetted and, perhaps, a tighter seal than critoseal, which shrinks upon cooling. Heating the straw may raise the temperature of the cryosolution to a point which reduces embryo survival and equipment for heat or ultrasonic sealing may not

be cost-effective. After glass ampules are sealed, the heated area
can be checked visually, with the aid of a dissection microscope, for
a complete seal. If a hole in the seal is observed, the embryo must
be removed from that ampule and placed into a fresh one. An alternate method of sealing glass ampules is to place a plastic cap over
the open neck (a 0.5 ml microfuge tube is adequate). The screw in
plug of plastic ampules should not be overtightened because the
rubber gasket becomes distorted and allows nitrogen access to the
container.

In general, the French 0.25 and 0.5 cc semen straws are most
popular due to their small size and optimal volume to surface area
ratios for cooling/warming. Problems are inherent with straws as
with any of the other container types. It cannot be stressed enough
that a thorough understanding and familiarity with the container
type, in addition to other components, must be achieved before
embarking on a human cryopreservation program. This understanding
will be invaluable when problems arise because solutions to the
problems will be more easily reached.

5.2. Cryoprotectants

Chemical deterioration of a cryoprotectant will lead to
poor embryo survival. The only method of checking cryoprotectant
toxicity is with laboratory animal embryos. Because cryoprotectants
such as DMSO are readily oxidized, they should be stored in air tight
containers at 4°C in the dark. Small containers are ideal since the
air space is minimized, and the chemical would be used quickly after
the storage container is opened. Additionally, it is good practice
to purchase small amounts of cryoprotectant, which insures the
rotation of fresh chemical.

5.3. Freezing Machine

In most cases representatives from the company will be
willing to set up and demonstrate their machines and equipment.
However, it is the user's responsibility to become proficient at
operation as well as detecting areas of potential malfunction. The
only valid check on machine function is to measure the actual
temperature of a solution in a dummy container. By placing a
thermister into a container and relaying the output to a strip
recorder, a chart of change in temperature over time can be produced.
Even the process of seeding or failure of the seed can be recorded in
this manner for future reference. Deviations in machine function,
which may indicate a major malfunction, can be detected.

5.4. Cooling Program

We have presented three basic cooling programs which were
optimized for specific cell-stage embryos. Embryo survival rates are

minimized when these protocols are not followed, that is, substituting glycerol for DMSO, changing final temperature, or changing warming rate. Thus, a strict technique for cryopreservation must be established within a laboratory so that technique will not confound the process of troubleshooting.

Two minor variations with cryoprotectant usage may decrease embryo viability. They are: shortening cryoprotectant equilibration times and failure to measure and mix the cryoprotectant thoroughly. When the equilibration process is shortened, protection of the embryo may be impaired due to inadequate penetration of the cryoprotectant. The embryo may be damaged by osmotic shock due to inadequate equilibration at lower cryoprotectant concentrations. In the case of glycerol, inadequate mixing or poor pipetting technique may result in erroneous cryoprotectant dilutions. Again when a embryo is subjected to the cooling process, inadequate cryoprotectant concentrations may result and viability may decrease. Thus, care must be taken when cryoprotectant solutions are constructed so that problems associated with cryoprotectants are minimized.

Failure to seed the sample properly is the most common error in the cryopreservation program. As discussed earlier, without seeding embryos have little chance of surviving. In the programs presented, seeding has been programmed to occur 2 to 3 degrees below the freezing point of the cryosolution. We recommend that manual seeding be employed, that is to touch the container at the meniscus of the cryosolution containing the embryo with an extremely cold object. The cold object is usually forceps placed in liquid nitrogen for a few minutes before seeding. A 10- to 20-min hold has been programmed at the seed point; thus, the sample can be seeded early in this period and rechecked near the end. One final caution, clear sample containers must be used since visualization of the solution is required to check for ice formation.

Cooling rate must be controlled and ideally recorded accurately. The freezing machine will control rates of change; however, a record must be made to determine whether or not the rate is appropriate. One method of recording temperature change is with a thermister inside a container containing cryosolution. Final temperature can be measured by the thermister method also.

5.5. Warming Rate

Two warming rates have been presented, slow (4–25°C/min) and rapid (100–500°C/min). Rapid warming rates are not difficult to achieve; however, the majority of IVF centers utilize 4- to 10-cell embryos so slow warming rates are required. To achieve these slow rates several methods can be employed: first, the freezing machine could be programmed to warm at a slow rate, 4 to 10°C/min, from −80°C to +10°C; second, an ice water bath can be employed; third, a

refrigerated water bath can be employed. With any of these methods,
the thermister and recorder should be employed to determine the
actual rate of warming.

5.6. Cryoprotectant Removal

Cryoprotectant removal can also destroy the embryo due to
excessive osmotic effects. Thus, careful preparation of solutions
and timing during the removal process is required for overall
success.

5.7. Final Caution

In summary, cryopreservation is the sum of several small,
but critical, steps. The process can be perfect up to the
cryoprotectant removal step when the embryos are rendered nonviable.
Unless rigid controls and records are kept, weeks or months of work
may be wasted troubleshooting the entire process when in fact a 30
sec dilution step at 6 to 4% glycerol was used instead of the
required 10 min. These problems must be worked out using animal
models before human embryos are employed in the procedure. That is
not to say other problems will not arise, but a familiarity with the
procedure will insure a faster solution for the problem.

6.   REPLACEMENT OF CRYOPRESERVED EMBRYOS AND PREGNANCY RATE AFTER
REPLACEMENT

6.1. Replacement Protocols

Early investigations with laboratory animals indicated that
close syncrony between the uterine environment and cell-stage
preimplantation embryo was important for the initiation of pregnancy
(Dickmann and Noyes, 1960; Noyes and Dickmann, 1960; Doyle et al.,
1963). If embryos were transferred to synchronous uterine environments, the highest pregnancy rates resulted; embryos of "older"
developmental stage than uterine environment rendered an acceptable
pregnancy rate, while embryos of "younger" age than uterine
environment resulted in few pregnancies. Thus, a 12-hr syncrony
between transferred embryo stage and uterine environment was found to
be desirable for cryopreserved embryos. Following are three
protocols for embryo replacement based on the cryopreservation
methods presented. Syncrony between embryo stage and expected
uterine age is the goal of each protocol.

The protocol for replacement of blastocyst-stage embryos is
presented in Cohen et al. (1985) and Fehilly et al. (1985). Cohen et
al. (1985) thawed and replaced embryos 4, 5, or 6 days after the
luteinizing hormone (LH) surge or, in the case of clomiphene citrate
(CC) stimulated cycles, after the administration of 500 IU human

chorionic gonadotropin (hCG). Pregnancies resulted when blastocysts
were transferred into the uterus 5 days after the LH surge. Follicular growth was monitored by ultrasound. However, the method employed
for detection of LH (urinary or serum) was not noted in this report.
Fehilly et al. (1985) thawed and replaced blastocyst stage embryos
118 to 132 hr after the LH surge was detected (urinary), approximately 4 days after ovulation, or 132 to 135 hr after the injection
of 500 IU hCG (CC stimulated cycles). Recommendation: twice daily
LH (serum or urine) measurements beginning day 11 of the replacement
cycle; ultrasound as necessary to confirm follicle collapse; transfer
embryos 4 days after ovulation, determined by LH surge.

        Both Trounson (1986) and Mohr et al. (1985) use the method
reported by Freemann et al. (1986) to calculate the 95% confidence
interval for day of ovulation based on previous menstrual cycle
lengths. Twice daily blood samples are then obtained beginning on
the day before predicted ovulation; serum is analyzed for LH,
estradiol, and progesterone. Ovulation is confirmed by a rise in
progesterone in addition to a rise in LH. Embryos are thawed 4 to 8
hr before exact uterine syncrony, approximately 40 hr after the onset
of the LH surge, and transferred into the uterus in exact syncrony 48
hr after the LH surge. Recommendation: twice daily LH measurements
beginning day 12 of the replacement cycle; progesterone documentation
of ovulation, or ultrasound to confirm follicle rupture; transfer
embryos 48 hr after the onset of the LH surge.

        Testart et al. (1986) employ daily plasma LH levels beginning on
the day a 15 mm follicle is visualized with ultrasound. To determine
the day of ovulation, an LH surge is defined by a rise in LH of 2x
the mean LH value calculated from the preceding days. Embryos are
thawed 1 to 4 days after ovulation and replaced 0 to 3 hr postthaw.
Recommendation: daily ultrasound starting day 10 of the cycle; daily
serum LH starting day that one 15 mm follicle is visualized; LH-surge
when LH concentration reaches 2x mean baseline value; replace embryos
in syncrony with LH surge (i.e., pronuclear 1 day after ovulation,
etc.).

        These protocols have resulted in pregnancies after the replace-
ment of cryopreserved embryos. As observed with laboratory animals,
higher pregnancy rates correlate with syncronous transfers. Thus,
twice daily LH levels and serum progesterone levels after the LH
rise, which identify ovulation more accurately than ultrasound and
single LH values, should be considered for monitoring the replacement
cycle.

        6.2. Expected Pregnancy Rates

        Two recent publications (Trounson, 1986, and Freemann et
al., 1986) have addressed factors which can be correlated to
pregnancy rate after the transfer of cryopreserved embryos. Freemann

et al. (1986) found that the number of excess embryos for cryopreservation was related to pregnancy rate and that embryos which
exhibited faster cleavage rates were associated with higher pregnancy
rates. Trounson (1986) found several interesting correlates to
pregnancy rate. Embryo morphology, as with livestock embryos, was
directly correlated to pregnancy rate. Pregnancy rate varied with
embryo classification as follows: 15%, regular appearing embryos
containing blastomeres of equal size; 12%, embryos containing minor
fragments but equal blastomere size; 5%, embryos of irregular
blastomere size and some fragmentation; 0%, embryos with severe
fragmentation. Also a higher pregnancy rate was associated with more
advanced cell-stage embryos, 32% for 8-cell versus 8% for 4-cell
embryos. Finally, an interesting observation, concerning either
embryo quality or uterine environment, that women who conceived on
the fresh IVF cycle were more likely to become pregnant when
cryopreserved embryos were replaced. Only 6% of women who failed to
become pregnant with fresh embryos at IVF became pregnant with
cryopreserved embryos compared to 41% of women who became pregnant
with cryopreserved embryos after a pregnancy with fresh IVF embryos.
In this case, the embryos for IVF and cryopreservation were harvested
during the same IVF cycle.

To date the publication with the largest sample size is Freemann
et al. (1986). This study reported the findings of 205 patients and
402 embryos; cryopreservation was accomplished with DMSO and the slow
program. Three hundred ninety-six (396) of the 402 embryos were
thawed, 229 (58%) of these were transferred to 144 (70%) of the
patients, and 16 pregnancies resulted. The pregnancy rate was 8%
based on cycles of cryopreservation or 11% based on transfer cycles.
An average of 1.6 embryos were replaced per cryopreservation cycle.

Other reports are not as easy to interpret; thus, for ease of
comparison, pregnancy rate is based on number of transfer cycles and
chemical pregnancies are not included. Trounson and Mohr (1983)
reported 2 pregnancies in 12 transfers (17%) with DMSO and the slow
program. Mohr et al. (1985) using DMSO reported a 15% pregnancy rate
(7/45). Cohen et al. (1985) reported 2 pregnancies from 16 transfers
(12%) of blastocyst stage embryos. Fehilly et al. (1985) compared
DMSO/cleavage-stage with glycerol/blastocyst stage embryos and found
a 15% (4/26) pregnancy rate with DMSO and a 34% (8/23) pregnancy rate
with glycerol; the figure for blastocyst stage embryos is inflated
since 70% of the extra embryos did not reach blastocyst in culture.
This report points to the need for a more accurate method of
reporting pregnancy data, such as a per embryo basis. Testart et al.
(1986) reported an 18% (6/32) pregnancy rate employing propanediol as
cryoprotectant.

In summary, reported pregnancy rates vary from 11% to 30%, based
on transfer data. When these data are standardized such that
pregnancy rate is calculated on a per embryo basis, pregnancy rate is

8% to 11% per embryo transferred. Pregnancy rate for blastocyst and early cleavage stage embryos is similar, because 50% of early cleavage stage embryos are lost during the freeze/thaw cycle and 50% or more of the early cleavage stage embryos do not progress to blastocyst in vitro. As Trounson (1986) stated, any pregnancies resulting from cryopreserved embryos are an advantage, because those embryos would not have contributed to pregnancy rate if they were not cryopreserved. Overall, an IVF team can expect an 8% to 10% increase in pregnancy rate when cryopreserved embryos are added to the program (Trounson, 1986).

7.   REFERENCES

Bernard, A., Imoedemhe, D. A., Shaw, R. W., and Fuller, B., 1985, Effects of cryoprotectants on human oocytes, Lancet 1:632.

Chang, M. C., 1947, Normal development of fertilized rabbit ova stored at low temperature for several days, Nature 159:602.

Chang, M. C., 1954, Development of parthenogenetic rabbit blastocysts induced by low temperature storage of unfertilized ova, J. Exp. Zool. 125:127.

Chen, C., 1986, Pregnancy after human oocyte cryopreservation, Lancet 1:884.

Cohen, J., Simons, R. F., Edwards, R. G., Fehilly, C. B., and Fishel, S. B., 1985, Pregnancies following the frozen storage of expanding human blastocysts, J. In Vitro Fert. Embryo Transfer 2:59.

Cummins, J. M., Breen, T. M., Harrison, K. L., Shaw, J. M., Wilson, L. M., and Hennessey, J. F., 1986, A formula for scoring human embryo growth rates in in vitro fertilization: its value in predicting pregnancy and in comparison with visual estimates of embryo quality, J. In Vitro Fert. Embryo Transfer 3:284.

Dickmann, Z. and Noyes, R. W., 1960, The fate of ova transferred into the uterus of the rat, J. Reprod. Fertil. 1:197.

Doyle, L. L., Gates, A. H., and Noyes, R. W., 1963, Asynchronous transfer of mouse ova, Fertil. Steril. 14:215.

Elsden, R. P., Nelson, L. D., and Seidel, G. E., 1978, Superovulating cows with follicle stimulating hormone and pregnant mare's serum gonadotropin, Theriogenology 9:17.

Fahy, G. M., MacFarlane, D. R., Angell, C. A., and Meryman, H. T., 1984, Vitrification as an approach to cryopreservation, Cryobiology 21:407.

Fehilly, C. B., Cohen, J., Simons, R. F., Fishel, S. B., and Edwards, R. G., 1985, Cryopreservation of cleaving embryos and expanded blastocysts in the human: a comparative study, Fertil. Steril. 44:638.

Fishel, S. B., Cohen, J., Fehilly, C., Purdy, J. M., Walters, D. E., and Edwards, R. G., 1985, Factors influencing human embryonic development in vitro, Ann. N.Y. Acad. Sci. 442:342.

Fleming, A. D., Yanagimachi, R., and Yanagimachi, H., 1979, Fertilizability of cryopreserved zona-free hamster ova, Gamete Res. 2:357.

Freemann, L., Trounson, A., and Kirby, C., 1986, Cryopreservation of human embryos: progress on the clinical use of the technique in human in vitro fertilization, J. In Vitro Fert. Embryo Transfer 3:53.

Hoagland, H. and Pincus, G., 1942, Revival of mammalian sperm after immersion in liquid nitrogen, J. Gen. Physiol. 25:337.

Jahnel, F., 1938, Uber die Wiederstands fahigkeit von menschlichen spermatozoen gegeneber starker kalte, Klin. Wochenschr. 17:1273.

Kane, M. T., 1975, Bicarbonate requirements for culture of one-cell rabbit ova to blastocysts, Biol. Reprod. 12:552.

Kerin, J. F., Warnes, G. M., Quinn, P. J., Jeffrey, R., Kirby, C., Matthews, C. D., Seamark, R. F., and Cox, L. W., 1983, Incidence of multiple pregnancy after in-vitro fertilisation and embryo transfer, Lancet 2:537.

Lassalle, B., Testart, J., and Renard, J.-P., 1985, Human embryo features that influence the success of cryopreservation with the use of 1, 2 propanediol, Fertil. Steril. 44:645.

Leibo, S. P., 1981, Introduction to embryo freezing, in: Frozen Storage of Laboratory Animal Embryos (G. H. Zeilmaker, ed.), Gustav Fischer Verlag, Stuttgart, p. 1.

Leibo, S. P., Mazur, P., and Jackowski, S. C., 1974, Factors affecting survival of mouse embryos during freezing and thawing, Exp. Cell Res. 89:79.

Leibo, S. P., McGrath, J. J., and Cravalho, E. G., 1978, Microscopic observation of intracellular ice formation in unfertilized mouse ova as a function of cooling rate, Cryobiology 15:257.

Lutjen, P., Trounson, A., Leeton, J., Findlay, J., Wood, C., and Renou, P., 1984, The establishment and maintenance of pregnancy using in vitro fertilization and embryo donation in a patient with primary ovarian failure, Nature 307:174.

Luyet, B. J. and Hodapp, E. L., 1938, Revival of frog's spermatozoa vitrified in liquid air, Proc. Soc. Exp. Biol. Med. 39:433.

Lyon, M., Gleinster, P., and Whittingham, D. G., 1981, Long-term viability of embryos stored under irradiation, in: Frozen Storage of Laboratory Animal Embryos (G. H. Zeilmaker, ed.), Gustav Fischer Verlag, Stuttgart, p. 139.

Mazur, P., 1963, Kinetics of water loss from cells at subzero temperatures and the likelihood of intracellular freezing, J. Gen. Physiol. 47:347.

Mohr, L. R., 1984, Assessment of human embryos, in: In Vitro Fertilization and Embryo Transfer (A. Trounson and C. Wood, eds.), Churchill Livingstone, New York, p. 159.

Mohr, L. R. and Trounson, A. O., 1980, The use of fluorescein diacetate to assess embryo viability in the mouse, J. Reprod. Fertil. 58:189.

Mohr, L. R., Trounson, A., and Freemann, L., 1985, Deep-freezing and transfer of human embryos, J. In Vitro Fert. Embryo Transfer 2:1.

Morris, G. J. and Farrant, J., 1972, Interactions of cooling rate and protective additive on the survival of washed human erythrocytes frozen to -196°C, Cryobiology 9:173.

Noyes, R. W. and Dickmann, Z., 1960, Relationship of ovular age to endometrial development, J. Reprod. Fertil. 1:186.

O'Neill, C., 1985, Examination of the causes of early pregnancy-associated thrombocytopenia in mice, J. Reprod. Fertil. 73:567.

O'Neill, C., Gidley-Baird, A. A., Pike, I. L., Porter, R. N., Sinosich, M. J., and Saunders, D. M., 1985, Maternal blood platelet physiology and luteal-phase endocrinology as a means of monitoring pre- and postimplantation embryo viability following in vitro fertilization, J. In Vitro Fert. Embryo Transfer 2:87.

Parkes, A. S., 1945, Preservation of human spermatozoa at low temperatures, Br. Med. J. 2:212.

Peura, A., Trounson, A. O., and Freemann, L., 1985, Ultra-rapid embryo freezing, Proceedings of the Fourth World Conference on In Vitro Fertilization, Melbourne, Abstract #244.

Polge, C., Smith, A. U., and Parkes, A. S., 1949, Revival of spermatozoa after vitrification and dehydration at low temperatures, Nature 164:666.

Quinn, P. and Kerin, J. F. P., 1986, Experience with the cryopreservation of human embryos using the mouse as a model to establish successful techniques, J. In Vitro Fert. Embryo Transfer 3:40.

Quinn, P., Barros, C., and Whittingham, D. G., 1982, Preservation of hamster oocytes to assay the fertilizing capacity of human spermatozoa, J. Reprod. Fertil. 66:161.

Quinn, P., Kerin, J. F., and Warnes, G. M., 1985, Improved pregnancy rate in human in vitro fertilization with the use of a medium based on the composition of human tubal fluid, Fertil. Steril. 44:493.

Rall, W. F. and Fahy, G. M., 1985, Ice-free cryopreservation of mouse embryos at -196°C by vitrification, Nature 313:573.

Rall, W. F. and Polge, C., 1984, Effect of warming rate on mouse embryos frozen and thawed in glycerol, J. Reprod. Fertil. 70:285.

Rall, W. F., Reid, D. S., and Polge, C., 1984, Analysis of slow-warming injury of mouse embryos by cryomicroscopical and physiochemical methods, Cryobiology 21:106.

Renard, J.-P., 1985, The cryopreservation of mammalian embryos, in: Human In Vitro Fertilization (J. Testart and R. Frydman, eds.), Elsevier Science Publishers, INSERM Symposium, 24:201.

Renard, J.-P., Bui-Xuan-Nguyen, and Garnier, V., 1984, Two-step freezing of two-cell rabbit embryos after partial dehydration at room temperature, J. Reprod. Fertil. 71:573.

Rogers, P. A. W., Milne, B. J., and Trounson, A. O., 1984, The effect
    of stimulation to produce multiple follicular development on
    embryo viability and uterine receptivity in human IVF, Proc.
    Fertil. Soc. Aust. 3:32.
Schneider, U. and Mazur, P., 1984, Osmotic consequences of cryopro-
    tectant permeability and its relation to the survival of frozen-
    thawed embryos, Theriogenology 21:68.
Sherman, J. K., 1973, Synopsis of the use of frozen human semen since
    1964: state of the art of human semen banking, Fertil. Steril.
    24:397.
Shettles, L. B., 1940, The respiration of human spermatozoa and their
    response to various gases and low temperatures, Am. J. Physiol.
    128:408.
Smith, A. U., 1952, Behavior of fertilized rabbit eggs exposed to
    glycerol and to low temperature, Nature 170:374.
Spallanzani, L., 1776, Opuscoli di fisca animale, e vegatabile.
    Opuscolo II. Osservazioni, e sperienze intorno ai vermicelli
    spermatiei dell'uomo edegli animali, Moderna.
Stiles, W., 1930, On the cause of cold death of plants, Protoplasm
    9:459.
Testart, J., Lassalle, B., Berthe, V., Belaisch-Allart, J., and
    Frydman, R., 1985, Cryopreservation of human embryos,
    Proceedings of the Fourth World Conference on In Vitro
    Fertilization, Melbourne, Abstract #248.
Testart, J., Lassalle, B., Belaisch-Allart, J., Hazout, A., Forman,
    R., Rainhorn, J. D., and Frydman, R., 1986, High pregnancy rate
    after early human embryo freezing, Fertil. Steril. 46:268.
Thorpe, P. E., Knight, S. C., and Farrant, J., 1976, Optimal
    conditions for the preservation of mouse lymph node cells in
    liquid nitrogen using cooling rate techniques, Cryobiology
    13:126.
Trounson, A. and Mohr, L., 1983, Human pregnancy following cryopres-
    ervation, thawing and transfer of an eight-cell embryo, Nature
    305:707.
Trounson, A. O., 1985, Clinical progress and new research develop-
    ments in human embryo and egg cryopreservation, Proceedings of
    the Fourth World Conference on In Vitro Fertilization,
    Melbourne, Abstract #217A.
Trounson, A. O., 1986, Preservation of human eggs and embryos,
    Fertil. Steril. 46:1.
Veeck, L. L., 1985, Extracorporeal maturation: Norfolk, 1984, Ann.
    N.Y. Acad. Sci. 442:357.
Whittingham, D. G., 1980, Principles of embryo preservation, in: Low
    Temperature Preservation in Medicine and Biology (M. J. Ashwood-
    Smith and J. Farrant, eds.), Pitman Medical, Tunbridge Wells,
    pp. 65-83.
Whittingham, D. G., Leibo, S. P., and Mazur, P., 1972, Survival of
    mouse embryos frozen to -196° and -269°C, Science 178:411.
Willadsen, S., Polge, C., and Rowson, L. E. A., 1978, The viability
    of deep-frozen cow embryos, J. Reprod. Fertil 52:391.

Wilmut, I., 1972, The effect of cooling rate, warming rate cryopro-
    tective agent and stage of development on survival of mouse
    embryos during freezing and thawing, Life Sci. 2:1071.
Wood, C., McMaster, R., Rennie, G., Trounson, A., and Leeton, J.,
    1985, Factors influencing pregnancy rates following in vitro
    fertilization and embryo transfer, Fertil. Steril. 43:245.
Wood, M. J., 1985, Recent progress in animal embryo cryopreservation,
    Proceedings of the Fourth World Conference on In Vitro
    Fertilization, Melbourne, Abstract #217.
Zeilmaker, G. H., Alberda, A. T., van Gent, I., Rijkmans,
    C. M. P. M., and Drogendijk, A. C., 1984, Two pregnancies
    following transfer of intact frozen-thawed embryos, Fertil.
    Steril. 42:293.

17

CONTROL MECHANISMS REGULATING MEIOTIC MATURATION OF MAMMALIAN OOCYTES

Richard Schultz

1.   INTRODUCTION

Meiotic maturation of mammalian oocytes, which is initiated in vivo by a hormonal stimulus, is characterized by dissolution of the nuclear membrane (germinal vesicle breakdown, GVBD), condensation of diffuse chromatin into distinct bivalents, separation of homologous chromosomes, and emission of the first polar body with arrest at metaphase II; fertilization triggers the second meiotic division. The ability of oocytes to mature properly is critical for development, since only oocytes that have successfully completed meiotic maturation are capable of being fertilized and giving rise to normal development. This discussion will focus on cellular events that occur during maturation, the basis for maintenance of meiotic arrest, and current ideas regarding the mechanism of gonadotropin-induced resumption of meiosis of oocytes enclosed in preovulatory antral follicles.

2.   ROLE OF PROTEIN SYNTHESIS IN COMPLETION OF MEIOTIC MATURATION

Although protein synthesis is not required for GVBD (the oocytes
arrest at the circular bivalent stage), dramatic changes in the
pattern of protein synthesis occur during meiotic maturation
(McGaughey and Van Blerkom, 1977; Schultz and Wassarman, 1977a,b; Van
Blerkom and McGaughey, 1978; Schultz et al., 1979). Temporally,
these changes occur concomitant with or subsequent to GVBD, and are
regulated at the post-transcriptional level, since they occur in
enucleated oocytes or oocytes treated with α-amanitin (Schultz et
al., 1978). These changes are not coupled with nuclear progression,
since drugs that inhibit nuclear progression at specific stages, but
that do not inhibit GVBD (Wassarman et al., 1976), do not inhibit
these changes in the pattern of protein synthesis in mouse oocytes
(Schultz and Wassarman, 1977b). For example, colcemid allows GVBD to
occur but inhibits spindle formation, and cytochalasin B arrests
maturation at metaphase I. Neither of these compounds prevents the
maturation-associated changes in the pattern of protein synthesis.

Some of these changes in types of proteins synthesized, however,
appear required for nuclear progression through polar body emission.
When protein synthesis is allowed to occur for a brief period subsequent to GVBD and during a time when the changes in the patterns of
protein synthesis occur and then transferred to medium containing a
protein synthesis inhibitor, these oocytes emit polar bodies. In
contrast, if the inhibitor of protein synthesis is added prior to the
time when the changes in pattern of protein synthesis occur, the
oocytes do not emit polar bodies (Wassarman et al., 1979). Thus,
while these changes in the pattern of protein synthesis must occur
for completion of meiotic maturation, they are not dependent on
nuclear progression beyond GVBD.

3.   ROLE OF cAMP AND PROTEIN PHOSPHORYLATION IN MAINTENANCE OF
     MEIOTIC ARREST AND RESUMPTION OF MEIOSIS

Spontaneous meiotic maturation in vitro accurately reflects in
vivo maturation, since oocytes matured and fertilized in vitro give
rise to viable offspring to a similar extent as oocytes matured in
vivo and then fertilized in vitro (Schroeder and Eppig, 1984). Thus,
oocyte maturation in vitro provides a valid system to study cellular
events associated with resumption of meiosis.

The observation that meiotically competent oocytes removed from
antral follicles resume spontaneously maturation in vitro (Pincus and
Enzmann, 1935) led to the concept that the follicle exerts an inhibitory influence on oocyte maturation that is relieved by gonadotropin.
A likely candidate for a molecule involved in maintenance of meiotic
arrest is cAMP, since membrane permeable cAMP analogs, such as
dibutyryl cAMP (dbcAMP) or 8-bromo cAMP, but not the corresponding

cGMP analogs, and phosphodiesterase inhibitors, such as theophylline
or 3-isobutyl-1-methyl-xanthine (IBMX), reversibly inhibit maturation
in vitro of oocytes derived from either mice, rats, or pigs (Cho et
al., 1974; Wassarman et al., 1976; Magnusson and Hillensjo, 1977;
Dekel and Beers, 1978; Rice and McGaughey, 1981). These results have
led to the model for oocyte maturation depicted in Figure 1.

Results of several additional lines of experimentation
strengthen the conclusions drawn from the cAMP analog experiments.
Forskolin, a reversible activator of adenylate cyclase in every
mammalian system tested to date (Seamon et al., 1981), induces a
concentration-dependent increase in mouse oocyte cAMP levels and
inhibition of GVBD (Schultz et al., 1983; Urner et al., 1983; Sato

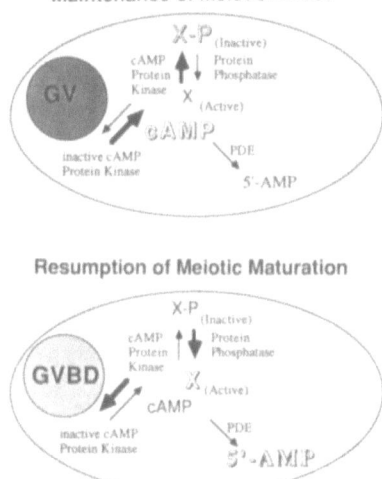

Fig. 1.   Model for regulation of meiotic arrest and resumption of
           meiosis. The model proposes that a phosphoprotein, XP,
           maintains meiotic arrest and that the dephosphorylated
           form, X, promotes resumption of meiosis. X is phospho-
           rylated by cAMP-dependent protein kinase and XP is
           dephosphorylated by a phosphoprotein phosphatase. Resump-
           tion of meiosis is initiated by a decrease in oocyte cAMP
           that leads to a decrease in cAMP-dependent protein kinase
           activity. Assuming that phosphoprotein phosphatase
           activity does not decrease, this decrease in cAMP-dependent
           protein kinase activity would induce a net dephosphoryla-
           tion of XP, triggering resumption of meiosis. It should be
           noted that although X is depicted as actively promoting
           germinal vesicle breakdown, XP could actively maintain
           meiotic arrest. The bold face type indicates the
           predominant species.

and Koide, 1984). In addition, GVBD is transiently inhibited in mouse oocytes microinjected with cAMP, whereas GVBD is not inhibited in oocytes microinjected with 2'-deoxy cAMP, which does not activate cAMP-dependent protein kinase (Bornslaeger et al., 1986); the transient nature of inhibition of GVBD is not surprising, since oocytes contain a high level of phosphodiesterase activity (Bornslaeger et al., 1984). These results also suggest that a cAMP-dependent protein kinase mediates the inhibitory effect of cAMP in maintenance of meiotic arrest. If such is the case, then inhibition of the catalytic subunit of cAMP-dependent protein kinase should induce maturation in the presence of high cAMP levels, since X-P can no longer be generated, and the protein phosphatase should shift the equilibrium to X. Oocyte microinjection with increasing amounts of protein kinase inhibitor, which specifically interacts with free catalytic subunit, and not with C in the holoenzyme, induces a concentration-dependent increase in resumption of meiosis in oocytes incubated in a totally inhibiting concentration of dbcAMP. This result strongly suggests that cAMP mediates inhibition of meiotic arrest through the activation of a cAMP-dependent protein kinase. In addition, injecting oocytes with increasing amounts of purified catalytic subunit of cAMP-dependent protein kinase inhibits spontaneous maturation in a concentration-dependent manner (Bornslaeger et al., 1986).

Both in vitro and in vivo oocyte maturation are associated with a decrease in mouse oocyte cAMP (Schultz et al., 1983; Vivarelli et al., 1983). Moreover, this decrease occurs during a period of time in which oocytes become committed to resume meiosis as defined by the observation that these oocytes undergo GVBD when returned to medium containing either IBMX or dbcAMP (Schultz et al., 1983). This decrease in oocyte cAMP is likely to be causally related to resumption of meiosis, since agents that block this decrease, such as IBMX, also inhibit maturation (Schultz et al., 1983), and microinjection of purified calmodulin-modulated sheep brain cyclic nucleotide phosphodiesterase into oocytes incubated in IBMX results in resumption of meiosis (Bornslaeger et al., 1986).

Implicit in the model for mammalian oocyte maturation is that the decrease in oocyte cAMP leads to a decrease in cAMP-dependent protein kinase activity, which in turn results in dephosphorylation of specific phosphoproteins prior to GVBD. Two dimensional gel electrophoresis of oocyte phosphoproteins reveals that specific proteins undergo dephosphorylation during, but not prior to, the commitment period (Schultz et al., 1983; Bornslaeger et al., 1986). In addition, an increase occurs during this time in spot intensity of several phosphoproteins. These maturation-associated changes in phosphoprotein metabolism are tightly coupled with maturation, since they occur in oocytes induced to resume maturation by injection of protein kinase inhibitor (Bornslaeger et al., 1986).

4.   MECHANISM OF GONADOTROPIN-INDUCED RESUMPTION OF MEIOSIS

    During the past decade, two models have emerged to account for
the mechanism of gonadotropin-induced resumption of meiosis of
oocytes present in preovulatory antral follicles. One is based on
the observation that follicular fluid can inhibit spontaneous meiotic
maturation in vitro (Chang, 1955; Tsafriri and Channing, 1975;
Tsafriri et al., 1976; Stone et al., 1978). This model proposes that
follicular fluid contains a low molecular weight polypeptide that is
synthesized by granulosa cells and called oocyte maturation inhibitor
(OMI). OMI is proposed to inhibit oocyte maturation and its amount
is reduced by LH. The other model (Dekel and Beers, 1978, 1980) is
based upon the well-documented observation that cAMP is involved, at
least in part, in maintenance of meiotic arrest. This model proposes
that LH terminates the flux of follicle cell cAMP to the oocyte,
which can no longer maintain levels of cAMP sufficient to maintain
meiotic arrest. This results in the maturation-associated decrease
in cAMP and resumption of meiosis.

    Results of recent studies cast serious doubt on OMI as being a
low molecular weight polypeptide. The inhibitory activity present in
porcine follicular fluid has been purified by HPLC and unequivocally
demonstrated to be hypoxanthine (Downs et al., 1985). The concentration of hypoxanthine in porcine follicular fluid is about 1.4 mM
(Downs et al., 1985) and that in murine follicular fluid about 2-4 mM
(Eppig et al., 1985). This concentration of hypoxanthine added to
medium totally mimics the effects of porcine follicular fluid on
oocyte maturation. In addition, serum, which does not inhibit
maturation, contains very low levels of hypoxanthine. The physiological role that hypoxanthine plays in maintenance of meiotic arrest
is not known.

    The original termination of intercellular communication model
postulated that gap junction-mediated transmission of follicle cell
cAMP to the oocyte inhibited maturation and that LH terminated the
flux of follicle cell cAMP to the oocyte, which was no longer able to
maintain inhibitory levels of cAMP, since the oocyte lacked an
adenylate cyclase but contained a cyclic nucleotide phosphodiesterase. This model explains the paradoxical situation in which cAMP
can act directly on the oocyte to inhibit maturation, but can act in
the follicle to induce maturation.

    Results of numerous experiments bolster many aspects of this
model. For example, a network of gap junctions exists between
granulosa cells, cumulus cells, and the oocyte, which results in a
syncytium, such that spatially separated compartments in the follicle
could communicate with each other (Anderson and Albertini, 1976;
Gilula et al., 1978). This syncytium is likely to be essential for
maintenance of meiotic arrest, since oocytes that are present in
cumulus cell-oocyte complexes grafted to follicle hemisections do not

resume spontaneous meiotic maturation, whereas they do if physical
contact with the follicle hemisection is prevented (Foote and
Thibault, 1969; Tsafriri and Channing, 1975; Leibfried and First,
1980).

The role of cAMP in maintenance of meiotic arrest and
resumption of meiosis is fairly well established. Consistent with
follicle cell cAMP transmission to the oocyte is the observation that
elevating cumulus cell cAMP results in both an increase in oocyte
cAMP and inhibition of maturation (Eppig et al., 1983; Racowsky,
1984, 1985; Bornslaeger and Schultz, 1985a). Oocytes possess cyclic
nucleotide phosphodiesterase and adenylate cyclase. Although the
original model postulated that oocytes not contain adenylate cyclase,
the level of activity appears quite low (Bornslaeger and Schultz,
1985a,b). The transmission of follicle cell cAMP to the oocyte is
probably necessary for generating sufficient levels to maintain
meiotic arrest. Spontaneous meiotic maturation of oocytes removed
from their follicles probably results, in part, from termination of
transfer of follicle cell cAMP to the oocyte, whose adenylate cyclase
activity is insufficient to maintain inhibitory levels of cAMP.
Lastly, examination of intercellular communication by electrical
coupling, dye transfer, or metabolic coupling, reveals that exposure
of follicles to LH results in termination of intercellular communication between cumulus cells and the oocyte (Gilula et al., 1978;
Moor et al., 1980; Eppig, 1982; Salustri and Siracusa, 1983). In
response to elevated cAMP levels, cumulus cells mucify--a cAMP-
induced process that involves synthesis and secretion of hyaluronic
acid (Eppig, 1979)--which results in the physical disruption of
contact of cumulus cells with the oocyte, and thus terminates
intercellular communication between the two cell types.

The major experimental finding apparently inconsistent with the
model is the temporal relationship between resumption of meiosis and
termination of intercellular communication. Results of experiments
using either sheep (Moor et al., 1980), mouse (Eppig, 1982; Salustri
and Siracusa, 1983), or hamster (Racowsky and Satterlie, 1985)
oocytes indicate that resumption of meiosis, which was induced in
vivo by administration of hCG, occurs prior to a reduction in the
extent of intercellular communication between cumulus cells and the
oocyte.

Results of a quantitative morphometric study of freeze-fractured
rat cumulus cell complexes reveal that a 10-20 fold reduction in the
extent of gap junctional surface area between cumulus cells occurs
during resumption of meiosis in vivo, and that this decrease
correlates well with resumption of meiosis (Larsen et al., 1986).
These results most likely provide a resolution to the apparent
paradox that the maturation-associated decrease in oocyte cAMP occurs
during a period of time when both follicle and cumulus cell cAMP are
increasing, and intercellular communication in the complex is not

reduced. Intercellular communication between cumulus cells and the
oocyte is probably mediated by the inner layers of cumulus cells,
which are the last cells to undergo the mucification reaction and
become uncoupled from one another and the oocyte (Eppig, 1982; Freter
and Schultz, 1984). A reduction in the extent of intercellular
communication amongst the outer cumulus cells of the complex,
therefore, would not result in a noticeable reduction in the extent
of communication between cumulus cells and the oocyte, using the
metabolite transfer assay to monitor communication, but would
effectively isolate the oocyte from communicating with the entire
follicle cell mass. Thus, if transfer of follicle cell cAMP to the
oocyte is vital for maintenance of meiotic arrest, such a reduction
in communication between cumulus cells could result in the
maturation–associated decrease in oocyte cAMP at a time when follicle
cell cAMP levels are increasing. It should be pointed out that even
if a reduction in intercellular communication preceded or was
concurrent with GVBD, uncoupling need not be the trigger event, since
the commitment event, e.g., inhibition of oocyte adenylate cyclase or
activation of oocyte cyclic nucleotide phosphodiesterase, could occur
prior to the reduction in communication.

In summary, cAMP participates in maintenance of meiotic arrest.
A decrease in oocyte cAMP triggers commitment to undergo GVBD, which
in turn initiates a series of changes in patterns of protein synthesis that are necessary for the successful completion of meiosis. It
remains to be demonstrated that the gonadotropin–induced reduction of
communication between cumulus cells initiates resumption of meiosis
by inducing the maturation–associated decrease in oocyte cAMP.

5.   REFERENCES

Anderson, E. and Albertini, D. F., 1976, Gap junctions between the
      oocyte and companion follicle cells in the mammalian ovary, J.
      Cell Biol. 71:680.
Bornslaeger, E. A. and Schultz, R. M., 1985a, Regulation of mouse
      oocyte maturation: Effect of elevating cumulus cell cAMP on
      oocyte cAMP levels, Biol. Reprod. 33:698.
Bornslaeger, E. A. and Schultz, R. M., 1985b, Adenylate cyclase
      activity in zona–free mouse oocytes, Exp. Cell Res. 156:277.
Bornslaeger, E. A., Wilde, M. W., and Schultz, R. M., 1984, Regula-
      tion of mouse oocyte maturation: Involvement of cyclic AMP
      phosphodiesterase and calmodulin, Dev. Biol. 105:488.
Bornslaeger, E. A., Mattei, P., and Schultz, R. M., 1986, Involvement
      of cAMP–dependent protein kinase and protein phosphorylation in
      regulation of mouse oocyte maturation, Dev. Biol. 114:453.
Chang, M. C., 1955, The maturation of rabbit oocytes in culture and
      their maturation, activation, fertilization, and subsequent
      development in the fallopian tubes, J. Exp. Zool. 128:378.

Cho, W. K., Stern, S., and Biggers, J. D., 1974, Inhibitory effect of dibutyryl cAMP on mouse oocyte maturation in vitro, J. Exp. Zool. 187:383.

Dekel, N. and Beers, W. H., 1978, Rat oocyte maturation in vitro: Relief of cyclic AMP inhibition by gonadotropins, Proc. Natl. Acad. Sci. U.S.A. 75:4369.

Dekel, N. and Beers, W. H., 1980, Development of the rat oocyte in vitro: Inhibition and induction of maturation in the presence or absence of the cumulus oophorus, Dev. Biol. 75:247.

Downs, S. M., Coleman, D. L., Ward-Bailey, P. F., and Eppig, J. J., 1985, Hypoxanthine is the principal inhibitor of murine oocyte maturation in a low molecular weight fraction of porcine follicular fluid, Proc. Natl. Acad. Sci. U.S.A. 82:454.

Eppig, J. J., 1979, Gonadotropin stimulation of the expansion of cumulus oophori isolated from mice: General conditions for expansion in vitro, Exp. Zool. 208:111.

Eppig, J. J., 1982, The relationship between cumulus cell–oocyte coupling, oocyte meiotic maturation, and cumulus expansion, Dev. Biol. 89:268.

Eppig, J. J., Freter, R. R., Ward-Bailey, P. F., and Schultz, R. M., 1983, Inhibition of oocyte maturation in the mouse: Participation of cAMP, steroid hormones, and a putative maturation-inhibitory factor, Dev. Biol. 100:39.

Eppig, J. J., Ward-Bailey, P. F., and Coleman, D. L., 1985, Hypoxanthine and adenosine in murine ovarian follicular fluid: Concentrations and activity in maintaining oocyte meiotic arrest, Biol. Reprod. 33:1041.

Foote, W. D. and Thibault, C., 1969, Recherches experimentales sur la maturation in vitro des oocytes de trui et de veau, Ann. Biol. Anim. Biochem. Biophys. 9:329.

Freter, R. R. and Schultz, R. M., 1984, Regulation of murine oocyte meiosis: Evidence for a gonadotropin-induced, cAMP-dependent reduction in a maturation inhibitor, J. Cell Biol. 98:1119.

Gilula, N. B., Epstein, M. L., and Beers, W. H., 1978, Cell-to-cell communication and ovulation. A study of the cumulus cell–oocyte complex, J. Cell Biol. 78:58.

Larsen, W. J., Wert, S. E., and Brunner, G. D., 1986, A dramatic loss of cumulus cell gap junctions is correlated with germinal vesicle breakdown in rat oocytes, Dev. Biol. 113:517.

Leibfried, L. and First, N. L., 1980, Follicular control of meiosis in the porcine oocyte, Biol. Reprod. 23:705.

Magnusson, C. and Hillensjo, T., 1977, Inhibition of maturation and metabolism in rat oocytes by cyclic AMP, J. Exp. Zool. 201:139.

McGaughey, R. W. and Van Blerkom, J., 1977, Patterns of polypeptide synthesis of porcine oocytes during maturation, Dev. Biol. 56:241.

Moor, R. M., Smith, M. W., and Dawson, R. M. C., 1980, Measurement of intercellular coupling between oocytes and cumulus cells using intracellular markers, Exp. Cell Res. 126:15.

Pincus, G. and Enzmann, E. V., 1935, The comparative behaviour of mammalian eggs in vivo and in vitro, J. Exp. Med. 62:665.

Racowsky, C., 1984, Effect of forskolin on the spontaneous maturation and cyclic AMP content of rat oocyte-cumulus complexes, J. Reprod. Fertil. 72:107.

Racowsky, C., 1985, Effect of forskolin on the spontaneous maturation and cyclic AMP content of hamster oocyte-cumulus complexes, J. Exp. Zool. 234:87.

Racowsky, C. and Satterlie, R. A., 1985, Metabolic, fluorescent dye and electrical coupling between hamster oocytes and cumulus cells during meiotic maturation in vivo and in vitro, Dev. Biol. 108:191.

Rice, C. and McGaughey, R. W., 1981, Effect of testosterone and dibutyryl cAMP on the spontaneous maturation of pig oocytes, J. Reprod. Fertil. 62:245.

Salustri, A. and Siracusa, G., 1983, Metabolic coupling, cumulus expansion, and meiotic resumption in mouse cumuli oophori cultured in vitro in the presence of FSH or dbcAMP, or stimulated in vivo by hCG, J. Reprod. Fertil. 68:335.

Sato, E. and Koide, S. S., 1984, Forskolin and mouse oocyte maturation in vitro, J. Exp. Zool. 230:125.

Schroeder, A. C. and Eppig, J. J., 1984, The developmental capacity of mouse oocytes that matured spontaneously in vitro is normal, Dev. Biol. 102:493.

Schultz, R. M. and Wassarman, P. M., 1977a, Biochemical studies of mammalian oogenesis: Protein synthesis during oocyte growth and meiotic maturation in the mouse, J. Cell Sci. 24:167.

Schultz, R. M. and Wassarman, P. M., 1977b, Specific changes in the pattern of protein synthesis during meiotic maturation of mammalian oocytes in vitro, Proc. Natl. Acad. Sci. U.S.A. 74:538.

Schultz, R. M., Letourneau, G. E., and Wassarman, P. M., 1978, Meiotic maturation of mouse oocytes in vitro: Protein synthesis in nucleate and anucleate oocyte fragments, J. Cell Sci. 30:251.

Schultz, R. M., Letourneau, G. E., and Wassarman, P. M., 1979, Program of early development in the mammal: Changes in the patterns and absolute rates of tubulin and total protein synthesis during oogenesis and early embryogenesis in the mouse, Dev. Biol. 68:341.

Schultz, R. M., Montgomery, R. R., and Belanoff, J. R., 1983, Regulation of mouse oocyte maturation: Implication of a decrease in oocyte cAMP and protein dephosphorylation in commitment to resume meiosis, Dev. Biol. 97:264.

Seamon, K. B., Padgett, W., and Daly, J. W., 1981, Forskolin: A unique diterpene activator of adenylate cyclase in membrane and intact cells, Proc. Natl. Acad. Sci. U.S.A. 78:3363.

Stone, S. L., Pomerantz, S. H., Schwartz-Kripner, A., and Channing, C. P., 1978, Inhibitor of oocyte maturation from porcine follicular fluid: Further purification and evidence for reversible action, Biol. Reprod. 19:585.

Tsafriri, A. and Channing, C. P., 1975, An inhibitory influence of granulosa cells and follicular fluid upon porcine oocyte meiosis in vitro, Endocrinology 96:922.

Tsafriri, A., Pomerantz, S. H., and Channing, C. P., 1976, Inhibition of oocyte maturation by porcine follicular fluid: Partial characterization of the inhibitor, Biol. Reprod. 14:511.

Urner, F., Herrmann, W. L., Baulieu, E. E., and Schorderet-Slatkine, S., 1983, Inhibition of denuded mouse oocyte meiotic maturation by forskolin, an activator of adenylate cyclase, Endocrinology 113:1170.

Van Blerkom, J. and McGaughey, R. W., 1978, Molecular differentiation of the rabbit ovum. I. During oocyte maturation in vivo and in vitro, Dev. Biol. 63:139.

Vivarelli, E., Conti, M., De Felici, M., and Siracusa, G., 1983, Meiotic resumption and intracellular cAMP levels in mouse oocytes treated with compounds which act on cAMP metabolism, Cell Differ. 12:271.

Wassarman, P. M., Josefowicz, W. J., and Letourneau, G. E., 1976, Meiotic maturation of mouse oocytes in vitro: Inhibition of maturation at specific stages of nuclear progression, J. Cell Sci. 22:431.

Wassarman, P. M., Schultz, R. M., Letourneau, G. E., LaMarca, M. J., Josefowicz, W. J., and Bleil, J. D., 1979, Meiotic maturation of mouse oocyte in vitro, in: Ovarian, Follicular, and Corpus Luteum Function (C. P. Channing, J. M. Marsh, and W. A. Sadler, eds.), Plenum Press, New York, pp. 251-268.

18

REGULATION OF EMBRYONIC DEVELOPMENT BY ENVIRONMENTAL FACTORS

Michael T. Kane

1.   INTRODUCTION

Growth and development of the human embryo from fertilization to implantation normally occurs in the environment of the oviduct and

uterus. Much of our knowledge concerning the importance of environment comes from studies of embryos cultured in vitro. For ethical
reasons most of these experimental studies involve animal rather than
human embryos. The information derived from these studies is
obviously relevant to the culture of human embryos after IVF.
However, it may also be relevant to problems of human infertility due
to very early embryonic mortality.

One problem that arises with much of this information is that it
comes from studies with the mouse embryo and the requirements of the
mouse embryo for development from 1-cell to blastocyst are extremely
limited. The reason for this is probably related to the fact that
the mouse blastocyst contains less protein than the 1-cell mouse
embryo (Brinster, 1967). Therefore, information from other species
such as the rabbit, hamster, and nonhuman primates whose embryos have
more complex culture requirements may be particularly relevant to the
human.

The present review will consider systematically available
results from embryo culture studies concentrating on information
obtained using defined and semidefined media. The term "defined
medium" will be used for medium prepared from bench chemicals and
whose only macromolecular component (if present) is composed of
synthetic macromolecules such as polyvinylalcohol (PVA). The term
"semidefined medium" will be used for medium prepared from bench
chemicals to which albumin or other purified protein has been added
as a macromolecular source. A medium containing any proportion of
serum or plasma will be considered merely as a dilution of serum or
plasma. While this approach may seem overly rigorous, it is perhaps
the only way to obtain reliable, reproducible information on the
effects of environmental components of media.

2.   PRESENT STATUS OF EMBRYO CULTURE ON A COMPARATIVE SPECIES BASIS

Before considering available information on the requirements of
mammalian embryos in culture, it is useful to summarize the present
status of embryo culture in the very limited number of species in
which success has been achieved. In most species, it is not possible
with current technology to culture embryos through the entire preimplantation period, i.e., from 1-cell just after fertilization to
blastocyst just before implantation. Blocks to development appear at
different stages in different species and it is probable that these
blocks are telling us something fundamental about embryo development,
i.e., that present media are inadequate even for the embryos of those
few species which can be cultured throughout the entire preimplantation period. Table I summarizes the degree of development obtainable
in culture with embryos of different mammalian species.

Table I. Success of embryonic development in vitro in mammals

Species	1-cell to blastocyst	4-cell to blastocyst	8-cell to blastocyst	Hatching in vitro	References
Mouse	+[a]			+	Whitten and Biggers, 1968 Cross and Brinster, 1973 Whittingham, 1975
Rabbit	+			+	Maurer et al., 1969 Kane, 1972
Hamster			+	+	Yanagimachi and Chang, 1964 Bavister et al., 1983 Kane et al., 1986
Rat			+		Folstad et al., 1969 Mayer and Fritz, 1974 Wood and Whittingham, 1981
Cattle			+[b]	+	Tervit et al., 1972 Wright et al., 1976 Wright and Bondioli, 1981
Sheep			+[c]		Tervit et al., 1972 Tervit and Rowson, 1974 Tervit and Goold, 1978
Pig		+		+	Lindner and Wright, 1978 Davis and Day, 1978 Niemann et al., 1983
Cat	+				Bowen, 1977
Ferret	+				Whittingham, 1975
Human	+			+	Steptoe et al., 1971 Fishel et al., 1984 Edwards, 1985
Rhesus monkey	+			+	Morgan et al., 1984
Baboon			+	+	Pope et al., 1982

[a]Development from 1-cell to blastocyst occurs readily for certain hybrid and (to a lesser extent) certain inbred strains of mice.
[b]A few 1-cell embryos developed to blastocysts (Wright et al., 1976) but this has not been repeated.
[c]Development of a 1-cell embryo to blastocyst reported (Tervit et al., 1972). Betterbed and Wright (1985) reported development of two 1-cell embryos to "compacted blastocysts."

2.1. Laboratory Animals

One-cell and 2-cell mouse embryos can be cultured to the blastocyst stage in simple semidefined media (Whitten, 1957; Whitten and Biggers, 1968). There is, however, a type of early block in the mouse embryo because culture of 1-cell embryos of certain strains of mice to blastocysts is only possible with great difficulty (Cross and Brinster, 1973; Whittingham, 1975). One-cell rabbit embryos can be cultured to blastocysts in a semidefined complex culture medium but

growth and development of the blastocysts are usually much less than
growth in vivo (Kane and Foote, 1971; Kane, 1972; Kane, 1987).

Eight-cell rat embryos can be grown to the blastocyst stage in a
simple semidefined medium (Folstad et al., 1969) but 1-cell, 2-cell
and 4-cell stages do not usually develop beyond the 8-cell stage
(Mayer and Fritz, 1974; Whittingham, 1975; Wood and Whittingham,
1981). The situation is similar for the hamster embryo. Eight-cell
hamster embryos can be readily cultured to the blastocyst stage in a
simple semidefined medium (Yanagimachi and Chang, 1964; Whittingham
and Bavister, 1974; Bavister et al., 1983), but 2- and 4-cell stages
rarely develop in culture (Bavister, 1987). Expansion and hatching
of the hamster blastocyst in culture is also limited (Kane et al.,
1986).

2.2. Farm Animals and Carnivores

Cattle embryos can be readily cultured from the late 8-cell
or early morula stages to blastocysts in media supplemented with
fetal calf or bovine serum (Wright and Bondioli, 1981). They can
also be grown from 1-cell to 8- or 16-cell in similar media (Brackett
et al., 1982; Sirard and Lambert, 1985) but in general, with the
exception of one report to the contrary with a limited number of
embryos (Wright et al., 1976), continuous development in culture from
1-cell or 2-cell to the blastocyst stage is almost impossible with
present media. One-cell sheep embryos were cultured to late morula
in a relatively simple semidefined medium (Tervit and Rowson, 1974)
but did not survive to term after transfer. Embryos cultured from
the 8-cell stage to morula and early blastocysts did, however, result
in the birth of young after transfer. Continuous development from 1-
cell to blastocyst in vitro does not appear to be possible (Lindner
et al., 1983). Four- to 8-cell pig embryos can be cultured to
blastocysts in semidefined medium (Lindner and Wright, 1978).
Development of 1- to 2-cell embryos is, however, very limited (Davis
and Day, 1978).

On the other hand, 1-cell ferret embryos (Whittingham, 1975) and
1-cell cat embryos (Bowen, 1977) can be cultured to blastocysts in
semidefined medium.

2.3. Primates

It is perhaps surprising that the first primate embryos to
be cultured from 1-cell to blastocyst were human embryos (Steptoe et
al., 1971; Edwards, 1985) using Ham's F-10 supplemented with human or
fetal calf serum. Some human embryos hatch in vitro and there is one
report of a human blastocyst developing trophoblastic outgrowths and
synthesizing large amounts of hCG (Fishel et al., 1984). In vitro
fertilized rhesus monkey embryos have been cultured to hatched
blastocysts in medium supplemented with human cord serum (Morgan et

al., 1984). In vitro fertilized cynomolgus monkey embryos have been cultured to morulae in medium supplemented with fetal calf serum (Kreitman et al., 1982). Squirrel monkey embryos fertilized in vitro have been cultured to 8-cell stages in medium with newborn calf serum (Kuehl and Dukelow, 1979; Dukelow et al., 1983) and chimpanzee embryos to 4-cell stages in medium with human serum (Gould, 1983). Eight-cell baboon embryos collected from the uterus have been cultured to hatched blastocysts in medium with fetal calf serum (Pope et al., 1982).

3. EFFECTS OF SPECIFIC ENVIRONMENTAL FACTORS ON EMBRYOS

The known requirements for growth of mammalian embryos will be considered under the following headings: water quality; major inorganic ions; osmolarity; pH, HCO_3/CO_2 and oxygen levels; energy sources; amino acids; vitamins, cofactors and trace elements; nucleic acid precursors; macromolecules; hormonal and other growth factors.

3.1. Water Quality

Water constitutes over 98% of the content of any culture medium. The importance of water quality is emphasized by Whittingham (1971) who found that triple distillation of water increased the percent of 2-cell mouse embryos developing to blastocysts to 92.5% as compared with 36.1% for double distillation. In principle, water which is purified to a very high specification in a water "polishing" system consisting of carbon filtration and ion exchange cartridges followed by membrane filtration should give adequate water quality. However, in some circumstances, there may be problems with these systems possibly due to production of endotoxins by bacteria growing in the cartridges. The automatic recirculation of water in the cartridges for 5 min in every hour may help avoid this problem (Mather et al., 1986).

3.2. Major Inorganic Ions - K, Ca, Mg, PO_4

Most of the information available on the role and optimal concentrations of major ions such as K, Ca, Mg, and PO_4 is from Wales' (1970) work with the mouse embryo (see Figures 1 and 2). A striking fact to emerge from this work is the wide range of ion concentrations to which the 2-cell mouse embryo is tolerant. Two-cell mouse embryos grew to blastocysts in concentrations of K varying from 0.4 to 48 mM, of Ca from 0.1 to 10.2 mM, of Mg from 0 to 9.6 mM and of PO_4 from 0 to 7.2 mM. The optimal range for each of these ions was, of course, narrower than this and there is evidence that the 1-cell mouse embryo may be much more sensitive than the 2-cell stage to concentrations of K (Whittingham, 1975). The embryos of other species including the human may not be quite as tolerant but the mouse data suggests that in general small variations in ion concentrations are not crucial to development.

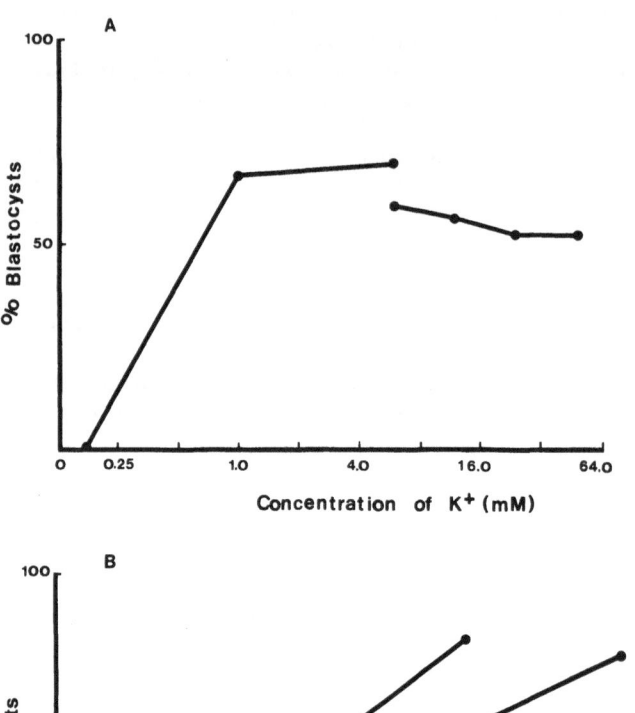

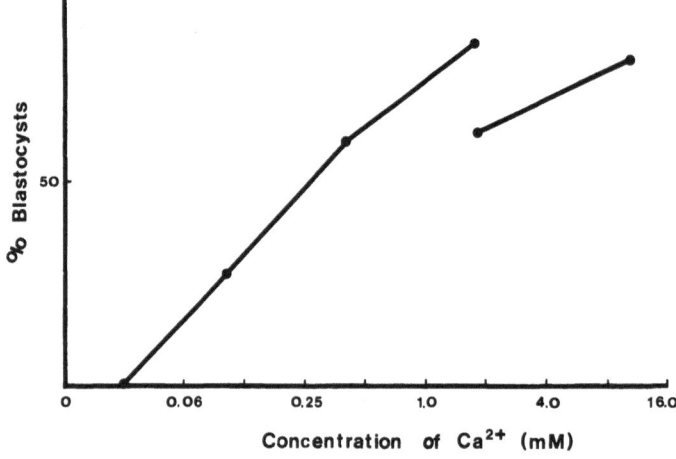

Fig. 1.   The effect of varying concentrations of K^+ (A) and Ca^{2+} (B)
in the culture medium on development of 2-cell mouse
embryos to blastocysts. Drawn from the data of Wales
(1970). Data in each case are from two experiments.

Calcium may be an exception in that Wales' data does suggest
that the level normally present in the mouse medium (1.71 mM) may be
suboptimal. Fishel (1980) found that both Ca and Mg are required for
mouse blastocysts to respond maximally to serum stimulation. This is
given added importance because Ca is known to be a major controller

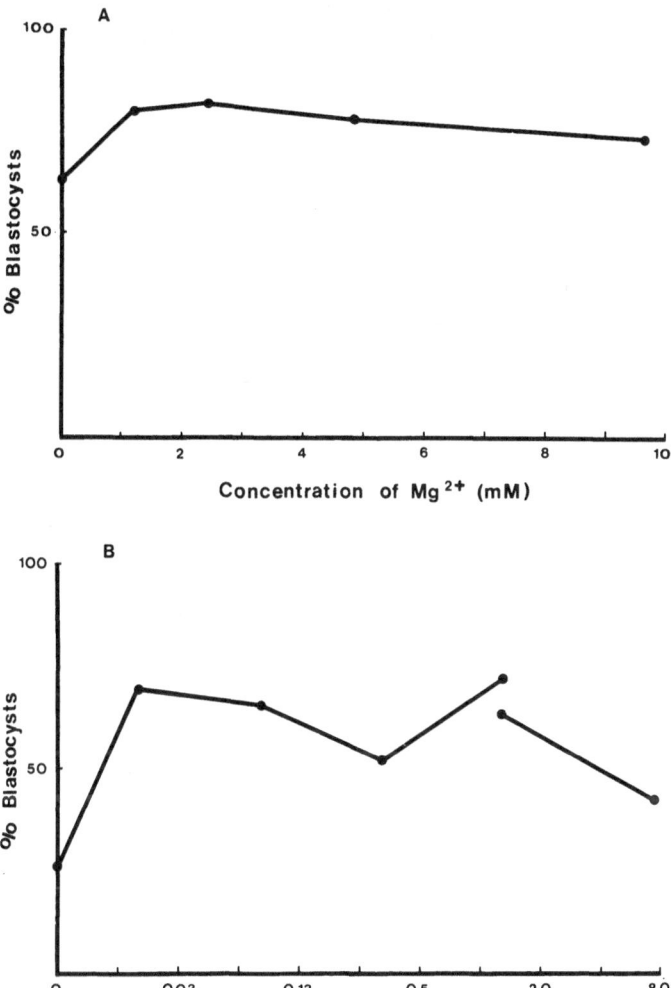

Fig. 2.   The effect of varying concentrations of Mg^{2+} (A) and PO_4^{3-} (B) on development of 2-cell mouse embryos to blastocysts. Drawn from the data of Wales (1970). Data on PO_4^{3-} is from two experiments.

of intracellular processes and of cell to cell adhesion. Embryos do not compact in the absence of Ca (Wales, 1970).

Even though Mg and PO_4 were not essential for development to the blastocyst stage, their absence did depress development (Wales,

1970). Also the possibility cannot be excluded that trace contamination of other salts in the medium allowed development in the absence of added Mg and PO_4.

3.3. Osmolarity

Brinster (1965a) examined the effects of osmolarity on the growth of 2-cell mouse embryos to blastocysts and found that 276 mOsM was optimal. Embryos developed to blastocysts over the range 216 to 330 mOsM (Figure 3). Naglee et al. (1969) found a very similar picture for the rabbit embryo with an optimum at about 270 mOsM. It is interesting that while changes in sodium chloride concentration are usually used to study osmolarity effects there is little if any work on the necessity for the sodium ion per se.

3.4. pH, HCO_3/CO_2 and Oxygen Levels

The usual buffer used for culturing mammalian embryos is a HCO_3/CO_2 system based on $NaHCO_3$ (25 mM) in the culture medium and CO_2 (5%) in the gas phase. Mouse embryos fail to cleave more than once or twice in the absence of HCO_3/CO_2 (Quinn and Wales, 1973). Rabbit embryos will cleave to the morula stage but fail to become blastocysts in a Hepes-buffered medium without HCO_3/CO_2 (Kane, 1975). The situation is even more critical in the hamster embryo in that Farrell and Bavister (1984) found that exposure of 2-cell hamster embryos to a Hepes-buffered modified Tyrode's solution for as little as 20 min drastically reduced subsequent embryo survival in vivo. Later work (Carney and Bavister, 1985) confirmed that collection of hamster embryos in low bicarbonate medium (2 mM) as compared with a high bicarbonate medium (25 mM) impaired later development in culture even though the culture medium contained 25 mM bicarbonate. Recently, these workers reported that increasing the CO_2 concentration in the gas phase from 5 to 10% improved development of 8-cell hamster embryos in culture (Carney and Bavister, 1986).

This need for HCO_3/CO_2 in an otherwise properly buffered medium may be partly related to its well known role in one-carbon metabolism and the formation of Krebs cycle intermediates. Also CO_2 may contribute to acidification of the intracellular environment since Carney and Bavister (1986) found that high CO_2 effects are mimicked by the addition of weak acids. It is unfortunate that embryos of many species have a requirement for CO_2 since the pH of bicarbonate-buffered media changes rapidly on exposure to air (Chetkowski et al., 1985).

The pH of embryo culture media is usually within the range of 7.3 to 7.4. Brinster (1965a) reported that 2-cell mouse embryos would grow to blastocysts in the range of 5.87 to 7.78. Interestingly, he found an optimum pH of 6.82 but later found that this depended on the energy substrate used and that, with optimal levels

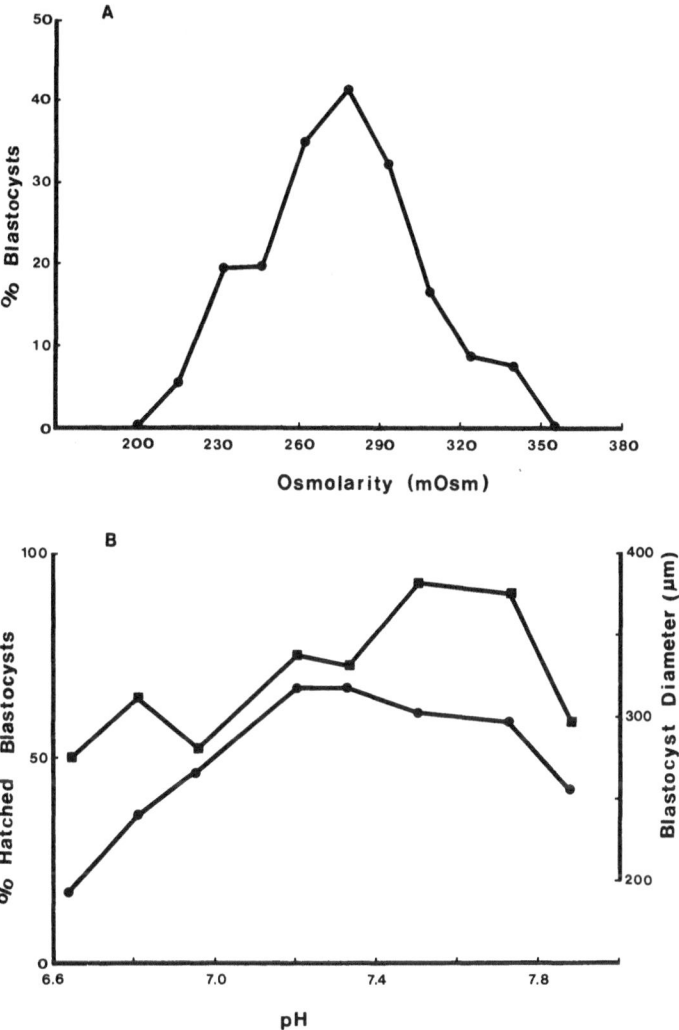

Fig. 3.   (A) Effect of osmolarity of the culture medium on development of 2-cell mouse embryos to blastocysts. Drawn from
the data of Brinster (1965a). (B) Effect of pH on development of 1-cell rabbit embryos to hatched blastocysts (●)
and on blastocyst expansion (■). Drawn from the data of
Kane (1974).

of pyruvate, pH 7.38 was superior to 6.82 (Brinster, 1965b). Kane
(1974) found that 1-cell rabbit embryos grew to blastocysts in
bicarbonate-buffered medium over the pH range 6.64 to 7.91. The

optimal pH for development to the blastocyst stage was 7.3 but
blastocyst expansion was greatest at 7.6 corresponding to a HCO_3
concentration of about 45 mM (see Figure 3). There is evidence that
the pH of rabbit uterine fluid may be as high as 7.69 (McLachlan et
al., 1970) or 7.9 (Vishwakarma, 1962).

The gas phase most commonly used for the culture of mammalian
embryos (5% CO_2 in air) contains about 20% oxygen. However, Auerbach
and Brinster (1968) have demonstrated that mouse embryos can develop
at very low O_2 concentrations (1%) and very high concentrations
(close to 100%) are deleterious (Whitten, 1957). A concentration of
5 to 10% would provide an oxygen tension close to that (40 to 75 mm
Hg) found in oviducal fluid (Mastroianni and Jones, 1965). Whitten
(1971) reported that 5% but not 20% would allow development of 1-cell
mouse embryos to blastocysts. In contrast, Brinster and Troike
(1979) found only a marginal advantage in using 5% instead of 10% for
culture of 1-cell mouse embryos. Although there have been some
further reports of the beneficial effect of using 5% O_2, e.g., Wright
et al. (1976), its usefulness has not been clearly and repeatedly
demonstrated.

3.5. Energy Sources

The discovery by Whitten (1957) that lactate allowed
development of 2-cell mouse ova to blastocysts in a simple salt
solution supplemented with bovine serum albumin (BSA) had a
revolutionary effect on embryo culture. It focused attention on the
mouse embryo as a model and on the simple type of semidefined medium
used by Whitten. This work was taken up by Brinster, Biggers and
their coworkers (Brinster, 1963; Brinster, 1965b; Biggers et al.,
1967). Examination of a wide range of energy substrates showed that
only pyruvate and oxaloacetate supported cleavage of the 1-cell mouse
embryo and in addition lactate and phosphoenolpyruvate supported
cleavage of the 2-cell embryo. Glucose is necessary for hatching,
outgrowth and attachment of the mouse blastocyst (Wordinger and
Brinster, 1976), but they suggest that its function in this situation
may not be that of an energy substrate but as a source of carbon for
a unique polysaccharide necessary for differentiation. This concept
is supported by work with a specific inhibitor of glucose utilization, 2-deoxyglucose, showing that utilization of glucose by the
embryo becomes essential at the blastocyst stage in both mice
(Thomson, 1967) and rabbits (Kane and Buckley, 1977). The requirement for glucose was not for energy production since, while the
inhibition was overcome by exogenous glucose, exogenous pyruvate had
no effect.

The requirement of the rabbit embryo for particular energy
substrates is less acute than that of the mouse embryo. Kane (1972)
showed that 1-cell rabbit embryos grew to blastocysts in the presence
of amino acids, vitamins and BSA but in the absence of any

carbohydrate-type energy sources. The addition of pyruvate or glucose did, however, improve growth. Later work showed that the addition of any one of a number of long- or short-chain fatty acids to a simple salt solution supplemented with defatted charcoal-treated BSA supported growth of 1-cell rabbit embryos to viable morula (Kane, 1979). More recently Kane (1986a) found that rabbit embryos will cleave from the 1-cell stage to a mean of about 10 cells per embryo over about 48 hr in a simple defined medium consisting of a physiological salt solution with 0.1% PVA but without any energy substrates in the medium. This result was surprising and indicates that the 1-cell rabbit embryo has considerable endogenous energy sources. The addition of pyruvate or amino acids to the medium significantly increased the cleavage rate by doubling the number of cells per embryo after 48 hr in culture.

Thus, the early rabbit embryo stands in striking contrast to the early mouse embryo. The 1-cell rabbit embryo makes considerable use of endogenous energy substrates and also has the capacity to use a wider range of exogenous energy substrates such as fatty acids, amino acids, and glucose. Relevant knowledge is almost entirely lacking for species other than the mouse and the rabbit, primarily because of difficulty of culturing the respective embryos in semidefined or defined media.

The 1-cell mouse embryo may have a requirement for fatty acids for purposes other than energy production, most probably for lipid synthesis. The results of Flynn and Hillman (1980) show that mouse embryos can utilize fatty acids in the medium both for energy production and for incorporation into embryo lipids. Quinn and Whittingham (1982) found that fatty acid-free albumin in a medium containing pyruvate supported development of 1-cell mouse embryos to blastocysts only when exogenous fatty acids (oleic and/or palmitic) were also added to the medium.

3.6. Amino Acids

One- or 2-cell mouse embryos will grow to blastocysts in a semidefined medium containing BSA without free amino acids (Whitten, 1957; Whitten and Biggers, 1968; Cross and Brinster, 1973). Two-cell mouse embryos have been cultured to blastocysts in a defined medium without any source of fixed nitrogen (Cholewa and Whitten, 1970). Outgrowth of mouse blastocysts in vitro does, however, require the presence of a number of amino acids in the medium (Gwatkin, 1966; Spindle and Pedersen, 1973; Sellens and Sherman, 1980; Spindle, 1980).

Amino acids or a fixed nitrogen source are not essential for cleavage of 1-cell rabbit embryos as they cleave to over 20 cells per embryo in a simple defined medium without amino acids or protein and with PVA as the only macromolecule (Kane, 1986a).

The situation changes drastically at later developmental stages
as amino acids are essential for progression of cleavage-stage rabbit
embryos to blastocysts in vitro. The experiment of Kane and Foote
(1970) is important in highlighting the requirements for development
to the blastocyst stage in the rabbit. Two- to 4-cell embryos were
cultured for four days either in a complex medium based on Ham's F-10
containing amino acids, vitamins, trace elements and nucleic acid
precursors or in media with one of these nutrient groups omitted.
The amino acid group of nutrients was the only group whose omission
completely prevented blastocyst formation (Table II). This difference between rabbit and mouse embryos is not surprising if one
considers that the mouse blastocyst at implantation has only about
100 cells (Bowman and McLaren, 1970) and a protein content of about
20 ng (less than at the 1-cell stage). This compares with a cell
population of about 80,000 for the implanting rabbit blastocyst
(Daniel, 1964a) and a protein content of about 200 µg (Lutwak-Mann,
1971).

Most of the twenty amino acids tested by Kane and Foote (1970)
were beneficial to blastocyst growth with methionine, serine and
threonine being the most important. Daniel and Krishnan (1967) found
that ten amino acids (methionine, serine, threonine, arginine,
lysine, histidine, tryptophane, phenylalanine, leucine and valine)
were essential for expansion of 5-day rabbit blastocysts.

The amino acid requirements for growth of hamster embryos are
extremely interesting. Bavister et al. (1983) found that a group of
four amino acids (glutamine, phenylalanine, methionine and
isoleucine) had a marked effect in promoting development of early 8-
cell hamster embryos to blastocysts. Carney and Bavister (1985) then
showed that glutamine was the only amino acid necessary for this
effect. Subsequently, Kane et al. (1986) found that amino acids were
necessary for hatching of hamster blastocysts in vitro and that
glutamine was as active in this regard as a full complement of 20
amino acids. As well as being a component of protein, glutamine may
be functioning here as a source of amino groups and as a readily
available energy source (Meister, 1984).

3.7. Vitamins, Cofactors, and Trace Elements

One- and 2-cell mouse embryos develop to hatched blastocysts in vitro in the absence of added vitamins and cofactors in the
culture medium (Brinster, 1963; Whitten and Biggers, 1968). Early
cleavage-stage rabbit embryos do not require vitamins to develop as
far as the start of blastocyst formation but further growth and
expansion is extremely limited in the absence of vitamins (Kane and
Foote, 1970) - see Table II. More recently Kane (1986b) has examined
the question of which vitamins are necessary for expansion of rabbit
blastocysts cultured from the morula stage. Preliminary results
indicate that inositol, pyridoxine, riboflavin, niacinamide and

Table II.   Effect of the omission of different nutrient groups from a
            complex culture medium on growth of 2- to 4-cell rabbit
            embryos to blastocysts[a]

| | | Basic medium minus | | |
	Basic medium	Amino acids	Vitamins	Trace elements	Nucleic acid precursors
No. of embryos	57	57	57	57	57
% Early blastocysts	83	0	54	79	81
% Expanding blastocysts	42	0	5	39	63

[a]Modified from Kane and Foote (1970)

thiamine are most important. Daniel (1967) found that four of these
five vitamins (inositol, pyridoxine, riboflavin and thiamine)
together with folic acid caused increased expansion of 5-day rabbit
blastocysts.

The potential importance of vitamins for the growth of preimplantation embryos is highlighted by the recent finding (Kane et al.,
1986) that vitamins are necessary for hatching of hamster blastocysts
cultured from the 8-cell or morula stages. This effect was clearly
visible after about 36 hr in culture.

These reports all refer to water soluble vitamins. There is an
almost total absence of reports on the effects of the fat-soluble
vitamins on preimplantation embryos. This is obviously related to
the difficulty of solubilizing these vitamins in culture media.

There is very little information available on the requirements
of mammalian embryos for trace elements. The question of which trace
elements may be necessary or beneficial to embryo growth is complicated by the fact that macromolecules such as BSA and even the
highest grades of salts such as NaCl are contaminated. Mouse embryos
are routinely cultured from the 1- or 2-cell stage to form blastocysts in the absence of added trace elements (Brinster, 1963; Whitten
and Biggers, 1968). Kane and Foote (1970) found that the omission of

Fe, Cu and Zn from the culture medium did not interfere with the
development of 2- and 4-cell embryos to expanding blastocysts. In
contrast, Daniel and Millward (1969) reported that the omission of
ferrous ion from Ham's F-12 medium completely abolished cleavage of
rabbit embryos and caused collapse of rabbit blastocysts. This
difference in results may reflect the use of macromolecule-free F-12
by Daniel and Millward. Eight-cell hamster embryos have also been
cultured to the blastocyst stage without the addition of trace
elements (Bavister et al., 1983). Further work needs to be carried
out in this area, particularly in regard to possible beneficial
effects on embryos of some of the newer trace elements being
introduced into culture media such as selenium and vanadium, etc.
(Nielsen, 1981). Requirements for these factors may become evident
as BSA or other serum protein levels in the medium are reduced
(McKeehan et al., 1981).

3.8. Nucleic Acid Precursors

        Mouse embryos can be cultured to the blastocyst stage in
the absence of added nucleic acid precursors such as thymidine or
hypoxanthine (Brinster, 1963; Whitten and Biggers, 1968). Tenbroeck
(1968) found no beneficial effect of adding various nucleosides and
bases to culture medium for mouse embryos. Thymidine and uridine
were toxic at levels of $> 10^{-5}$ M. Thymidine is toxic to the growth
of 2- to 4-cell rabbit embryos to blastocysts at 3×10^{-6} M (Kane and
Foote, 1970, 1971). Daniel (1967) found that expansion of 5-day
rabbit blastocysts was promoted by hypoxanthine. Even though nucleic
acid precursors are not generally beneficial to preimplantation
embryo growth, it is quite clear that this is not because of failure
to incorporate these molecules into DNA. Incorporation of tritiated
nucleosides into the DNA and RNA of embryos has been shown by various
workers for the mouse embryo (e.g., Mintz 1965; Barlow et al., 1972;
Streffer et al., 1980) and for the rabbit embryo (e.g., Gulyas et
al., 1969; Manes, 1969; El-Banna and Daniel 1972; Fischer, 1986).
Thus, the general failure of added nucleosides and bases to exert a
beneficial effect on embryo growth indicates that the normal de novo
pathway for nucleic acid synthesis is adequate for the requirements
of the embryo.

3.9. Macromolecules

3.9.1.   General Role of Macromolecules

        In cell culture generally and in particular in
embryo culture, it is normal practice to add a source of macromolecules. The source of macromolecules may be serum or dialyzed serum
protein, Fraction V or crystallized BSA or a synthetic macromolecule
such as polyvinylpyrollidone (PVP), ficoll or polyvinylalcohol (PVA).
Any use of whole serum involves the addition of a very wide range of
undefined proteins together with micromolecules both bound to the

serum proteins and free in the medium. Dialysis of the serum merely
removes the micromolecules that are free in the serum and one is
still left with the undefined proteins and their bound micromolecules. The use of BSA involves the addition of a relatively pure
protein which has a range of undefined micromolecules attached.
These small molecules may include fatty acids, hormones and vitamins,
etc. (Goodman, 1958; Westphal, 1970). Normally the level of BSA used
is 1-4 mg/ml for mouse embryos (Brinster, 1965c; Whitten and Biggers,
1968) and 5-15 mg/ml for rabbit embryos (Kane and Foote, 1970; Kane,
1985). Use of a synthetic macromolecule such as PVA, ficoll or PVP
provides a defined macromolecule and circumvents the problem of small
molecules.

The role played by macromolecules may vary with the type of
macromolecule, and the species and stage of embryo. Kane (1987) has
listed three possibilities: (1) A physical effect on the stickiness
of the embryo or blastomeres which may be due to a change in the
electrical surface charge. Embryos in macromolecule-free media are
extremely difficult to handle due to their stickiness and while this
difficulty is alleviated by PVP or ficoll, in our experience, it is
eliminated completely by PVA (Bavister, 1981) or protein. It is
difficult to find any report in the literature of culture of embryos
for any prolonged period in media completely free of macromolecules.
Cholewa and Whitten (1970) reported growth of 2-cell mouse embryos to
blastocysts in the presence of PVP and in the absence of any other
macromolecule or source of fixed nitrogen. Fewer embryos, however,
underwent hatching in the absence of albumin. They also found that
albumin was necessary for cleavage of 1-cell embryos beyond the 2-
cell stage. Kuzan et al. (1982) found that whereas PVA and BSA would
individually allow development of 2-cell mouse embryos to blastocysts, outcome was improved by the combined use of both compounds.
Kane (1986a) has found that 1-cell rabbit embryos could be cultured
to early morulae in the presence of PVA and in the absence of a fixed
nitrogen source but that development was improved by the addition of
BSA. Development to the blastocyst stage requires the presence of
either BSA (Kane and Foote, 1970) or uterine proteins (Krishnan and
Daniel, 1967).

(2) Macromolecules may protect against the effects of toxic
compounds by binding or chelation. Cholewa and Whitten (1970)
suggested that, even though a fixed nitrogen source is not normally
necessary for development of 2-cell mouse embryos to the blastocyst
stage, albumin, peptides or amino acids may be beneficial if toxic
metallic ions are present in the medium. This may explain the
difference between their results and those of Wales and Whittingham
(1973) who found that the substitution of PVP for albumin did not
maintain development of even 8-cell mouse embryos. If this is true,
it means that the requirements for macromolecules may vary depending
on the quality of the water and other constituents of the medium.
The report of Abramczuk et al. (1977), that the addition of a

chelating agent (EDTA) to the culture medium promoted in vitro
development of 1-cell embryos from inbred strains of mice, may be
relevant here.

(3) Protein macromolecules such as BSA may contain bound low
molecular-weight contaminants which promote embryo growth. Albumin
is known to bind fatty acids, steroid hormones, vitamins and many
other small molecules; one of its major functions in the blood is to
act as a carrier protein (Goodman, 1958; Westphal, 1970; Spector,
1975). Wales and Whittingham (1973) found that the use of fatty
acid-free albumin instead of normal albumin reduced the proportion of
8-cell mouse embryos developing to blastocysts. However, the
clearest evidence for the effects of low molecular weight contaminants of albumin is derived from work on the rabbit embryo. Kane
(1979) showed that fatty acids bound to albumin could act as energy
substrates for development of 1-cell rabbit embryos to morulae. Kane
and Headon (1980) later showed that this effect of albumin-bound
fatty acids also works at the blastocyst stage. Much more interestingly, however, they also found that a contaminant of BSA other than
a fatty acid appeared to cause increased blastocyst expansion and
hatching. Later work (Kane, 1985) showed that there is a low
molecular weight embryotropic factor attached to BSA (see Section
3.10.).

It is this author's opinion that much of the growth promoting
effect of albumin on embryos, particularly at the blastocyst stage,
is due to one or other of these low molecular-weight contaminants.
The fact that albumin is contaminated with low molecular weight
compounds which affect embryo growth means that BSA or any form of
albumin is a variable product and media containing it can at best be
termed semidefined. The importance of this type of variability is
emphasized by the report of Kane (1983a) that one batch of BSA
resulted in a high proportion of completely hatched blastocysts
whereas a second batch from the same supplier gave no hatched
blastocysts and less than half the number of cells per blastocyst.

3.9.2.   Uterine Proteins

The discovery of a distinctive rabbit uterine
protein called "blastokinin" or uteroglobin (Krishnan and Daniel,
1967; Beier, 1968) caused a tremendous surge of interest in the role
of uterine proteins in development of preimplantation embryos. The
concentration of this protein can be close to 50% of the total
uterine-fluid protein during the 5th and 6th days of pregnancy. The
initial claim that it functioned as a unique regulator of rabbit
blastocyst growth and development (Krishnan and Daniel, 1967) was not
sustained in later work (Daniel, 1971; El-Banna and Daniel, 1972).
Maurer and Beier (1976) compared the effects of BSA and both
fractionated and unfractionated uterine proteins on growth of rabbit
blastocysts and found that unfractionated uterine proteins had a

greater growth promoting effect than any specific uterine fraction
such as uteroglobin. They also found that BSA was only marginally
less effective than the unfractionated uterine protein. It could be
said that uteroglobin is still a protein in search of a function. A
possible function could be one of masking the antigenicity of the
developing trophoblast and embryo as suggested by Mukherjee et al.
(1980, 1982).

Because of the research effort that has been concentrated on the
role of uteroglobin, the fact that rabbit uterine fluid contains a
number of other unique proteins, glycoproteins and peptides
(Kirchner, 1969; Petry et al., 1970) has been ignored. An examination of such factors present in low and, therefore, possibly
hormonal levels might well indicate a potential role for growth
factors in uterine fluid. A recent report (Fischer, 1986) shows that
whole unfractionated uterine fluid has a definite growth promoting
effect on rabbit embryos as compared with BSA when added to a complex
culture medium.

3.9.3.   Enzymes as Environmental Factors

In all mammalian species, the ovulated egg is
surrounded by a noncellular mucoprotein layer, the zona pellucida.
In the rabbit, there is also an additional layer laid down in the
oviduct and uterus, usually called the mucin coat. Cleavage and
blastocyst formation take place within the zona pellucida and at the
time of implantation the zona (and mucin coat if present) must be
eroded and/or shed ("hatched") to allow the trophoblast access to the
uterine tissues during implantation (Boyd and Hamilton, 1952;
Bergstrom, 1972). Erosion of the blastocyst coverings in vivo could
be due to uterine enzymes as suggested for the mouse (McLaren, 1969)
and rat (Wu, 1973). In the case of the rabbit blastocyst, which does
not shed the zona as a discreet structure and so does not "hatch" in
vivo, Denker and Gerdes (1979) have provided evidence that blastocyst
enzymes may be involved in erosion and remodeling of the blastocyst
coverings. This type of information has stimulated workers to
examine the effects of various enzymes, both proteases and glycosidases, on cultured embryos. Another factor which has awakened
interest in possible effects of enzymes on embryos has been the
evidence that exogenous proteases added to certain types of cells in
culture are mitogenic (Kaplan and Bona, 1974; Chen and Buchanan,
1975; Keil, 1979; Cunningham, 1981).

There is extensive information on the effects of exogenous
enzymes on the mouse embryo. Proteases such as trypsin and pronase
cause lysis of the mouse zona in culture (Mintz, 1962; Gwatkin, 1964;
Bowman and McLaren, 1970). A number of workers have also shown that
short-term treatment of cultured mouse embryos with pronase or
trypsin to remove the zona pellucida stimulated growth of the embryos
(Pienkowski et al., 1974; Konwinski et al., 1978). Menino and

O'Claray (1986) found that plasmin and plasminogen increased the incidence of hatching and trophoblastic outgrowth in vitro. In the rabbit embryo, Onuma et al. (1968) showed that short-term treatment with pronase stimulated blastocyst expansion. However, in none of these reports is it clear that the effect of proteases in apparently stimulating cell division is a true mitogenic effect directly on the embryonic cells. It is more likely that partial or complete digestion of the zona with protease allows the blastocyst to expand and thus stimulates growth indirectly (Kane, 1983b).

Kane (1986c) carried out a survey of the effects of a wide range of proteases and glycosidases added in relatively low levels to the culture medium on culture of rabbit morulae to blastocysts. The only enzymes to cause zona lysis and shedding in low levels were trypsin and streptomyces griseus protease (pronase). Elastase and chymotrypsin were active only at high levels and all other proteases tested were inactive as were all glycosidases tested. None of the enzymes stimulated growth to any marked extent.

3.10. Hormonal and Other Growth Factors

It is clear that the embryo is at the center of reproductive endocrinology. This makes it extraordinary that there are no well-documented examples of a beneficial effect of a hormone directly on the preimplantation embryo. There are numerous examples of the deleterious effects of hormones on embryos, e.g., Daniel (1964b) and Daniel and Levy (1964) found that progesterone specifically inhibited cleavage of the rabbit embryo in culture. A general review of the effects of steroid hormones on preimplantation embryos found little evidence for beneficial direct effects (Warner, 1977). It appears that such effects in vivo are mediated indirectly via the reproductive tract.

The addition of a cocktail of hormones and growth factors including glucagon, thyrocalcitonin, transferrin, triodothyronine, insulin, parathyroid hormone, liver cell growth factor, LH releasing hormone, FSH and LH to the culture medium did not stimulate rabbit blastocyst growth (Kane, unpublished data). However, current work has produced evidence that a low molecular weight growth factor for rabbit embryos can be extracted from some samples of bovine serum albumin (Kane, 1985). This embryotropic factor is extracted using 5% formic acid and a membrane filter with a cut off of M_r 10,000 (subsequent unpublished work has shown that the factor will pass through a filter with an M_r of 1,000). Addition of the lyophilized extract to a complex semi-defined medium containing charcoal-treated BSA resulted in an 8-fold increase in cell number in blastocysts cultured from the morulae stage (See Table III). Subsequent chromatographic purification of the extract and amino acid analysis indicates that the active factor may be a peptide. These results are exciting because it seems the presence of the factor in the culture medium is

Table III.   Effect of a low molecular weight extract of BSA on rabbit
             blastocyst formation, expansion, and cell division[a]

| | Basic medium | Basic medium + low molecular weight extract (mg/ml) | | |
		0.04	0.2	1.0
No. of embryos	51	50	48	49
% Blastocysts	100	98	98	96
Embryo cell count (mean ± s.e.m.)	99 ± 7	210 ± 29	807 ± 65	806 ± 55
Embryo diameters (μm, mean ± s.e.m.)	155 ± 7	191 ± 9	349 ± 13	328 ± 14

[a]The basic medium was a complex culture medium supplemented with 0.5%
charcoal-treated BSA. Modified from Kane (1985).

essential for any marked degree of blastocyst growth and because it
may also act on the embryos of the hamster (Kane and Bavister,
unpublished data).

There is evidence indicating an important role for growth
factors in postimplantation mouse embryonic development (see review
by Rizzino, 1987). Factors resembling transforming growth factors
are apparently released by early postimplantation mouse embryos.
Another exciting development is the finding of Camous et al. (1984)
that trophoblastic vesicles from day-14 bovine embryos cocultured
with 1- to 8-cell cattle embryos stimulated their development to
morulae. This suggests that trophoblastic tissue secretes soluble
growth factor(s).

4.   GENERAL COMMENTS AND PROSPECTS FOR THE FUTURE

There are a number of obvious points that arise from this
review.

a.   The information based on studies with defined or semi-
     defined media is derived from an extremely limited number
     of species - mouse, rabbit, and hamster.

b.   It is clear from what has been said previously, that the
     media in current use for embryo culture are grossly
     inadequate. This is reflected in the numerous blocks to in
     vitro development recorded for most species studied. The
     use of various types of serum may partly compensate for
     these inadequacies but only at the cost of introducing
     other problems associated with serum variability. Sirard
     and Lambert (1985) provide a striking documentation of this
     problem. They found that one batch of bovine serum allowed
     19 of 42 four-cell bovine embryos to cleave to 8 cells
     whereas a second batch from a different animal prepared
     under the same conditions only allowed 2 of 20 four-cell
     embryos to do so.

c.   The mouse embryo is probably an inadequate model for most
     other species including primates. The mouse embryo has
     less protein at the blastocyst stage just before implanta-
     tion than at the 1-cell stage. It has minimal environmen-
     tal requirements for development from 1-cell to blastocyst
     and is clearly different from most other species. I would
     suggest that, of the laboratory animals, the hamster is
     perhaps the most demanding model for the cleavage stages
     (Bavister, 1987) and that for blastocyst formation the
     rabbit embryo may be the most suitable model (Kane 1987).
     It would seem reasonable that information from these
     species should first be tested on nonhuman primate species
     such as the rhesus monkey before considering its applica-
     bility to human embryos.

d.   It is likely that some of the current work on growth
     factors in embryos (Rizzino, 1987, with postimplantation
     mouse embryos; Camous et al., 1984, on growth factors from
     bovine trophoblast; Kane, 1985, with the BSA low molecular-
     weight growth factor) will shed new light on the control of
     preimplantation embryo growth.

e.   There are undoubtedly some very interesting biological
     questions underlying the blocks to early embryonic develop-
     ment occurring in vitro. Solutions to these problems
     should provide insights not only into the control of
     embryonic growth but also into the control of growth and
     differentiation in normal and malignant cells.

5.   REFERENCES

Abramczuk, J., Solter, D., and Koprowski, H., 1977, The beneficial
     effect of EDTA on development of mouse one-cell embryos in
     chemically defined medium, Dev. Biol. 61:378-383.

Auerbach, S. and Brinster, R. L., 1968, Effect of oxygen concentra-
     tion on the development of two-cell mouse embryos, Nature
     217:465-466.
Barlow, P., Owen, D. A. J., and Graham, C., 1972, DNA synthesis in
     the preimplantation mouse embryo, J. Embryol. Exp. Morphol.
     27:431-445.
Bavister, B. D., 1981, Substitution of a synthetic polymer for
     protein in a mammalian gamete culture system, J. Exp. Zool.
     217:45-51.
Bavister, B. D.,, 1987, Studies on the developmental blocks in
     cultured hamster embryos, in: The Mammalian Preimplantation
     Embryo: Regulation of Growth and Differentiation In Vitro
     (B. D. Bavister, ed.), Plenum Press, New York, pp. 219-249.
Bavister, B. D., Leibfried, M. L., and Lieberman, G., 1983, Develop-
     ment of preimplantation embryos of the golden hamster in a
     defined culture medium, Biol. Reprod. 28:235-247.
Beier, H. M., 1968, Uteroglobin: a hormone sensitive endometrial
     protein involved in blastocyst development, Biochim. Biophys.
     Acta 160:289-291.
Bergstrom, S., 1972, Shedding of the zona pellucida of the mouse
     blastocyst in normal pregnancy, J. Reprod. Fertil. 31:275-277.
Betterbed, B. and Wright, R. W., 1985, Development of one-cell ovine
     embryos in two-culture media under two gas atmospheres,
     Theriogenology 23:547-553.
Biggers, J. D., Whittingham, D. G., and Donahue, R. P., 1967, The
     pattern of energy metabolism in the mouse oocyte and zygote,
     Proc. Natl. Acad. Sci. U.S.A. 58:560-567.
Bowen, R. A., 1977, Fertilization in vitro of feline ova by spermato-
     zoa from the ductus deferens, Biol. Reprod. 17:144-147.
Bowman, P. and McLaren, A., 1970, The reaction of the mouse blasto-
     cyst and its zona pellucida to enzymes in vitro, J. Embryol.
     Exp. Morphol. 24:331-334.
Boyd, J. D. and Hamilton, W. J., 1952, Cleavage, early development
     and implantation of the egg, in: Marshall's Physiology of
     Reproduction, Volume 2 (A. S. Parkes, ed.), Longmans, London,
     pp. 1-126.
Brackett, B. G., Bousquet, D., Boice, M. L., Donawick, W. J., Evans,
     J. F., and Dressel, M. A., 1982, Normal development following in
     vitro fertilization in the cow, Biol. Reprod. 27:147-158.
Brinster, R. L., 1963, A method for in vitro cultivation of mouse ova
     from two-cell to blastocyst, Exp. Cell Res. 32:205-208.
Brinster, R. L., 1965a, Studies on the development of mouse embryos
     in vitro. I. The effect of osmolarity and hydrogen ion concen-
     tration, J. Exp. Zool. 158:49-58.
Brinster, R. L., 1965b, Studies on the development of mouse embryos
     in vitro. II. The effect of energy source, J. Exp. Zool.
     158:59-68.
Brinster, R. L., 1965c, Studies on the development of mouse embryos
     in vitro. III. The effect of fixed-nitrogen source, J. Exp.
     Zool. 158:69-77.

Brinster, R. L., 1967, Protein content of the mouse embryo during the first five days of development, J. Reprod. Fertil. 13:413-420.

Brinster, R. L. and Troike, D. E., 1979, Requirements for blastocyst development in vitro, J. Anim. Sci. 49(Suppl. 2):26-34.

Camous, S., Heyman, Y., Méziou, W., and Ménézo, Y., 1984, Cleavage beyond the block stage and survival after transfer of early bovine embryos cultured with trophoblastic vesicles, J. Reprod. Fertil. 72:479-485.

Carney, E. W. and Bavister, B. D., 1985, Development of hamster preimplantation embryos in vitro: Effect of bicarbonate and amino acids, Biol. Reprod. 32(Suppl. 1):98 (abstract).

Carney, E. W. and Bavister, B. D., 1986, Increased atmospheric carbon dioxide stimulates hamster embryo development, Biol. Reprod. 34(Suppl. 1):199 (abstract).

Chen, L. B. and Buchanan, J. M., 1975, Mitogenic activity of blood components. I. Thrombin and prothrombin, Proc. Natl. Acad. Sci. U.S.A. 72:131-135.

Chetkowski, R. J., Nass, T. E., Matt, D. W., Hamilton, F., Steingold, K. A., Randle, D., and Meldrum, D. R., 1985, Optimization of hydrogen-ion concentration during aspiration of oocytes and culture and transfer of embryos, J. In Vitro Fert. Embryo Transfer 2:207-212.

Cholewa, J. A. and Whitten, W. K., 1970, Development of two-cell mouse embryos in the absence of a fixed-nitrogen source, J. Reprod. Fertil. 22:553-555.

Cross, P. C. and Brinster, R. L., 1973, The sensitivity of one-cell mouse embryos to pyruvate and lactate, Exp. Cell Res. 77:57-62.

Cunningham, D. D., 1981, Proteases as growth factors, in: Tissue Growth Factors (R. Baserga, ed.), Springer-Verlag, New York, pp. 229-248.

Daniel, J. C., Jr., 1964a, Early growth of rabbit trophoblast, Am. Naturalist 98:85-97.

Daniel, J. C., Jr., 1964b, Some effects of steroids on cleavage of rabbit eggs in vitro, Endocrinology 75:706-710.

Daniel, J. C., Jr., 1967, Vitamins and growth factors in the nutrition of rabbit blastocysts in vitro, Growth 31:71-77.

Daniel, J. C., Jr., 1971, Uterine proteins and embryonic development, in: Schering Symposium on Intrinsic and Extrinsic Factors in Early Mammalian Development (G. Raspe, ed.), Pergamon Press, Oxford, pp. 191-203.

Daniel, J. C., Jr., and Krishnan, R. S., 1967, Amino acid requirements for growth of the rabbit blastocyst in vitro, J. Cell. Physiol. 70:155-160.

Daniel, J. C., Jr., and Levy, J. D., 1964, Action of progesterone as a cleavage inhibitor of rabbit ova in vitro, J. Reprod. Fertil. 7:323-329.

Daniel, J. C., Jr., and Millward, J. T., 1969, Ferrous ion requirement for cleavage of the rabbit egg, Exp. Cell Res. 54:135-136.

Davis, D. L. and Day, B. N., 1978, Cleavage and blastocyst formation by pig eggs in vitro, J. Anim. Sci. 46:1043-1053.

Denker, H.-W., and Gerdes, H.-J., 1979, The dynamic structure of rabbit blastocyst coverings. 1. Transformation during regular preimplantation development, Anat. Embryol. 157:15-34.

Dukelow, W. R., Chan, P. J., Hutz, R. J., DeMayo, F. J., Dooley, V. D., Rawlins, R. G., and Ridha, M. T., 1983, Preimplantation development of the primate embryo after in vitro fertilization, J. Exp. Zool. 228:215-221.

Edwards, R. G., 1985, Current status of human conception in vitro, Proc. R. Soc. Lond. [Biol.]. 223:417-448.

El-Banna, A. A. and Daniel, J. C., Jr., 1972, The effects of protein fractions from rabbit uterine fluids on embryo growth and uptake of nucleic acid and protein precursors, Fertil. Steril. 23:105-114.

Farrell, P. S. and Bavister, B. D., 1984, Short-term exposure of two-cell hamster embryos to collection media is detrimental to viability, Biol. Reprod. 31:109-114.

Fischer, B., 1986, Studien zur Thymidine-Inkorporation und Entwicklung von in vitro kultivierten und transferierten Kaninchenembryonen nach FSH-stimuliertem Follikel wachstum, Dissertation Rheinisch-Westfalische Technische Hochschule, Aachen.

Fishel, S. B., 1980, The role of divalent cations in the metabolic response of mouse blastocysts to serum, J. Embryol. Exp. Morphol. 58:217-229.

Fishel, S. B., Edwards, R. G., and Evans, C. J., 1984, Human chorionic gonadotrophin secreted by preimplantation embryos cultured in vitro, Science 223:816-818.

Flynn, T. J. and Hillman, N., 1980, The metabolism of exogenous fatty acids by preimplantation mouse embryos developing in vitro, J. Embryol. Exp. Morphol. 56:157-168.

Folstad, L., Bennett, J. P., and Dorfman, R. I., 1969, The in vitro culture of rat ova, J. Reprod. Fertil. 18:145-146.

Goodman, D. S., 1958, The interaction of human serum albumin with long-chain fatty acid anions, J. Am. Chem. Soc. 80:3892-3898.

Gould, K. G., 1983, Ovum recovery and in vitro fertilization in the chimpanzee, Fertil. Steril. 40:378-383.

Gulyas, B. J., Daniel, J. C., Jr., and Krishnan, R. S., 1969, Incorporation of labelled nucleosides in vitro by rabbit and mink blastocysts in the presence of blastokinin or serum, J. Reprod. Fertil. 20:255-262.

Gwatkin, R. B. L., 1964, Effect of enzymes and acidity on the zona pellucida of the mouse egg before and after fertilization, J. Reprod. Fertil. 7:99-105.

Gwatkin, R. B. L., 1966, Amino acid requirements for attachment and outgrowth of the mouse blastocyst in vitro, J. Cell. Physiol. 68:335-344.

Kane, M. T., 1972, Energy substrates and culture of single cell rabbit ova to blastocysts, Nature (London) 238:468-469.

Kane, M. T., 1974, The effects of pH on culture of one-cell rabbit
    ova to blastocysts in bicarbonate-buffered medium, J. Reprod.
    Fertil. 38:477-480.
Kane, M. T., 1975, Bicarbonate requirements for culture of one-cell
    rabbit ova to blastocysts, Biol. Reprod. 12:552-555.
Kane, M. T., 1979, Fatty acids as energy sources for culture of one-
    cell rabbit ova to viable morulae, Biol. Reprod. 20:323-332.
Kane, M. T., 1983a, Variability in different lots of commercial
    bovine serum albumin affects cell multiplication and hatching of
    rabbit blastocysts in culture, J. Reprod. Fertil. 69:555-558.
Kane, M. T., 1983b, Evidence that protease action is not specifically
    involved in the hatching of rabbit blastocysts caused by
    commercial bovine serum albumin in culture, J. Reprod. Fertil.
    68:471-475.
Kane, M. T., 1985, A low molecular weight extract of bovine serum
    albumin stimulates rabbit blastocyst cell division and expansion
    in vitro, J. Reprod. Fertil. 73:147-150.
Kane, M. T., 1986a, Minimal nutrient requirements for cleavage of
    one-cell rabbit embryos in culture, Biol. Reprod.
    34(Suppl. 1):199 (abstract).
Kane, M. T., 1986b, The effects of vitamins on blastocyst growth of
    cultured rabbit embryos, Ir. J. Med. Sci. 155:321-322.
Kane, M. T., 1986c, A survey of the effects of proteases and
    glycosidases on culture of rabbit morulae to blastocysts, J.
    Reprod. Fertil. 78:225-230.
Kane, M. T., 1987, In vitro growth of preimplantation rabbit embryos,
    in: The Mammalian Preimplantation Embryo: Regulation of Growth
    and Differentiation in Vitro (B. D. Bavister, ed.), Plenum
    Press, New York, pp. 193-217.
Kane, M. T. and Buckley, N. J., 1977, The effects of inhibitors of
    energy metabolism on the growth of one-cell rabbit ova to
    blastocysts, J. Reprod. Fertil. 49:261-266.
Kane, M. T. and Foote, R. H., 1970, Culture of two- and four-cell
    rabbit embryos to the expanding blastocyst stage in synthetic
    media, Proc. Soc. Exp. Biol. Med. 133:921-925.
Kane, M. T. and Foote, R. H., 1971, Factors affecting blastocyst
    expansion of rabbit zygotes and young embryos in defined media,
    Biol. Reprod. 4:41-47.
Kane, M. T. and Headon, D. R., 1980, The role of commercial bovine
    serum albumin preparations in the culture of one-cell rabbit
    embryos to blastocysts, J. Reprod. Fertil. 60:469-475.
Kane, M. T., Carney, E. W., and Bavister, B. D., 1986, Vitamins and
    amino acids stimulate hamster blastocysts to hatch in vitro, J.
    Exp. Zool. 239:429-432.
Kaplan, J. G. and Bona, C., 1974, Proteases as mitogens. The effect
    of trypsin and pronase on mouse and human lymphocytes, Exp. Cell
    Res. 88:388-394.
Keil, B., 1979, Proteolysis - one of the major tools of physiological
    regulation, in: Proteases and Hormones (M. K. Agarwal, ed.),
    Elsevier/North Holland, Amsterdam, pp. 1-17.

Kirchner, C., 1969, Untersuchungen an uterusspezifischen Glykoproteinen während der frühen Gravidität des Kaninchens Oryctolagus cuniculus, Wilhelm Roux Archiv. 164:97-133.

Konwinski, M., Solter, D., and Koprowski, H., 1978, Effect of removal of the zona pellucida on subsequent development of mouse blastocysts in vitro, J. Reprod. Fertil. 54:137-143.

Kreitman, O., Lynch, A., Nixon, W. E., and Hodgen, G. D., 1982, Ovum collection, induced luteal dysfunction, in vitro fertilization, embryo development and low tubal ovum transfer in primates, in: In Vitro Fertilization and Embryo Transfer (E. S. E. Hafez and K. Semm, eds.), MTP Press, Lancaster, U.K., pp. 303-324.

Krishnan, R. S. and Daniel, J. C., Jr., 1967, "Blastokinin": Inducer and regulator of blastocyst development in the rabbit uterus, Science 158:490-492.

Kuehl, T. J. and Dukelow, W. R., 1979, Maturation and in vitro fertilization of follicular oocytes of the squirrel monkey (Saimiri sciureus), Biol. Reprod. 21:545-556.

Kuzan, F. B., Pomeroy, K. O., and Seidel, G. E., Jr., 1982, Polyvinyl alcohol as a macromolecular substitute for bovine serum albumin in mouse embryo culture medium, Biol. Reprod. 26(Suppl. 1):65A (abstract).

Lindner, G. M. and Wright, R. W., Jr., 1978, Morphological and quantitative aspects of the development of swine embryos in vitro, J. Anim. Sci. 46:711-718.

Lindner, G. M., Dickey, J. F., and Hill, J. R., Jr., 1983, Effect of bovine serum albumin concentration on the development of ovine embryos in vitro, J. Anim. Sci. 57:466-472.

Lutwak-Mann, C., 1971, The rabbit blastocyst and its environment: Physiological and biochemical aspects, in: The Biology of the Blastocyst (R. J. Blandau, ed.), University Chicago Press, Chicago, pp. 243-260.

Manes, C., 1969, Nucleic acid synthesis in preimplantation rabbit embryos. I. Quantitative aspects, relationship to early morphogenesis and protein synthesis, J. Exp. Zool. 172:303-310.

Mastroianni, L. and Jones, R., 1965, Oxygen tension within the rabbit fallopian tube, J. Reprod. Fertil. 9:99-102.

Mather, J., Kaczarowski, F., Gabler, R., and Wilkins, F., 1986, Effects of water purity and addition of common water contaminants on the growth of cells in serum-free media, Biotechniques 4:56-63.

Maurer, R. R. and Beier, H. M., 1976, Uterine proteins and development in vitro of rabbit preimplantation embryos, J. Reprod. Fertil. 48:33-41.

Maurer, R. R., Whitener, R. H., and Foote, R. H., 1969, Relationship of in vivo gamete aging and exogenous hormones to early embryo development in rabbits, Proc. Soc. Exp. Biol. Med. 131:882-885.

Mayer, J. F., Jr., and Fritz, H. I., 1974, The culture of preimplantation rat embryos and the production of allophenic rats, J. Reprod. Fertil. 39:1-9.

McKeehan, W. L., McKeehan, K. A., and Ham, R. G., 1981, The relationship between defined low-molecular weight substances and undefined serum-derived factors in the multiplication of untransformed fibroblasts, in: The Growth Requirements of Vertebrate Cells In Vitro (C. Waymouth, R. G. Ham, and P. J. Chapple, eds.), Cambridge University Press, Cambridge, pp. 223-243.

McLachlan, J. A., Sieber, S. M., Cowherd, C. M., Straw, J. A., and Fabro, S., 1970, The pH values of the uterine secretions and preimplantation blastocyst of the rabbit, Fertil. Steril. 21:84-87.

McLaren, A., 1969, A note on the mouse zona pellucida, Adv. Reprod. Physiol. 4:207-210.

Meister, A., 1984, Enzymology of glutamine, in: Glutamine Metabolism in Mammalian Tissues (D. Haussinger and H. Sies, eds.), Springer-Verlag, Berlin, pp. 3-15.

Menino, A. R., Jr., and O'Claray, J. L., 1986, Enhancement of hatching and trophoblastic outgrowth by mouse embryos cultured in Whitten's medium containing plasmin and plasminogen, J. Reprod. Fertil. 77:159-167.

Mintz, B., 1962, Experimental study of the developing mammalian egg: removal of the zona pellucida, Science 138:594-595.

Mintz, B., 1965, Synthetic processes and early development in the mammalian egg, J. Exp. Zool. 157:85-100.

Morgan, P. M., Boatman, D. E., Collins, K., and Bavister, B. D., 1984, Complete preimplantation development in culture of in vitro fertilized rhesus monkey oocytes, Biol. Reprod. 30(Suppl. 1):96A (abstract).

Mukherjee, A. B., Laki, K., and Agrawal, A. K., 1980, Possible mechanism of success of an allotransplantation in nature: Mammalian pregnancy, Med. Hypotheses 6:1043-1051.

Mukherjee, A. B., Ulane, R. E., and Agrawal, A. K., 1982, Role of uteroglobin and transglutaminase in masking the antigenicity of implanting rabbit embryos, Am. J. Reprod. Immunol. 2:135-141.

Naglee, D. L., Maurer, R. R., and Foote, R. H., 1969, Effect of osmolarity on in vitro development of rabbit embryos in a chemically defined medium, Exp. Cell Res. 58:331-333.

Nielsen, F. H., 1981, Consideration of trace element requirements for preparation of chemically defined media, in: The Growth Requirements of Vertebrate Cells In Vitro (C. Waymouth, R. G. Ham, and P. J. Chapple, eds.), Cambridge University Press, Cambridge, pp. 68-81.

Niemann, H., Illera, M. J., and Dziuk, P. J., 1983, Developmental capacity, size and number of nuclei in pig embryos cultured in vitro, Anim. Reprod. Sci. 5:311-321.

Onuma, H., Maurer, R. R., and Foote, R. H., 1968, In vitro culture of rabbit ova from early cleavage stages to the blastocyst stage, J. Reprod. Fertil. 16:491-493.

Petry, G., Kühnel, W., and Beier, H.M., 1970, Untersuchungen zur hormonellen Regulation der Praimplantationsphase der Gravidität. I. Histologische, topochemische und biochemische Analysen am normalenKaninchenuterus, Cytobiologie 2:1-32.

Pienkowski, M., Solter, D., and Koprowski, H., 1974, Early mouse embryos: growth and differentiation in vitro, Exp. Cell Res. 85:285-290.

Pope, C. E., Pope, V. Z., and Beck, L. R., 1982, Development of baboon preimplantation embryos to post-implantation stages in vitro, Biol. Reprod. 27:915-923.

Quinn, P. and Wales, R. G., 1973, Growth and metabolism of preimplantation mouse embryos cultured in phosphate-buffered medium, J. Reprod. Fertil. 35:289-300.

Quinn, P. and Whittingham, D. G., 1982, Effect of fatty acids on fertilization and development of mouse embryos in vitro, J. Androl. 3:440-444.

Rizzino, A., 1987, Defining the roles of growth factors during early mammalian development, in: The Mammalian Preimplantation Embryo: Regulation of Growth and Differentiation in Vitro (B. D. Bavister, ed.), Plenum Press, New York, pp. 151-174.

Sellens, M. H. and Sherman, M. I., 1980, Effects of culture conditions on the developmental programme of mouse blastocysts, J. Embryol. Exp. Morphol. 56:1-22.

Sirard, M. A. and Lambert, R. D., 1985, In vitro fertilization of bovine follicular oocytes obtained by laparoscopy, Biol. Reprod. 33:487-494.

Spector, A. A., 1975, Fatty acid binding to plasma albumin, J. Lipid Res. 16:165-179.

Spindle, A., 1980, An improved culture medium for mouse blastocysts, In Vitro 16:669-674.

Spindle, A. I. and Pedersen, R. A., 1973, Hatching, attachment and outgrowth of mouse blastocysts in vitro: Fixed nitrogen requirements, J. Exp. Zool. 186:305-318.

Steptoe, P. C., Edwards, R. G., and Purdy, J. M., 1971, Human blastocysts grown in culture, Nature (London) 229:132-133.

Streffer, C., Van Beuningen, D., Molls, M., Zamboglou, N., and Schultz, S., 1980, Kinetics of cell proliferation in the preimplanted mouse embryo in vivo and in vitro, Cell Tissue Kinet. 13:135-143.

Tenbroeck, J. T., 1968, Effect of nucleosides and nucleoside bases on the development of pre-implantation mouse embryos in vitro, J. Reprod. Fertil. 17:571-573.

Tervit, H. R. and Goold, P. G., 1978, Culture of sheep embryos in either a bicarbonate-buffered medium or a phosphate-buffered medium enriched with serum, Theriogenology 9:251-257.

Tervit, H. R. and Rowson, L. E. A., 1974, Birth of lambs after culture of sheep ova in vitro for up to 6 days, J. Reprod. Fertil. 38:177-179.

Tervit, H. R., Whittingham, D. G., and Rowson, L. E. A., 1972,
    Successful culture in vitro of sheep and cattle ova, J. Reprod.
    Fertil. 30:493–497.
Thomson, J. L., 1967, Effect of inhibitors of carbohydrate metabolism
    on the development of preimplantation mouse embryos, Exp. Cell
    Res. 46:252–262.
Vishwakarma, P., 1962, The pH and bicarbonate-ion content of the
    oviduct and uterine fluids, Fertil. Steril. 13:481–485.
Wales, R. G., 1970, Effects of ions on the development of the
    preimplantation mouse embryo in vitro, Aust. J. Biol. Sci.
    23:421–429.
Wales, R. G. and Whittingham, D. G., 1973, Development of eight-cell
    mouse embryos in substrate-free medium, J. Reprod. Fertil.
    32:316–317.
Warner, C. M., 1977, RNA polymerase activity in preimplantation
    mammalian embryos, in: Development in Mammals, Volume I (M. H.
    Johnson, ed.), North-Holland, Amsterdam, pp. 99–136.
Westphal, U., 1970, Corticosteriod-binding globulin and other steroid
    hormone carriers in the blood stream, J. Reprod. Fertil.
    (Suppl.) 10:15–38.
Whitten, W. K., 1957, Culture of tubal ova, Nature (London) 179:1081–
    1082.
Whitten, W. K., 1971, Nutrient requirements for the culture of
    preimplantation embryos in vitro, in: Advances in the
    Biosciences, Volume 6 (G. Raspe, ed.), Pergamon Press, Oxford,
    pp. 129–141.
Whitten, W. K. and Biggers, J. D., 1968, Complete development in
    vitro of the preimplantation stages of the mouse in a simple
    chemically defined medium, J. Reprod. Fertil. 17:399–401.
Whittingham, D. G., 1971, Culture of mouse ova, J. Reprod. Fertil.
    (Suppl.) 14:7–21.
Whittingham, D. G., 1975, Fertilization, early development and
    storage of mammalian ova, in: The Early Development of Mammals
    (M. Balls and A. E. Wild, eds.), Cambridge University Press,
    Cambridge, pp. 1–24.
Whittingham, D. G. and Bavister, B. D., 1974, Development of hamster
    eggs fertilized in vitro or in vivo, J. Reprod. Fertil. 38:489–
    492.
Wood, M. J. and Whittingham, D. G., 1981, Low temperature storage of
    rat embryos, in: Frozen Storage of Laboratory Animals (G. H.
    Zeilmaker, ed.), Gustav Fischer Verlag, Stuttgart, pp. 119–128.
Wordinger, R. J. and Brinster, R. L., 1976, Influence of reduced
    glucose levels on the in vitro hatching, attachment and
    trophoblast outgrowth of the mouse blastocyst, Dev. Biol.
    53:294–296.
Wright, R. W., Jr., and Bondioli, K. R., 1981, Aspects of in vitro
    fertilization and embryo culture in domestic animals, J. Anim.
    Sci. 53:702–729.

Wright, R. W., Jr., Anderson, G. B., Cupps, P. T., and Drost, M., 1976, Successful culture in vitro of bovine embryos to the blastocyst stage, Biol. Reprod. 14:157-162.

Wu, J. T., 1973, Precocious shedding of the zona pellucida in rats treated with progesterone, J. Reprod. Fertil. 33:331-335.

Yanagimachi, R. and Chang, M. C., 1964, In vitro fertilization of golden hamster ova, J. Exp. Zool. 156:361-376.

19

CLINICAL PARAMETERS INFLUENCING SUCCESS IN IVF

Sander S. Shapiro

1. INTRODUCTION

    Building an IVF program from the ground up is a difficult
undertaking. Equipment must be obtained, set up and tested, protocols developed and personnel trained. In choosing equipment, methods
and protocols, the experience of the pioneer IVF groups can be
helpful. Review of their carefully enunciated experiences will save
the neophyte time, energy and hopefully limit the potential for poor
results. To these ends, this chapter will review the available data

up to early 1986 concerned with several of the more important
clinical variables encountered during in vitro fertilization and
embryo transfer.

IVF-ET in human beings was originally conceived primarily as a
means of bypassing nonfunctional fallopian tubes to achieve
pregnancy. Since immature ova from secondary follicles cannot be
hormonally matured in vitro or fertilized and successfully transferred, present day IVF involves the harvest of mature or near-mature
preovulatory ova. Through the readily combined use of ultrasound,
luteinizing hormone and estrogen monitoring, this can now be
accomplished with considerable success in natural cycles.

2.   SUPERSTIMULATION

Despite the fact that the first successful IVF-ET, resulting in
a live birth, occurred after spontaneous follicular maturation,
active clinical treatment programs now universally resort to pharma-
cologic stimulation (super-stimulation, hyperstimulation, supraovula-
tion) for follicular development. In part, this is because of the
difficulty encountered when monitoring natural cycles and the
inconvenience that accrues to the hospital and personnel involved in
such efforts. However, it is to a greater extent because of a
demonstrated increase in the rate of pregnancy when multiple embryos
are transferred that superstimulation is so widely used (Edwards and
Steptoe, 1983; Leeton et al., 1983; Jones et al., 1984).

Much of our present knowledge concerning superstimulation stems
from experience with the drugs used to treat anovulation. There the
goal is to induce maturation of a single follicle, the antithesis of
current stimulation protocols. The occurrence of multiple ovulations
during treatment of anovulation is a complication to be avoided. In
addition to the difference in goals between ovulation and superstimulation protocols, there is the fact that, in the one case, ovulation
is coaxed from a semiquescent ovary while, in the other, stimulation
is superimposed upon an active cycling system. It is not surprising,
then, that considerable evolution in the use of ovulatory drugs has
occurred during the development of stimulation protocols for IVF.

There are a plethora of stimulation protocols described in the
published literature. From the number currently in use, it should be
evident that an ideal regimen has not been found. Among the various
regimens it is difficult to assess which, if any, demonstrates
significant superiority. Part of the problem in making a comparative
judgement is the complexity of the IVF process. The widely different
success rates (defined as pregnancies/laparoscopy, pregnancies/stimu-
lation cycle, pregnancies/transfer) quoted by various institutions
may be the result of different stimulatory protocols but could also
be due to any number of other variables inherent in the IVF process.

In an effort to isolate stimulation methods from the overall IVF
process, data on intermediate functions has been collected to assess
and compare protocols. Estradiol levels, number of mature follicles,
fertilizability of ova and frequency of spontaneous LH surges have
been evaluated. A summary of the reported findings will be given
here (Table I). The reader should be cautious when interpreting
these data as none of these intermediate parameters relates strongly
to the frequency of clinical pregnancy.

The human female is usually monotocous. However, from the many
follicles that begin the maturational process, there appear to be a
number that remain candidates for complete maturation at the
beginning of an ovarian cycle. The exact number and its variability
from cycle to cycle are not known but may be roughly estimated as
from four to twelve. The optimum time to stimulate these follicles
so that they may mature along with the dominant one is not known.
Experience has shown that stimulation beginning on day three of a
cycle will provide for the development of more follicles than similar
efforts begun on day five. When natural cycles are short (< 27
days), it has been the custom of most therapists to begin stimulation
on day two. Controlled studies to show that this is effective have
not been published. It may be presumed that initiation of hyperstimulation at even earlier times might result in the development of
still larger numbers of follicles. The harvesting of large numbers
of eggs may be counter productive. Lopata has stated that when his
group could recover more than eight eggs, they found a significantly
lower fertilization and cleavage rate (Lopata, 1983).

Ultimately, the success of every regimen must be judged by the
rate that it will produce live-born, healthy infants. However, a
multiplicity of variables (both recognized and unrecognized, control-
lable and uncontrollable) in other parts of the IVF procedure make it
difficult to isolate and evaluate variations in follicular stimulation protocols using pregnancy as the sole criteria. Thus, we must
often resort to data on intermediate factors as mentioned above. The
number of mature follicles and the frequency of ovum fertilization
and/or cleavage are such criteria. Follicular fluid contents, serum
estrogen concentrations and frequency of spontaneous LH elevations
have also been evaluated in this regard.

2.1. Clomiphene Citrate

Clomiphene citrate, begun on days three to seven at doses
of 50 to 150 mg per day, has been used extensively for IVF (Lopata,
1983; Quigley et al., 1983). Lopata (1983) reported the number of
recovered oocytes to be considerably greater (1.7 vs. 2.6) when the
earlier cycle day was used to initiate therapy. Marrs et al. (1984),
on the other hand, found that starting clomiphene citrate on day five
provided for maximal oocyte recovery. Little or no difference in the
number of eggs recovered was found when 50-mg- and 150-mg-per-day

Table I. Stimulation protocols employed in human IVF-ET[a]

Regimen	Reference	# Large follicles	Mature eggs	Embryos/transfer	Pregnancy/laparoscopy (%)	Pregnancy/transfer (%)
CC 100/d 3-9 hMG: 300 IU/d 7 hCG: foll > 18 mm	Lopata, 1983	6.4	2.9	2.2	14.9	18.5
CC 100/d 3-7, 9 or d 5-9 hMG: 150 or 225 IU hCG: foll > 18 mm	Lopata, 1983	5.6	2.1	2.4	23.2	38.1
CC 100/d 4-8 hMG: 75 IU d 4-8 150-225 IU d 9 hCG: after 6 d E2	Quigley et al., 1985	6.0	4.9	3.2	13	18
CC 100 mg/d 5-9 hMG: 150-225 IU d 6, 8, 10 hCG: E2 > 300 pg/foll/foll > 18 mm	Plachot et al., 1985	3.4	2.8	1.5	5	
CC 150 mg/d 3-7 hMG: 150 IU d 8 hCG: 2 foll > 18 mm	Vargyas et al., 1984	4.4	3.0	2.2	16	16
CC 100-150 mg/d 5-9 hCG: when E2 > 500 pg/ml or foll > 20 mm	Lopata, 1983	3	1.2	1.5	9.3	14.8
CC 100 or 150 mg/d d 3 until foll > 14 mm hCG: E2 > 500 pg/ml or foll > 20 mm	Lopata, 1983	4.1	1.7	2.1	11.6	15.5

Protocol	Reference					
CC 50 to 200 mg/d 2-6 hCG: Urine E2 > 150 ug/d or spont. LH	Fishel et al., 1984		0.8	1.5	37	47
CC 50 mg/d 5-9 hCG: > 20 mm	Quigley et al., 1983	2.3	1.6	1.7	12	19
hMG: 150 IU/d 3 hCG: E2 > 600 or E2 > 300 + biologic shaft	Garcia et al., 1981	4.0	1.7	2.0	20	25
hMG: 75 IU/d 3, 5 hCG: > 15 mm	Yuen et al., 1985		0.7	1.4	0	0
hMG: 225 IU/d 4 hCG: > 15 mm + E2 > 1000 pg/ml	Yuen et al., 1985		2.5	1.4	11	11
hMG: 225 IU/d 3-7 then by 1 amp hCG: 2 foll > 1.6 mm	Laufer et al., 1983	4.3	3.1	2.3	16	17
hMG: 150 IU/d hCG: > 17 mm	Vargyas et al., 1984	4.2	2.5	1.4	30	37

[a]This table contains representative published stimulation protocols. It is not meant as a complete list of published accounts and should not be used to strenuously compare various protocols.

protocols were compared although higher serum estradiol levels developed in the women receiving 150 mg/day (Quigley et al., 1984a). The length of the cycle and the ultrasound determined size of the follicle at ovulation were not found to be different when stimulated cycles were compared to natural cycles (Kerin et al., 1983). It, therefore, seems to make little difference which dose of drug is used or which day (3 or 5) drug administration is initiated. The relatively low numbers of follicles obtained by the clomiphene-only regimens has caused most IVF groups to abandon them in favor of regimens that include menotropins (hMG).

2.2. Pergonal

Follicle-stimulating hormone (FSH) and luteinizing hormone (LH) have been used in a commercial combination (Pergonal) for super-stimulation (Jones, 1984). Two ampules (150 IU) per day beginning on day three has been the most extensively explored regimen. It has produced about 4.4 mature follicles per cycle without spontaneous LH surges in the experience of at least one group (Mantzavinos et al., 1983). When three ampules were used each day in another IVF program, approximately the same number of follicles were aspirated (Laufer et al., 1983). The number of oocytes recovered was higher in the three-ampule protocol, but the pregnancy rate was higher in the two-ampule program. Significant differences in reporting and in program methods make it hazardous to derive conclusions from these comparisons. In a small study from a single institution comparing a one- to two-ampule regimen with a three-ampule regimen, considerable difference was found in the number of oocytes recovered (1.8 vs. 2.9) but fertilization rates were similar (Yuen et al., 1985). In the only study thus far published in which the number of ampules was reduced from three to two after two days, there was an improvement in the frequency with which mature eggs were obtained. This did not result, however, in an improved pregnancy rate (Jones, 1984). There may be a limit to the degree of stimulation that can be attained by increasing amounts of Pergonal administration. High levels of estradiol developed in a series where four ampules a day was given. This resulted in a shortening of the luteal phase and a lowered frequency of pregnancy (Quigley, 1984).

To date, there has not been a comparative study contrasting a menotropin regimen that starts high and is lowered as follicular development dictates with one in which low doses at the beginning of a cycle are increased to gain multiple follicular development.

2.3. Combinations

The use of clomiphene citrate and Pergonal sequentially or in combination has been studied by several groups (Leeton et al., 1983; Lopata, 1983; Quigley et al., 1984b). Most frequently, 100 mg of clomiphene is begun on day three or five and is followed by daily doses of menotropins (Lopata, 1983). The day of clomiphene initiation in this type of regimen is usually varied with women who have short natural cycles beginning earlier than those with long natural cycles. One variant of this regimen alters the daily dose of Pergonal according to the size of the most advanced follicle on the last day of clomiphene. Results from this kind of protocol include an average of 5.5 follicles aspirated and four eggs recovered per laparoscopy. The frequency of fertilization can be lower with these combinations than with clomiphene citrate alone (58% vs. 75%), but the pregnancy rate has been higher (23% vs. 9%). Combination protocols in which clomiphene citrate and Pergonal are given simultaneously (clomiphene citrate day three to day nine, Pergonal day seven onward or clomiphene citrate day three-nine, + hMG days six, eight, and ten) followed by Pergonal daily or every other day have given similar results in terms of follicle development and ovum recovery by aspiration (Lopata, 1983; Plachot et al., 1985). As might be expected by extrapolating from the data of Quigley (Quigley et al., 1984a), increasing the clomiphene citrate dosage from 100 mg to 150 mg/day did not significantly raise the pool of follicles that were recruited (Plachot et al., 1985). Doubling the dosage of Pergonal, on the other hand, allowed for the recruitment of about one additional follicle. When a straight Pergonal protocol was compared with a clomiphene-Pergonal regimen in a single study, the combination proved superior by several criteria: more large follicles, fertilized ova, and embryos transferred. Yet, this same study reported a higher but statistically nonsignificant pregnancy rate in the menotropin-only group (Vargyas et al., 1984).

2.4. Follicle-Stimulating Hormone

It has long been recognized that one may vary the success of ovulation induction cycles by altering the FSH to LH ratio of the menotropin preparation (Jacobson and Marshall, 1969). This has prompted the use of Pergonal plus FSH for IVF cycles in an effort to improve upon the pregnancy rate established using Pergonal only. When 150 IU of FSH was given on days three and four of a standard two amp per day Pergonal regimen, there was a dramatic increase in the number of large follicles that developed and the numbers of ova retrieved (3 vs. 1.5) (Muasher et al., 1985). Unfortunately, the increase in recovered preovulatory ova from these cycles was not accompanied by a significant increase in pregnancy rate. The FSH-Pergonal combination protocol also produced more large follicles than a regimen that used three ampules of menotropins each day (Laufer et al., 1983) and one in which three ampules were given on days three and four followed by two ampules on each of the remaining stimulation days (Jones et al., 1984). Here again, however, the pregnancy rate was not significantly different. An attempt to use FSH alone over the whole stimulation period has produced encouraging results in a small cohort (Jones et al., 1985). In twelve cycles, a mean of 2.7 preovulatory and 4.0 immature ova were recovered. From these ova,

there was a transfer rate of three embryos per attempt and five women
became pregnant.

The relatively high rates of follicular maturation and ovum
transfer in FSH-Pergonal and pure FSH cycles rival those attained
using clomiphene citrate and Pergonal. This may be the result of the
high serum FSH levels developed with both types of regimen at an
early and possible critical time in the cycle.

After surveying the available published data, it would be
difficult for someone who was constructing an IVF program to
enthusiastically choose one particular ovulation regimen. There is
simply no single regimen that is significantly superior to all other
currently available regimens. Available data do, however, suggest
that protocols using clomiphene citrate alone are at considerable
disadvantage to those that embody menotropins or clomiphene plus
menotropins. These give greater numbers of large follicles, more
preovulatory ova and higher pregnancy rates. None of them seems,
however, to be universally optimal.

It has been the experience of many IVF therapists that once a
patient responds favorably to a particular stimulation protocol, she
tends to do so in repeated cycles. It also seems that when a woman
responds poorly to a regimen, she is likely to do so repeatedly.
Thus, it is advantageous to be familiar with and have available
several regimens, using one for all first attempts and reserving
another for those women who fail to respond favorably to the
"standard" stimulation regimen.

In the near future, there will inevitably be new and improved
stimulation protocols developed. Experience may also provide the
means by which we can identify the regimen most likely to produce
optimal stimulation in a given individual. It may then become even
more requisite that each practitioner be facile with several
stimulation protocols.

3.   CRITERION FOR hCG

Equally important to success of a stimulation protocol is the
manner in which patients are monitored and how the appropriate time
for final maturation with human chorionic gonadotropin (hCG) is
identified. The practice of stimulating follicular maturation but
then waiting for a spontaneous LH surge has been almost totally
abandoned (Trounson and Calabrese, 1984). Three parameters have been
studied in this regard: follicular size, serum estradiol concentrations and end-organ response to estrogen. Various IVF centers have
chosen to emphasize, to different degrees, the data they obtained
about each of these parameters.

3.1. Ultrasound

As the definition of ultrasound scanning has improved, it has become possible to measure the size of individual ovarian follicles with considerable accuracy (Jones et al., 1985). In natural cycles, the diameters of tertiary follicles increase in a linear fashion during the week prior to ovulation--the rate being about 2 mm per day (Bryce et al., 1982). Ovulation takes place when the dominant follicle reaches a diameter of 20 to 24 mm (Buttery et al., 1983). A slightly faster growth rate with a similar size at ovulation has been recorded for clomiphene-stimulated cycles (O'Herlihy et al., 1982; Leerentveld et al., 1985). Menotropin-induced folliculogenesis seems to progress at about the same rate and ovulation most often results in pregnancy if hCG is administered when the mean follicular diameter is greater than 15 mm (Seibel et al., 1981).

There is a general belief that menotropin-induced follicles are smaller than clomiphene-menotropin follicles at the time they reach maturity. This is not yet firmly substantiated by data from published studies. And, in fact, it may be a self-fulfilling injunction, in that most groups currently using ultrasound for timing determinations pick a smaller diameter as a target when using menotropins alone as compared to when clomiphene is used. In the one instance in which ultrasound was reported but not used to determine the time of hCG administration after Pergonal, follicle size was actually smaller than that reported by others for clomiphene cycles (Mantzavinos et al., 1983). Most centers with published descriptions of their protocols use a target of greater than 15 mm for the lead follicle in menotropin-stimulated cycles and 18 mm for clomiphene-menotropin-stimulated cycles.

3.2. Estradiol

Serum estradiol levels have been shown to rise in exponen-tial fashion during the late follicular phase (Wilson et al., 1982). Almost all of the circulating steroid is produced by the lone preovulatory follicle in natural cycles. This rise in estrogen levels correlates well with both the size and degree of follicular maturation. At the time of the LH surge, serum estradiol levels are in the 250 to 400 pg/ml range. When multiple follicles develop, the correlation between size of the most advanced follicle and serum estradiol concentrations is poor. This negates the use of estradiol levels as sole predictors of follicular maturation. Multiple follicle maturation beyond 10 mm shows a correlation of estradiol levels with total follicular volume (Haning et al., 1982). Thus, in stimulated cycles, the maturity of the advanced cohort may be judged best by the estradiol level only if one also knows the total number and volume of those follicles. Using the figures for estradiol concentrations in natural cycles, one might estimate that maturity is

reached when each large follicle is producing enough steroid to
provide 250 to 400 pg/ml in peripheral blood. There is, however,
evidence that pharmacologically-stimulated follicles may produce more
estrogen than those of natural cycles (Smith et al., 1980). It has,
therefore, become common to expect a contribution of 300 to 500 pg
from each follicle over 14 mm in diameter. Because multiple small
follicles can produce as much steroid as one or two large, more
mature follicles, it is advantageous to use ultrasound when evaluating the significance of serum estrogen levels and for determining
the time of hCG administration.

There are several estrogen-dependent changes in the genital
tract that can be monitored during follicular maturation:
karyopicnotic indices, cervical dilatation and the character of the
cervical mucus. These changes relate to the rising levels of
estrogen but it is unclear just how they may relate to the maturity
of a group of synchronously ripening follicles. Biological changes
that occur when serum estrogen levels reach 200 to 300 pg/ml will not
distinguish between one or two mature follicles and five smaller,
less mature ones.

While most protocols for determining the timing of hCG administration depend on serum levels of estradiol, there is one that uses
the length of time that estradiol levels have been elevated over
baseline values (Quigley et al., 1985). In that protocol, hCG is
given after serum estradiol levels have shown a sustained rise for
six days, provided there are at least two large follicles present.
The rate of ovum fertilization and embryo cleavage following this
protocol is similar to that of the more frequently used methods. It
has, however, been used to show that waiting for a seventh day
results in a significant deterioration in ovum quality. Recognition
of this phenomenon in isolation seems sufficient to justify the
effort of daily monitoring from an early cycle day and the aborting
of all cycles requiring stimulation for more than six days beyond the
initial estradiol rise.

3.3. Biological Changes

Among the many published descriptions of IVF programs,
there are only a small number of protocols for determining when to
administer hCG. A few groups rely entirely on the follicular size of
the lead follicle (> 18-20 mm for clomiphene-menotropin cycles and
> 15 mm for menotropin only cycles) or the two most advanced
follicles (Laufer et al., 1983; Lopata, 1983; Quigley et al., 1984a).
Others require that in addition to size, each follicle produce more
than 300 pg/ml of estradiol (Vargyas et al., 1984; Dirnfeld et al.,
1985; Plachot et al., 1985; Yuen et al., 1985). One program has
relied heavily on changes in biological parameters as well as
estrogen levels to determine the time of induced ovulation (Jones et
al., 1982). In that group, menotropins are given until clinical

parameters have shown an estrogen-induced change and estradiol levels are over 300 pg/ml or estradiol has reached 600 pg/ml without a biological change. Attention to the, so called, biological shift has proven helpful with a Pergonal-only stimulation protocol that seldom induces more than three large follicles. It remains to be seen whether this parameter can be of help in stimulation programs that routinely involve the maturation of four to six follicles. The reader should remember that each of the indicators used to determine the time for hCG administration is related to a particular type of follicular stimulation. When a program alters its stimulation protocol it will, in all probability, find it necessary to make some changes in the criteria used for determining the time of hCG administration. As it has been shown that administering hCG, when follicles have not yet reached an advanced stage of maturity, can impede ovulation, more attention should be paid to this aspect of the IVF process (Tamada and Matsumoto, 1969; Williams and Hodgen, 1980).

4.   hCG DOSE

The completion of follicular maturation and ovulation is induced by the LH surge during natural cycles. Continued LH secretion is then necessary to sustain luteal activity. Human chorionic gonadotropin has long been used as a substitute for LH during induced ovulation in anovulatory women. The choice of a surrogate was originally affected because of the nonavailability of pure LH preparations. It turned out to be fortuitous as the short half life of LH would have required repeated injections to maintain the corpus luteum for its full 14-day life span. A 10,000 IU injection of hCG is retained in a bioactive state for sufficient time to stimulate the corpus luteum for the whole of its natural life span. In IVF cycles, with the patient's capacity to secrete LH intact, only the follicle maturing component of hCG appears to be needed. Therefore, a dose of 10,000 IU is not required and may not even be optimal. The amount of hCG given to IVF patients varies from group to group. Four, five, six and ten thousand international units given in a single intermuscular injection are the most common dosages used. One group has reported using 3000 IU twice in a twelve-hour interval, but without a rational explanation for the schema (Kerin et al., 1983). There has been, essentially, no work done to determine the merits of the different hCG doses used. To date, concerns that excess hCG might evoke a down regulation of hCG granulosa receptors has not been experienced in clinical practice.

5.   hCG TIMING

The interval between last menotropin injection and hCG administration has not been evaluated exhaustively. Most IVF groups monitor daily serum estrogens in the morning and administer menotropins in

the evening. They have the results of their assays by early afternoon and make decisions as to continuing menotropins or giving hCG at
that time. hCG is then administered so as to have 35 hr elapse
between injection and the planned time of oocyte recovery. The most
common procedure is to give hCG approximately 24 hr after the last
menotropin injection. One group has been notable for consistently
delaying (coasting) this injection for 48 hr (Jones, 1984). They
have not stated a concise rationale for choosing the delay, but have
noted in a retrospective evaluation that there is a higher pregnancy
rate when hCG is given at 50 hr than at less than 30 hr. Their data
conflicts with the one small prospective, randomized study designed
to evaluate the value of delaying hCG (Laufer et al., 1984). In that
study, a significantly lower frequency of fertilization was found
when hCG was administered more than 47 hr after the last menotropin
injection. To date, an evaluation of data involving the simultaneous
injection of the last menotropin dose and hCG has not been published.

6.   LUTEAL PHASE SUPPORT

        The majority of IVF cycles fail after transfer of one or more
multicelled embryos into the uterus. This may be due to poor embryo
quality, problems generated by the transfer technique or by
inadequacy of the luteal phase. Since oocyte recovery involves the
aspiration of large numbers of granulosa cells (Feichtinger et al.,
1982) and because a classic luteal phase defect has been found in
some ovulation induced cycles (Frydman et al., 1982), the possibility
of a luteal progesterone deficiency has been evaluated (Cohen et al.,
1984); either a defect does not occur or it seldom occurs after
follicular aspiration for IVF. Nevertheless, many programs support
the luteal phase with progesterone. Suppositories of 25 mg twice a
day or single injections of 12.5 and 25 mg are used most frequently
(Smith et al., 1980; Jones, 1984; Laufer et al., 1984). Some
successful programs do not use progesterone (Smith et al., 1980;
Kerin et al., 1983; Laufer et al., 1984). At present, there are no
published studies comparing outcome with or without progesterone
support. The amounts used seldom delay the onset of menses, and
probably cause no harm. One theoretic objection to this therapy is
the possibility that excess progesterone might depress intracellular
progesterone receptor concentrations and thereby inhibit progesterone
induced endometrial maturation. Without extensive understanding of
human endometrial receptor turnover and the temporal sequence for
each alteration in the progesterone domain, the likelihood of a
harmful effect cannot be predicted.

7.   ABORTING TREATMENT CYCLES

        One facet of IVF programs that varies widely is the frequency
with which cycles are aborted before ovum retrieval. Figures in the

literature vary between 7 and 20 percent (Johnston et al., 1981; Leerentveld et al., 1985) and seem to be getting higher as therapists gain greater insight into what characterizes a good stimulation cycle and what does not. The reasons for abandoning a cycle include: failure to develop more than one accessible follicle; low levels or aberrant patterns of serum estrogen; and the occurrence of a spontaneous LH surge. When one ovary is obscured by adhesions, it may be necessary to initiate several cycles before obtaining a consentient stimulation and proceeding to laparoscopy. Ultrasound directed aspiration can obviate this constraint, but when neither the facilities nor skills are available, waiting for the right situation may be the only workable alternative. To increase the likelihood that several good follicles will develop on the accessible side, a more vigorous stimulation protocol than is usually used may be undertaken.

Many groups have advanced theories on just how high estradiol concentrations must be and what the shape of the preovulatory estradiol concentration curve ought to take for optimal egg retrieval (Johnston et al., 1981; Jones et al., 1983). Most investigators think that estradiol levels of at least 300 pg/ml for each large follicle are desirable. At Norfolk, using a fairly fixed menotropin schedule, some patients experienced a slow rise or premature fall in serum estradiol concentrations. These women were found to have a significantly lower frequency of pregnancy (Jones et al., 1983). Another unfavorable sign reported from Norfolk was a down turn in estradiol during menotropin stimulation. Early in the general experience with IVF, some authors suggested that the best time to give hCG was when the estrogen level was beginning to plateau. More recently, in agreement with the Norfolk group, it has become dogma to assume that a high pregnancy rate derives from cycles in which hCG is administered while estradiol levels are logarithmically ascending. Therefore, it seems appropriate to consider canceling cycles in which either the estrogen level remains low or plateaus before hCG can be given (Figure 1).

8.   THE SPONTANEOUS LH SURGE

The occurrence of a spontaneous LH surge can be used to target the time of ovum retrieval. It can also be a disconcerting event when it develops prior to full follicular maturation. In natural cycles, a single mature follicle signals its state of readiness by the amount of estrogen that it produces. However, when several follicles mature, their cumulative estrogen production may trigger a surge before the proper degree of maturity is attained. This is said to be a relatively rare event where monotropins are used, but is seen in almost half of the cycles in which clomiphene is the sole agent of induction (Fishel et al., 1984). However, in at least one well-

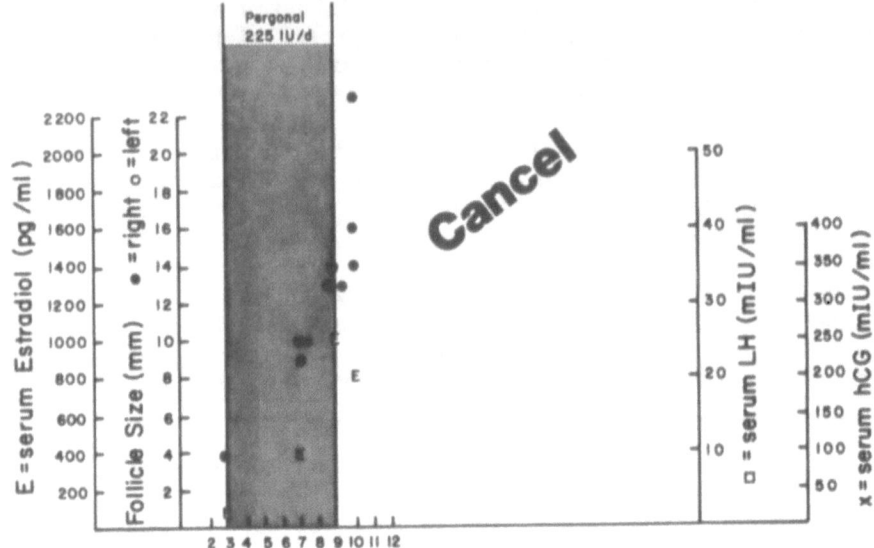

Fig. 1.   This 29-year-old woman with a history of two ectopic
           pregnancies and one ovary underwent her first IVF cycle
           using hMG. With the lead follicles at 13 mm diameters on
           day 9, the patient received her P.M. injection of hMG. On
           day 10, one follicle became cystic with a fall in serum
           estradiol. The cycle was then canceled.

monitored hMG series, the frequency of an LH surge was 33% (Vargyas
et al., 1984).

Should an LH surge occur when both estrogen and follicular size
suggest maturity but before hCG is given, the time set for aspiration
must be rescheduled. This can only be done when LH monitoring has
been performed frequently enough to pinpoint the beginnings of the
surge. Then, a 24-hr interval to retrieval will produce reasonably
healthy ova (Fishel et al., 1984). Should early recovery result in
the harvest of relatively immature ova, a delay in the planned time
of insemination can be instituted. In those IVF programs where
frequent urine or serum monitoring is not carried out and the surge
onset cannot be identified, abandonment of the cycle becomes a
reasonable alternative (Figure 2).

9.   SUMMARY

In every IVF program, but especially in a new one, there is a
desire to attain pregnancies with respectable frequency--to have a
good batting average (Figure 3). Success will, of course, depend to
a large extent on a well maintained laboratory. Judicious selection

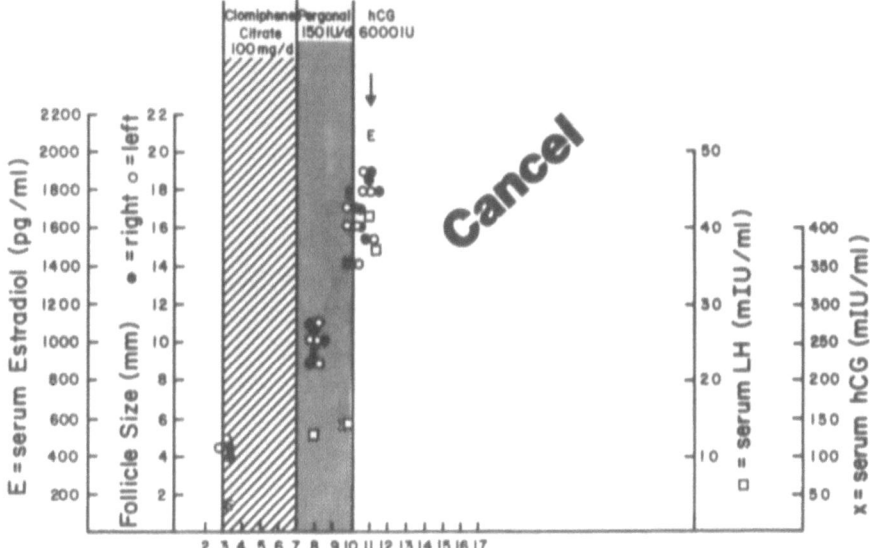

Fig. 2.   This 34-year-old woman with a past history of salpingitis
           isthmica nodosa underwent IVF stimulation for the first
           time with a combination regimen. hCG was given on the
           evening of the eleventh cycle day. On the twelfth day, the
           results of the LH assays became available and the cycle was
           canceled. The decision to cancel was made in concert with
           the patient based on the inability to closely identify the
           time at which LH began to surge.

of candidates and a willingness to abandon less than ideal cycles
will also help one to attain this goal. No new program can be
identical to one that has a long history of success. Circumstances
of time and place will make for differences in protocols and patient
characteristics. A certain amount of flexibility in planning and
execution will be required. This will, however, make it difficult to
compare and contrast locally derived data to that of published
series.

        The use of two or more stimulation protocols can provide alter-
natives when a patient fails to respond in an initial attempt at
stimulation. In the future, pure FSH, LRF and LRF agonists coupled
with menotropins may be used to advantage. At present, the use of
clomiphene-Pergonal and straight Pergonal protocols within the same
IVF program (an approach used at the University of Wisconsin) will
provide a tailored stimulation regimen to each patient.

        Judicious selection of candidates will also enhance the success
rate of any program. Tubal obstruction and endometriosis patients
seem to respond favorably to IVF while severe oligospermics have a

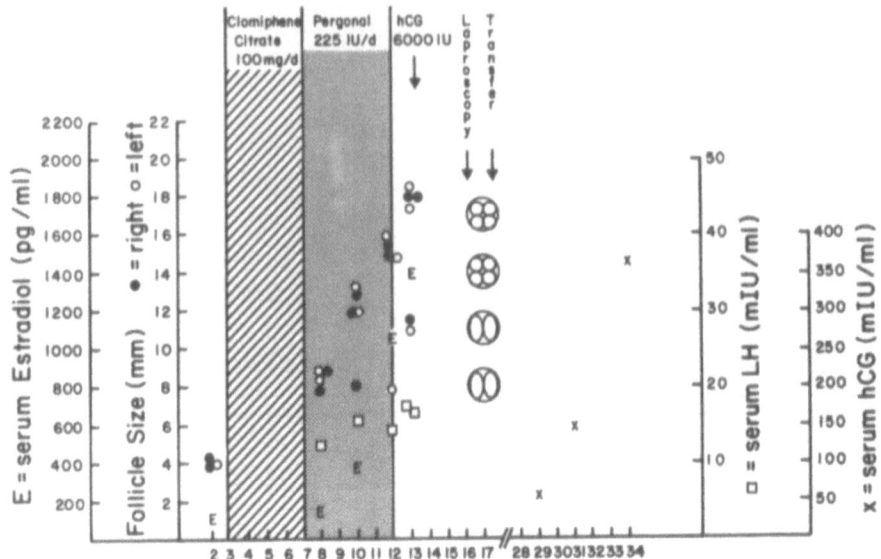

Fig. 3.   This 32-year-old woman with a past history of pelvic
           adhesions and hydrosalpinges failed to become pregnant
           after microsurgery. On her first attempt at IVF, she
           became pregnant with twins that delivered uneventfully at
           37 weeks.

much lower chance of successful outcome by this methodology.
Encouraging results have been obtained with a small cohort of DES
women, making it likely that this problem will be amenable to IVF
therapy (Muasher et al., 1984). Little experience has been gained
with polycystic women with tubal disease, but the problems involved
in inducing maturation of multiple follicles in such people should
give the neophyte pause. On the other hand, successes have been
noted with mild hyperprolactinemia by the addition of bromocriptine
to a standard stimulation regimen. Unless a program is large enough
to systematically evaluate specific patient or stimulation variables,
experimental and unproven protocols should be avoided.

       The importance of patient selection to the success of IVF
recommends that a thorough evaluation of each couple be undertaken
before inclusion into the program. Table II lists those studies that
are worth considering as part of a routine screening.

       Once an IVF program has been planned, assembled, and initiated,
careful attention must be paid on a frequent and repeated basis to
those parameters which can be evaluated. Good record keeping and the
ability to retrieve data for critical appraisal will help to identify

Table II. Pre-IVF laboratory tests

---

1. Semen analysis
2. Serum prolactin
3. Serum T_3 and T_4
4. Timed endometrial biopsy
5. Cervical ureaplasma culture
6. Pelvic ultrasound (absence of ovarian cysts)

---

problem areas early and point the way to workable solutions.
Continuous monitoring should help insure early and repeated success
in this very demanding area of infertility therapy.

10. REFERENCES

Bryce, R. L., Shuter, B., Sinosich, M. J., Stiel, J. N., Picker,
    R. H., and Saunders, D. M., 1982, The value of ultrasound,
    gonadotropin, and estradiol measurements for precise ovulation
    prediction, Fertil. Steril. 37:42.
Buttery, B., Trounson, A., McMaster, R., and Wood, C., 1983, Evalu-
    ation of diagnostic ultrasound as a parameter of follicular
    development in an in vitro fertilization program, Fertil.
    Steril. 39:458.
Cohen, J. J., Debache, C., Pigeau, F., Mandelbaum, J., Plachot, M.,
    and de Brux, J., 1984, Sequential use of clomiphene citrate,
    human menopausal gonadotropin and human chorionic gonadotropin
    in human in vitro fertilization. II. Study of luteal phase
    adequacy following aspiration of the preovulatory follicles,
    Fertil. Steril. 42:360.
Dirnfeld, M., Lejeune, B., Camus, M., Vekemans, M., and Leroy, F.,
    1985, Growth rate of follicular estrogen secretion in relation
    to the outcome of in vitro fertilization and embryo replacement,
    Fertil. Steril. 43:379.
Edwards, R. G. and Steptoe, P. C., 1983, Current status of in-vitro
    fertilisation and implantation of human embryos, Lancet ii:1265.
Feichtinger, W., Kemeter, P., Szalay, S., Beck, A., and Janisch, H.,
    1982, Could aspiration of the Graafian follicle cause luteal
    phase deficiency?, Fertil. Steril. 37:205.
Fishel, S. B., Edwards, R. G., and Purdy, J. M., 1984, Analysis of 25
    infertile patients treated consecutively by in vitro fertiliza-
    tion at Bourn Hall, Fertil. Steril. 42:191.
Frydman, R., Testart, J., Giacomini, P., Imbert, M. C., Martin, E.,
    and Nahoul, K., 1982, Hormonal and histological study of the
    luteal phase in women following aspiration of the preovulatory
    follicle, Fertil. Steril. 38:312.

Garcia, J., Jones, G. S., Acosta, A. A., and Wright, G. L., Jr., 1981, Corpus luteum function after follicle aspiration for oocyte retrieval, Fertil. Steril. 36:565.

Haning, R. V., Jr., Austin, C. W., Kuzma, D. L., Shapiro, S. S., and Zweibel, W. J., 1982, Ultrasound evaluation of estrogen monitoring for induction of ovulation with menotropins, Fertil. Steril. 37:627.

Jacobson, A. and Marshall, J. R., 1969, Ovulatory response rate with human menopausal gonadotropins of varying FSH/LH ratios, Fertil. Steril. 20:171.

Johnston, I., Lopata, A., Speirs, A., Hoult, I., Kellow, G., and du Plessis, Y., 1981, In vitro fertilization: the challange of the eighties, Fertil. Steril. 36:699.

Jones, G. S., 1984, Update on in vitro fertilization, Endocr. Rev. 5:62.

Jones, G. S., Acosta, A. A., Garcia, J. E., Bernardus, R. E., and Rosenwaks, Z., 1985, The effect of follicle-stimulating hormone without additional luteinizing hormone on follicular stimulation and oocyte development in normal ovulatory women, Fertil. Steril. 43:696.

Jones, H. W., Jr., Jones, G. S., Andrews, M. C., Acosta, A., Bundren, C., Garcia, J., Sandow, B., Veeck, L., Wilkes, C., Witmyer, J., Wortham, J. E., and Wright, G., 1982, The program for in vitro fertilization at Norfolk, Fertil. Steril. 38:14.

Jones, H. W., Jr., Acosta, A., Andrews, M. C., Garcia, J. E., Jones, G. S., Mantzavinos, T., McDowell, J., Sandow, B., Veeck, L., Whibley, T., Wilkes, C., and Wright, G., 1983, The importance of the follicular phase to success and failure in in vitro fertilization, Fertil. Steril. 40:317.

Jones, H. W., Jr., Acosta, A. A., Andrews, M. C., Garcia, J. E., Jones, G. S., Mayer, J., McDowell, J. S., Rosenwaks, Z., Sandow, B. A., Veeck, L. L., and Wilkes, C. A., 1984, Three years of in vitro fertilization at Norfolk, Fertil. Steril. 42:826.

Kerin, J. F., Warnes, G. M., Quinn, P., Jeffrey, R., Godfrey, B., Broom, T. J., McEvoy, M., Kirby, C., Johnson, M., and Cox, L. W., 1983, The effect of clomid induced superovulation on human follicular and luteal function for extracorporeal fertilization and embryo transfer, Clin. Reprod. Fertil. 2:129.

Laufer, N., DeCherney, A. H., Haseltine, F. P., Polan, M. L., Mezer, H. C., Dlugi, A. M., Sweeney, D., Nero, F., and Naftolin, F., 1983, The use of high-dose human menopausal gonadotropin in an in vitro fertilization program, Fertil. Steril. 40:734.

Laufer, N., DeCherney, A. H., Tarlatzis, B. C., Zuckerman, A. L., Polan, M. L., Dlugi, A. M., Graebe, R., Barnea, E. R., and Naftolin, F., 1984, Delaying human chorionic gonadotropin administration in human menopausal gonadotropin-induced cycles decreases successful in vitro fertilization of human oocytes, Fertil. Steril. 42:198.

Leerentveld, R. A., van Gent, I., van der Stoep, M., and Wladimiroff, J. W., 1985, Ultrasonographic assessment of Graafian follicle growth under monofollicular and multifollicular conditions in clomiphene citrate-stimulated cycles, Fertil. Steril. 43:565.

Leeton, J., Trounson, A., Wood, C., and Gianaroli, L., 1983, In vitro fertilization and embryo-transfer: The Monash group experiences 1981-83, Acta Eur. Fertil. 14:95.

Lopata, A., 1983, Concepts in human in vitro fertilization and embryo transfer, Fertil. Steril. 40:289.

Mantzavinos, T., Garcia, J. E., and Jones, H. W., Jr., 1983, Ultrasound measurement of ovarian follicles stimulated by human gonadotropins for oocyte recovery and in vitro fertilization, Fertil. Steril. 40:461.

Marrs, R. P., Vargyas, J. M., Shangold, G. M., and Yee, B., 1984, The effect of time of initiation of clomiphene citrate on multiple follicle development for human in vitro fertilization and embryo replacement procedures, Fertil. Steril. 41:682.

Muasher, S. J., Garcia, J. E., and Jones, H. W., Jr., 1984, Experience with diethylstilbestrol-exposed infertile women in a program of in vitro fertilization, Fertil. Steril. 42:20.

Muasher, S. J., Garcia, J. E., and Rosenwaks, Z., 1985, The combination of follicle-stimulating hormone and human menopausal gonadotropin for the induction of multiple follicular maturation for in vitro fertilization, Fertil. Steril. 44:62.

O'Herlihy, C., Pepperell, R. J., and Robinson, H. P., 1982, Ultrasound timing of human chorionic gonadotropin administration in clomiphene-stimulated cycle, Obstet. Gynecol. 59:40.

Plachot, M., Mandelbaum, J., Cohen, J. J., Debache, C., Pigeau, F., and Junca, A.-M., 1985, Sequential use of clomiphene citrate, human menopausal gonadotropin, and human chorionic gonadotropin in human in vitro fertilization. I. Follicular growth and oocyte suitability, Fertil. Steril. 43:255.

Quigley, M. M., 1984, The use of ovulation-inducing agents in in-vitro fertilization, Clin. Obstet. Gynecol. 27:983.

Quigley, M. M., Maklad, N. F., and Wolf, D. P., 1983, Comparison of two clomiphene citrate dosage regimens for follicular recruitment in an in vitro fertilization program, Fertil. Steril. 40:178.

Quigley, M. M., Berkowitz, A. S., Gilbert, S. A., and Wolf, D. P., 1984a, Clomiphene citrate in an in vitro fertilization program: Hormonal comparisons between 50- and 150-mg daily dosages, Fertil. Steril. 41:809.

Quigley, M. M., Schmidt, C. L., Beauchamp, P. J., Pace-Owens, S., Berkowitz, A. S., and Wolf, D. P., 1984b, Enhanced follicular recruitment in an in vitro fertilization program: Clomiphene alone versus a clomiphene human menopausal gonadotropin combination, Fertil. Steril. 42:25.

Quigley, M. M., Sokoloski, J. E., and Richards, S. I., 1985, Timing human chorionic gonadotropin administration by days of estradiol rise, Fertil. Steril. 44:791.

Seibel, M. M., McArdle, C. R., Thompson, I. E., Berger, M. J., and
    Taymor, M. L., 1981, The role of ultrasound in ovulation
    induction: a critical appraisal, Fertil. Steril. 36:573.
Smith, D. H., Picker, R. H., Sinosich, M., and Saunders, D. M., 1980,
    Assessment of ovulation by ultrasound and estradiol levels
    during spontaneous and induced cycles, Fertil. Steril. 33:387.
Tamada, T. and Matsumoto, S., 1969, Suppression of ovulation with
    human chorionic gonadotropin, Fertil. Steril. 20:840.
Trounson, A. O. and Calabrese, R., 1984, Changes in plasma proges-
    terone concentrations around the time of the luteinizing hormone
    surge in women superovulated for in vitro fertilization, J.
    Clin. Endocrinol. Metab. 59:1075.
Vargyas, J. M., Morente, C., Shangold, G., and Marrs, R. P., 1984,
    The effect of different methods of ovarian stimulation for human
    in vitro fertilization and embryo replacement, Fertil. Steril.
    42:745.
Williams, R. F. and Hodgen, G. D., 1980, Disparate effects of human
    chorionic gonadotropin during the late follicular phase in
    monkeys: normal ovulation, follicular atresia, ovarian
    acyclicity, and hypersecretion of follicle-stimulating hormone,
    Fertil. Steril. 33:64.
Wilson, E. A., Jawad, M. J., and Hayden, T. L., 1982, Rates of
    exponential increase of serum estradiol concentrations in normal
    and human menopausal gonadotropin-induced cycles, Fertil.
    Steril. 37:46.
Yuen, B. H., Pride, S. M., Rowe, T. C., Moon, Y. S., McComb, P. F.,
    Poland, B. J., and Gomel, V., 1985, Comparison of the outcome of
    ovulation induction therapy in an in vitro fertilization program
    employing a low-dose and an individually adjusted high-dose
    schedule of human menopausal gonadotropins, Am. J. Obstet.
    Gynecol. 151:172.

CONTRIBUTORS

Barry D. Bavister, Ph.D.
    Wisconsin Regional Primate Research Center, and Department of
    Veterinary Science, University of Wisconsin, Madison, Wisconsin

Albert S. Berkowitz, Ph.D.
    Department of Obstetrics, Gynecology and Reproductive Sciences,
    University of Texas Health Science Center, Houston, Texas

William Byrd, Ph.D.
    Department of Obstetrics and Gynecology, University of Texas
    Health Science Center, Dallas, Texas

Marybeth Gerrity, Ph.D.
    IVF and Andrology Programs, Evanston and Glenbrook Hospitals,
    Glenview, and Department of Clinical Obstetrics and Gynecology,
    Northwestern University School of Medicine, Chicago, Illinois

Michael Kane, Ph.D.
    Department of Physiology, University College of Galway, Galway,
    Ireland

Gregory S. Kopf, Ph.D.
    Department of Obstetrics and Gynecology, Division of
    Reproductive Biology, University of Pennsylvania, Philadelphia,
    Pennsylvania

Frank B. Kuzan, Ph.D.
    Department of Obstetrics and Gynecology, University of
    Washington, Seattle, Washington

Patrick Quinn, Ph.D.
    Department of Obstetrics and Gynecology, Cedars-Sinai Medical
    Center/UCLA, Los Angeles, California

409

John S. Rinehart, M.D., Ph.D.
    IVF and Andrology Programs, Evanston and Glenbrook Hospitals,
    Glenview, and Northwestern University School of Medicine,
    Chicago, Illinois

Richard Schultz, Ph.D.
    Department of Biology, University of Pennsylvania, Philadelphia,
    Pennsylvania

Sander S. Shapiro, M.D.
    Department of Obstetrics and Gynecology, University of Wisconsin
    Medical School, Madison, Wisconsin

Don P. Wolf, Ph.D.
    Division of Reproductive Biology and Behavior, Oregon Regional
    Primate Research Center, Beaverton, and Department of Obstetrics
    and Gynecology, Oregon Health Sciences University, Portland,
    Oregon

INDEX